H. G. Parsa
Francis A. Kwansa
Editors

Quick Service Restaurants, Franchising and Multi-Unit Chain Management

Quick Service Restaurants, Franchising and Multi-Unit Chain Management has been co-published simultaneously as *Journal of Restaurant & Foodservice Marketing*, Volume 4, Numbers 3/4 2001.

Pre-publication REVIEWS, COMMENTARIES, EVALUATIONS . . .

"FILLS AN IMPORTANT VOID, bringing together a much-needed unique collection of the latest research. This is important reading for experienced management personnel within the Quick Service Restaurant industry as well as FOR GRADUATE STUDENTS AND UPPER-LEVEL UNDERGRADUATES."

Audrey C. McCool, EdD, RD FADA, FMP
Assistant Dean for Research
William F. Harrah College
of Hotel Administration
University of Nevada, Las Vegas

More pre-publication
REVIEWS, COMMENTARIES, EVALUATIONS . . .

"A MUST-READ MANIFESTO for the entrepreneur contemplating opening QSR franchise/multichain operations, or for STUDENTS wanting to understand the phenomenon of this megastar within the restaurant industry galaxy. FASCINATING . . . examines the strategy and functional management areas (including finance, marketing, international management and human resource management) and technology composing the international QSR and multi-unit industry. CURRENT AND THOROUGH."

Frederick J. DeMicco, PhD FMP, RD
ARAMARK Chair and Professor of Hotel, Restaurant & Institutional Management
The University of Delaware
Newark

The Haworth Hospitality Press

Quick Service Restaurants, Franchising and Multi-Unit Chain Management

Quick Service Restaurants, Franchising and Multi-Unit Chain Management has been co-published simultaneously as *Journal of Restaurant & Foodservice Marketing*, Volume 4, Numbers 3/4 2001.

Journal of Restaurant & Foodservice Marketing Monographic "Separates"

Below is a list of "separates," which in serials librarianship means a special issue simultaneously published as a special journal issue or double-issue *and* as a "separate" hardbound monograph. (This is a format which we also call a "DocuSerial.")

"Separates" are published because specialized libraries or professionals may wish to purchase a specific thematic issue by itself in a format which can be separately cataloged and shelved, as opposed to purchasing the journal on an on-going basis. Faculty members may also more easily consider a "separate" for classroom adoption.

"Separates" are carefully classified separately with the major book jobbers so that the journal tie-in can be noted on new book order slips to avoid duplicate purchasing.

You may wish to visit Haworth's website at . . .

http://www.HaworthPress.com

. . . to search our online catalog for complete tables of contents of these separates and related publications.

You may also call 1-800-HAWORTH (outside US/Canada: 607-722-5857), or Fax 1-800-895-0582 (outside US/Canada: 607-771-0012), or e-mail at:

getinfo@haworthpressinc.com

Quick Service Restaurants, Franchising, and Multi-Unit Chain Management, edited by H. G. Parsa, PhD, FMP, and Francis A. Kwansa, PhD (Vol. 4, No. 3/4, 2001). *"Fills an important void, bringing together a much-needed unique collection of the latest research. This is important reading for experienced management personnel within the Quick Service Restaurant industry as well as for graduate students and upper-level undergraduates." (Audrey C. McCool, EdD, RD, FADA, FMP, Assistant Dean for Research, William F. Harrah College of Hotel Administration, University of Nevada, Las Vegas)*

Quick Service Restaurants, Franchising and Multi-Unit Chain Management

H. G. Parsa, PhD, FMP
Francis A. Kwansa, PhD
Editors

Quick Service Restaurants, Franchising and Multi-Unit Chain Management has been co-published simultaneously as *Journal of Restaurant & Foodservice Marketing*, Volume 4, Numbers 3/4 2001.

The Haworth Hospitality Press
An Imprint of
The Haworth Press, Inc.
New York • London • Oxford

Published by

The Haworth Hospitality Press®, 10 Alice Street, Binghamton, NY 13904-1580 USA

The Haworth Hospitality Press® is an imprint of The Haworth Press, Inc., 10 Alice Street, Binghamton, NY 13904-1580 USA.

Quick Service Restaurants, Franchising and Multi-Unit Chain Management has been co-published simultaneously as *Journal of Restaurant & Foodservice Marketing,* Volume 4, Numbers 3/4 2001.

Cover design by Thomas J. Mayshock Jr.

Library of Congress Cataloging-in-Publication Data

Quickservice restaurants, franchising, and multi-unit chain management / edited by H. G. Parsa and Francis A. Kwansa.

p. cm.

Includes bibliographical references and index. " Co-published simultaneously as Jouranl of Restaurant & Foodservice Marketing, Volume 4, Numbers 3/4 2001." ISBN 0-7890-1704-0 ((hard) : alk. paper)–ISBN 0-7890-1705-9 ((pbk) : alk. paper)

1. Fast food restaurants. I. Parsa, H. G. II. Kwansa, Francis A. III. Journal of Restaurant & Foodservice

TX945.3 .Q53 2002

647.95–dc21

2002003158

Indexing, Abstracting & Website/Internet Coverage

This section provides you with a list of major indexing & abstracting services. That is to say, each service began covering this periodical during the year noted in the right column. Most Websites which are listed below have indicated that they will either post, disseminate, compile, archive, cite or alert their own Website users with research-based content from this work. (This list is as current as the copyright date of this publication.)

Abstracting, Website/Indexing Coverage . . . . Year When Coverage Began

- *AGRICOLA Database <www.natl.usda.gov/ag98>* **1994**
- *BUBL Information Service, an Internet-based Information Service for the UK higher education community. <URL: http://bubl.ac.uk/>* **1997**
- *CIRET (Centre International de Recherches et d'Etudes Touristiques) <www.ciret-tourism.com>* **2001**
- *CNPIEC Reference Guide: Chinese National Directory of Foreign Periodicals* **1996**
- *ELMAR (Current Table of Contents Services), American Marketing* **2001**
- *Emerald Management Reviews (formerly known as Anbar Management Intelligence Abstracts) <www.emeraldinsight.comreviews/index.htm>* **1997**
- *FINDEX www.publist.com* **1999**
- *Food Institute Report "Abstracts Section"* **1996**
- *Food Management* **2000**
- *Food Market Abstracts* **1994**

(continued)

- *Food Science and Technology Abstracts (FSTA) Scanned, abstracted and indexed by the International Food Information Service (IFIS) for inclusion in Food Science and Technology Abstracts (FSTA) <www.ifis.org>* **1994**
- *Foods Adlibra* **1994**
- *Foodservice & Hospitality* **2000**
- *Herbal Connection, The "Abstracts Section" <http://www.herbnet.com>* **1994**
- *IBZ International Bibliography of Periodical Literature <www.saur.de>* **1996**
- *Index Guide to College Journals (core list compiled by integrating 48 indexes frequently used to support undergraduate programs in small to medium sized libraries)* **1999**
- *INSPEC <www.iee.org.uk/publish/>* **1994**
- *International Hospitality and Tourism Database, The* **1996**
- *Leisure, Recreation & Tourism Abstracts (c/o CAB International/CAB ACCESS) <www.cabi.org>* **1994**
- *Lodging, Restaurant & Tourism Index* **1994**
- *Management & Marketing Abstracts* **1994**
- *Restaurant Information Abstracts* **1997**
- *South African Association for Food Science and Technology (SAAFOST)* **2000**
- *Swets Net <www.swetsnet.com>* **2001**
- *"Travel Research Bookshelf" a current awareness service of the Journal of Travel Research "Abstracts from other Journals Section" published by the Travel & Tourism Association* **1997**
- *Worldwide Hospitality & Tourism Trends Database ("WHATT" Database), International CD-ROM research database for hospitality students, researchers, and planners with 3 international editorial bases, and functions with support of the Hotel & Catering International Management Association (HCIMA)* **1996**

Special Bibliographic Notes related to special journal issues (separates) and indexing/abstracting:

- indexing/abstracting services in this list will also cover material in any "separate" that is co-published simultaneously with Haworth's special thematic journal issue or DocuSerial. Indexing/abstracting usually covers material at the article/chapter level.
- monographic co-editions are intended for either non-subscribers or libraries which intend to purchase a second copy for their circulating collections.
- monographic co-editions are reported to all jobbers/wholesalers/approval plans. The source journal is listed as the "series" to assist the prevention of duplicate purchasing in the same manner utilized for books-in-series.
- to facilitate user/access services all indexing/abstracting services are encouraged to utilize the co-indexing entry note indicated at the bottom of the first page of each article/chapter/contribution.
- this is intended to assist a library user of any reference tool (whether print, electronic, online, or CD-ROM) to locate the monographic version if the library has purchased this version but not a subscription to the source journal.
- individual articles/chapters in any Haworth publication are also available through the Haworth Document Delivery Service (HDDS).

ALL HAWORTH HOSPITALITY PRESS
BOOKS AND JOURNALS ARE PRINTED
ON CERTIFIED ACID-FREE PAPER

Quick Service Restaurants, Franchising and Multi-Unit Chain Management

CONTENTS

RESEARCH NOTES

ABOUT THE EDITORS

H. G. Parsa, PhD, FMP, is Associate Professor in the Hospitality Management Program at the Ohio State University and Editor of the *Journal of Foodservice Business Research*. He teaches foodservice management, hospitality marketing, and cost management in hospitality. His published articles are included in the *Journal of Business Research*; *Cornell HRA Quarterly*; *Hospitality and Tourism Research;* and the *Journal of Restaurant & Foodservice Marketing*. His research interests include franchising in QSRs, brand management and consumer behavior in QSRs, and pricing strategies. Professor Parsa has master's degrees in natural sciences and food science and his doctorate in hospitality and tourism management. He has over thirteen years of management experience in QSRs with several international corporations. He is also a member of several national and international organizations and has presented papers in countries such as Australia, China, Canada, and India. He is also an active member of CHRIE.

Francis A. Kwansa, PhD, is Associate Professor in the Hotel, Restaurant and Institutional Management Department at the University of Delaware, where he teaches hospitality finance and managerial accounting. Professor Kwansa has edited one book and written several articles on hospitality management and tourism. He serves on the editorial boards of four academic journals in hospitality and tourism management. He is the former President of the Association of Hospitality Financial Management Educators, and a current member of the Financial Management Committee of the American Hotel and Motel Association. Professor Kwansa is the current Associate Editor of the *Journal of Hospitality Financial Management*. His graduate degree is in hospitality and tourism management from Virginia Tech with specialization in finance and statistics.

Foreword

As Counihan and Esterik (1997) stated, "Food is the foundation of every economy." Food was the driving force behind the great migration of the human race to the far-flung parts on the earth. In human history, food has been the source of the rise and fall of many political empires. The stability and advancement of every society depends on its ability to meet the food needs of its people. In fact, the advancement of a society is often measured in terms of abundance of available food, and the sophistication of a culture is measured in terms of evolution of culinary arts. Measuring on these two terms, in a short duration of two hundred years, the American society has advanced to become a giant in the free world by achieving the highest level of per capita consumption of food in the contemporary human society. From the culinary perspective, America has changed the food habits of the world like no other country before by making Coca Cola and McDonald's the truly global brands and two of the best-known brands in the world. This success of American society in food abundance is often attributed to the technological innovations in agriculture. But to convert the raw food produced by the agriculturists into a wholesome human meal America has turned to the foodservice industry. In other words, it is the foodservice industry that made America a great society by making it the food basket of the world.

The quick-service restaurant industry in the U.S. embraced the concept of globalization in the eighties, in part, because the domestic market was saturated and companies needed to expand and grow. Therefore, in the nineties many domestic chains became multinational and aggressively pursued strategies, such as franchising, designed to export their successful concepts overseas. This volume titled *Quick Service Restaurants, Franchising and Multi-Unit Chain Management* focuses on various issues relevant to the quick-service restaurant industry. In general, the presented articles also provide a window into the reasons behind choices made by the consumers in making their choice with quick-service res-

[Haworth indexing entry note]: "Foreword." Parsa, H. G., and Francis A. Kwansa. Published in Quick Service Restaurants, Franchising and Multi-Unit Chain Management (ed: H. G. Parsa and Francis A. Kwansa) The Haworth Hospitality Press, an imprint of The Haworth Press, Inc., 2001, pp. xvii-xxi. Single or multiple copies of this article are available for a fee from The Haworth Document Delivery Service [1-800-HAWORTH, 9:00 a.m. - 5:00 p.m. (EST). E-mail address: getinfo@haworthpressinc.com].

taurants in the marketplace. Individually, however, the articles tackle specific segments of quick-service customers, exploring trends in these groups, and how they impact on the products and services offered in the industry.

This volume begins with a review of the different strategies employed by the larger foodservice chains to compete for consumer patronage both domestically and internationally as illustrated by Michael D. Olsen and Jinlin Zhao (*"The Restaurant Revolution–Growth, Change and Strategy in the International Foodservice Industry"*). In the USA, food safety and sanitation has become a prominent public health issue, and it has become so for several reasons. The foodservice industry has grown phenomenally over the past two decades; today a greater percentage of meals is eaten away from home; and consumers are more health conscious than ever before. Kim Harris (*"Food Safety: A Public Crime"*) addresses the challenges state governments face in ensuring that restaurant food is handled safely for public consumption.

Then a series of articles converge on the marketplace examining consumption behavior trends in the quick-service industry as well as attitudes and perceptions that determine customer choices in selected segments of the industry. One article (*Changing Food Consumption Patterns and Their Impact on the Quick Service Restaurant Industry"* by Sung-Pil Hahm and Mahnood A. Khan) acknowledges that consumers have reduced their consumption of milk, eggs, pork and beef, and increased their intake of fish, poultry, fruits and vegetables. How will these changes influence the production, marketing and strategic decisions of quick- service restaurant companies in the long-term? Another article (*"Understanding Mature Customers in the Restaurant Business: Inferences from a Nationwide Survey"* by WonSeok Seo, Vivienne J. Wildes and Fred J. DeMicco) uses results from a nationwide survey of mature customers to identify their characteristics, service requirements, and the kinds of unpleasant service experiences that diminish their satisfaction in quick-service restaurants. Another area of the foodservice industry that has brought concern and frustration to consumers is airline food. There have been some serious attempts by the airlines to entertain, even experiment with, practical and feasible suggestions on ways to provide palatable in-flight meals to passengers (*"Consumer Attitude Toward Adding Quick Service Foods to In-Flight Meals"* by Juline Mills and Joan Marie Clay). This article examines what makes customers satisfied with quick-service foods, proposes their addition to airline in-flight menus, and investigates consumer attitudes toward this idea. Finally, in this series, another article identified and studied the factors that influence parents' choice to patronize quick-service restaurants (*"A Study of Factors Influencing Parental Patronage of Quick Service Restaurants"* by Lisa R. Kennon and Johnny Sue Reynolds).

This volume also includes a case study on White Castle Restaurants, the oldest restaurant chain in the world. Dave Hogan tracks the history of White Castle from the 1920s through the 1990s explaining the contributions of White Castle to the American food habits and the impact of socio-economic factors that influenced the growth and development of the oldest restaurant chain.

In the 21st Century, delivery of wholesome meal on demand and on site at an affordable price is the primary challenge faced by the American foodservice industry. An individual's choice of food selection and consumption of food is affected by various types of factors including physiological, psychological, cultural, sociological, nutritional, functional, political, economical, religious, medicinal, technological, ethnological, anthropological, communicational, recreational, vocational, managerial, legal, ecological, familial, situational, etc. Thus exploration of human need for food challenges the intellectual boundaries of many disciplines. Therefore, in the current volume we present a selection of interdisciplinary papers that discuss the role of foodservice industry. Roger D. Blackwell and Kristina S. Blackwell's opinion article (*"Creating Consumer-Driven Demand Chains in Food-Service"*) emphasizes the need to evaluate customers' demand as the primary force in growth and development of the foodservice industry. This article presents a generic consumer decision model with reference to foodservice. Roger D. Blackwell is involved in teaching/research of consumer behavior for over 30 years, authors of several books and articles and is actively consulted by the foodservice industry.

The first set of articles focus on management of restaurant chains in the 2lst century. These article emphasize the role of multiunit management and entrepreneurship in development of QSR chains. In their article, Jinlin Zhou and Lan Li (*"The Restaurant Industry in China: Assessment of the Current Status and Future Opportunities"*) bring their expertise on China and the foodservice industry. As ex-patriots and frequent visitors to China, Zhou and Lan Li recount the growth and development of QSR chains in China. Liberalization of political policies in China has allowed implementation of American-style chain restaurant management. As this article presents, development of QSR chains may be one of the answers to the food needs of Chinese society. In the next article, Jerrold Leong (*"Impact of Mentoring and Empowerment on Employee Performance and Customer Satisfaction in Chain Restaurants"*) emphasizes the importance of mentoring in enhancing the financial performance of a QSR chain.

The next two articles present the role information technology in QSR industry. Marsha Huber and Paul Pilmanis (*"Strategic Information Systems in the Quick Service Restaurant Industry: A Case Study Approach"*) discuss the importance of information technology in a medium-size QSR chain. She chose case study approach as the preferred method of research. This article illustrates how QSR chains could enhance financial performance by carefully selecting an appropriate information technology system. This research is based on "Stage Theory" of IT adaptation. Radesh Palakurthi (*"A Conceptual Model for Quantifying the Effects of a New Entrant in a Competitive Restaurant Market"*) took advantage of the Markov Model, a stochastic model, in quantifying effects of new entrant in given restaurant market. It is a good example of usage of modeling technique to address the issues facing the QSR industry.

In his article, W. Terry Umbreit (*"Study of the Changing Role of Multi-Unit Managers in Quick Service Restaurant Segment"*) discusses the role of multiunit managers in Quick Service Restaurant chains. This is a follow up article to his earlier work focusing on multiunit management. This study is based on interviews conducted with executives of 10 major QSR organizations. It offers rich content and provides good insights in multiunit management encouraging further empirical research on this topic. Sandy Changfeng Chen and John T. Bowen (*"An Exploratory Study of Determinant Factors for the Success of Chinese Quick Service Restaurant Chains"*) present results of their empirical research with Chinese QSR chains in the USA. Growth and development of Chinese QSR chains has been a real challenge in spite of the remarkable success achieved by individual entrepreneurs with Chinese restaurants. This article sheds some light on this topic.

Success of a Quick Service Restaurant, like any other business, primarily depends on the performance of its people, employees and managers. The importance of this human resource aspect is aptly illustrated in a conceptual paper by Dennis Reynolds (*"Integrating 'My Fair Lady' with Foodservice Management: A Theoretical Framework and Research Agenda"*). Reynolds discusses the Pygmalion effect and the Golem effect with reference to performance of foodservice employees. He offers rich discussion of human resource theories and ample opportunities for further empirical research.

This volume ends with a couple of research notes articles. The purpose of a research note is to bring forth a research idea for further discussion. These articles are neither strictly conceptual nor strictly empirical. We thought they present some interesting research ideas worth considering by the hospitality educators and the industry for further research. In some instances, they may be work-in-progress reports. In her article, Nancy Swanger (*"Quick Service Chain Restaurant Managers: Temperament and Profitability"*) makes an effort to suggest a more economical alternative to the well-established Myers and Briggs' psychological profile technique. The suggested alternative has certain limitations such as lack of popular support and absence of established reliability scores. Low response rate and smaller sample size reported by the author provide ample opportunities for other researchers to repeat the study with methodological refinements. Yae Sock Roh and Catherine Smestad (*"Strategic Alliances: The Economics of Dual Branding in Restaurant Franchising"*) discuss the various forms of co-branding and their importance in the QSR industry. It is a brief discussion of an important topic with some value to practitioners as well as educators. Franchising has been the backbone of the QSR industry for over four decades. Inclusion of a research note on this topic should promote further discussion. The selected set of articles present a wide array of research methods from modeling to case study. The articles also include conceptual papers and simple discussion of research topics as presented in research notes.

In summary, it was a challenge to bring forth this volume on the QSR industry for the first time ever. The presented collection offers a glimpse of the status of research on the topic of QSR industry at this time in the hospitality discipline. This volume, we hope, offers good starting material for further research relevant to the QSR industry. The editors' goal for this collection was to bring together research studies, both empirical and conceptual, that center around the quick-service restaurant industry. Together the articles have described the industry during its most recent decade providing some insight into corporate strategy and customer choice. We believe that the substance of these articles further offer valuable information to, on the one hand, restaurant companies with regard to product and service development and delivery, and on the other, to researchers regarding topics for future investigation in the foodservice pedagogy. We hope this volume may help the hospitality researchers in taking the research in QSR to the next level in the future.

H. G. Parsa
Francis A. Kwansa

REFERENCE

Counihan, C. and P. V. Esterik (1997), *Food and Culture–A Reader*. New York, NY Rutledge.

The Restaurant Revolution–Growth, Change and Strategy in the International Foodservice Industry

Michael D. Olsen
Jinlin Zhao

SUMMARY. This paper reports the results of a two-year investigation into the forces driving change in the global foodservice industry and the identification and analysis of competitive methods used by multinationals to respond to these forces. The five forces identified are: (1) globalization and economic change, (2) a knowledge based environment, (3) the future labor force and its entrants, (4) concern for well being, and (5) threatened natural resources. The competitive methods as identified by analyzing multinational foodservice firms are: (1) strategic expansion into the international marketplace, (2) investment in technological development, (3) internal competency development, (4) effective communication to target markets, and (5) competitive pricing strategies. Using the co-alignment model of strategy as the underpinning of the analysis it was determined that multinational firms have responded well to the forces of globalization and economic change, and the knowledge based environment. However, it appears that little competitive methods have been de-

Michael D. Olsen, PhD, is Professor, Department of Hotel, Restaurant and Institutional Management, Virginia Tech, Blacksburg, VA.

Jinlin Zhao, PhD, is Associate Professor, School of Hospitality Management, Florida International University, Miami, FL.

[Haworth co-indexing entry note]: "The Restaurant Revolution–Growth, Change and Strategy in the International Foodservice Industry." Olsen, Michael D., and Jinlin Zhao. Co-published simultaneously in *Journal of Restaurant & Foodservice Marketing* (The Haworth Hospitality Press, an imprint of The Haworth Press, Inc.) Vol. 4, No. 3, 2001, pp. 1-34; and: *Quick Service Restaurants, Franchising and Multi-Unit Chain Management* (ed: H. G. Parsa and Francis A. Kwansa) The Haworth Hospitality Press, an imprint of The Haworth Press, Inc., 2001, pp. 1-34. Single or multiple copies of this article are available for a fee from The Haworth Document Delivery Service [1-800-HAWORTH, 9:00 a.m. - 5:00 p.m. (EST). E-mail address: getinfo@ haworthpressinc.com].

veloped to address the forces such as labor force, well being and threatened natural resources. The implication here is that firms that are not in alignment with environmental forces will have to address this concern if they expect to continue to add value for all industry stakeholders.

KEYWORDS. Multinational, foodservice, international expansion, strategy, competition, environment

INTRODUCTION

The forthcoming millennium will provide new opportunities and challenges for the foodservice industry as,

- multinational chains seek to push into new markets
- pressure increases from global capital for economic performance
- technology impacts how the customer makes the eating out decision
- a shrinking labor supply threatens to limit growth potential
- threats to the food chain increase from man made and natural forces

Responding to these complex forces driving change will require new skills and more strategic thinking from foodservice managers.

Managers of foodservice must:

- identify the key forces driving change, their dynamics and complexity to help tomorrow's foodservice manager invest in the products and services that will put them ahead of the competition
- review the competitive methods adopted over the past five by multinational foodservice companies in response to these forces and assess their readiness to meet the challenges of the new millennium

GLOBALIZATION AND ECONOMIC CHANGE

The emergence of the global economic marketplace and the free market system is changing the criteria on which companies attract capital to fuel growth. In an environment where there are no serious impediments to the free movement of capital between nations, the foodservice industry is competing with other sectors for the capital needed to sponsor growth. Investors will fund those firms that offer the greatest returns.

Firms will be valued on their future prospects (i.e., anticipated cash flow) rather than current performance. To win investor confidence, foodservice managers must anticipate change and adopt future-oriented competitive methods. Given the risks involved in leading change, the temptation to wait for others to take the risk first is strong. But taking too long to adapt to change in the marketplace may involve a higher risk than being innovative to begin with.

Tomorrow's managers must therefore be able to respond to innovations and take advantage of emerging opportunities before others do. They must also anticipate the useful economic life of new competitive methods and project future cash flows according to the trends developing as a result of each force. For example, will new menu items, service experiences, or technological investments last one month or five years?

Investors in the new capital market system " . . . will search primarily for a promise of growth" (Davis & Meyer, 1998) and increasingly value foodservice companies on the abilities of management to think strategically with foresight. This runs counter to the way money has traditionally been made in the industry for the last forty years, that is, through the combination of inflation and real estate transactions. In the future, investments will be made in "what's moving, not what's standing still" (Davis & Meyer, 1998) making it imperative for industry executives to develop strategic skills focused on adding future value to the firm–a serious challenge for those schooled in the craft side of the industry over the last fifty years. Exhibit 1 summarizes the above issues.

A KNOWLEDGE-BASED ENVIRONMENT

Tomorrow's foodservice industry will face new challenges resulting from the transition to the information age, characterised by the convergence of computing, telecommunications and software programs which will shape the way people live and work (Tapscott, 1996; Gate, 1995). According to Davis and Meyer (1998), "almost instantaneous communication and computation are shrinking time and focusing us on speed," creating a fast-paced environment they refer to as "*Blur.*"

Above all, this convergence will change the nature of interaction between the employee and the customer. It will reshape the exchange process between providers of products and experiences and more informed customers with higher expectations of both the tangible and intangible elements of the exchange. Customer demands will not stop at high quality food; they will also

EXHIBIT 1. Globalization and Economic Change: Elements of the Force Driving Change

- A global capital market system is demanding more
- Valuing the enterprise based upon future earnings, not asset value
- A growing number of investor choices in a global market increases competition for available capital
- Demands for improving the response time to competitive threats
- Strategic financial skills are growing in importance

want to know if management has insured that the food has been carefully looked after "from farm to fork."

In an information age, the consumer will be able to get this information from several sources, many of them beyond the control of the foodservice operator. This information–the intangible part of the exchange–will grow in value to the customer.

Likewise, tomorrow's foodservice operator will be expected to know much more about the customer–to systematically accumulate information about the customers' desires and needs (another intangible element of the exchange process) and use it to customize the product/service to the guest's needs. For example, customers with specific dietary requirements will be able to request special menu items and preparation procedures in advance. It will transform foodservice operations from physical assets built to achieve standardized products and services (the McDonald's "cookie-cutter" approach) to units designed with in-built flexibility to meet customized consumer demands. This "segment of one" approach will continue to reshape how foodservice providers interact with the customer of tomorrow. The "segment of one" refers to a new way of segmenting the customer. Instead of grouping them in a large segments by artificial criteria, the customer is considered an individual segment and products and services are customized for the customer.

Customers will gain more power in the new exchange environment, as the convergence process creates numerous ways for them to find out whether a foodservice outlet makes good on its promises. Convergence will make it possible for everyone to be connected, allowing future customers to learn about the experience of past customers at any particular foodservice outlet. Customers who are not satisfied will be able to vent their feelings to an Internet-connected community that could represent thousands of current and future customers.

Consumers will also gain power over the pricing process. The traditional cost-of-food multiplied by a constant of 2.5 formula as the basis of selling price (assuming a 40% projected food cost) will give way to a pricing system similar to the stock market. The menu price will be posted on an electronic trading board in the restaurant or the Internet and change according to what the market (the customer) is willing to pay. As with stocks and shares, some customers may take options on menu items hoping the price will go up later in the day as more customers seek their special menu item. This trend will force managers to determine what the value drivers of the foodservice really are and how to price them in order to achieve the return on investment sought by demanding investors.

Much of the exchange process will be accomplished by personal software agents and "infomediaries" that match specific requests from the customer with various providers and act as filters insuring that the buyer obtains only what is wanted at the price desired. Such new intermediaries as Priceline.com, Expedia and Travelocity can be anticipated to gain in popularity as information providers.

They will also give consumers privacy protection and the option to earn revenue in return for the information they are willing to release about themselves (Anon. 1999). Permission marketing where, in exchange for a fee, the customer is willing to receive certain commercial messages that match his or her interests (Davis & Meyer, 1998) will challenge foodservice firms to get the customer's attention in order to offer the tangible and intangible products and services needed at at the point in time that s/he is seeking to buy. It will require the capability to operate a cyber-based "radar" twenty-four hours a day, seven days a week which can connect to the buyer at the right moment, with the right product at the right price, once the buyer has given permission to receive the message.

This will necessitate an entirely new approach to managing the hospitality enterprise (see Exhibit 2). It will place more emphasis on providing intangible experiences alongside the tangibles like food and beverage which will become more like commodities. Managers must learn how to build an experience portfolio (the growing importance of the intangibles) to satisfy the guest and meet the demands of the capital market players seeking returns on their investments (Olsen, 1999). Improving the intangibles of the exchange, while insuring that the tangibles remain of consistently high quality, will not be easy for an industry that has primarily relied on the tangible product in the past. Early attempts at this, such as those by Planet Hollywood and Rainforest Café, have not proved successful over time.

As the Internet phenomenon continues, the new exchange model between buyer and seller will be centered on a networked environment where service providers, suppliers, and customers will all be linked seamlessly. This connected environment will create opportunities for new inter-and intra-industry alliances as well as outsourcing. As hotels and restaurants build alliances with more members of the supply chain and their interdependency grows, they will

EXHIBIT 2. A Knowledge Based Environment: Elements of the Force Driving Change

- The accelerating rate of change
- A new model of the customer exchange process
- The need for more knowledge about the customers' individual needs
- Customers gaining more power
- The transparency of the foodservice enterprise
- Market pricing controlled by the customer, not the supplier
- The rise of infomediaries
- Adding value through intangibles
- A new industry and enterprise structure
- Your work is never complete
- The lifetime value of the customer
- Cyber marketing

inevitably relinquish some control over their destiny. Their management of these new relationships will determine how they compete and to what purpose they innovate.

In the connected environment, every aspect of business will operate in real time terms, anywhere, 24 hours per day, 7 days per week. Connectivity in the distribution chain will link the customer to the farmer, producer, supplier, regulator and maybe even the competitor.

The recent acquisition of cable provider MediaOne Group Inc. by AT&T (the US telecommunications giant) suggests that telephone and cable services will be combined to create a system with the potential to deliver interactive television. This could, for example, allow viewers with high-speed Internet access in the future to select their own camera angles of a foodservice operation, perhaps even the cleanliness of the kitchen (Schiesel, 1999). Microsoft has invested over $4 billion USD in this effort and also acquired a 30% stake in Telewest, the second largest cable operator in the United Kingdom (Parkes &Thal, 1999).

It is obvious from these examples that connectivity has arrived. While truly interactive television is still years away, the pattern is already clear when AT&T is spending in excess of $90 billion on the belief that the merging of communications and television technology will be the way the consumer will acquire all types of offerings of products and services in the near future (Davis & Meyer, 1998). To remain competitive, AOL Inc. recently announced it will provide interactive television service through satellite TV and set-top box suppliers to US homes in the near future (Price & Kehoe, 1999). As the Internet eventually penetrates everyday life, the possibilities are exciting but daunting to an industry unused to leading change.

As indicated, technology is changing the marketing practices of the industry, redefining the individual customer as a segment of one. Customer information will be systematically gathered and stored in data warehouses for use in customizing the offering. This will in turn encourage businesses to evaluate the lifetime value of that customer. McDonald's has for decades understood that each current customer contributes to the future cash flows of the firm and invested in winning customer loyalty from childhood on.

In addition to the complex process of reaching the "segment of one" customer and maximizing their lifetime value, marketers will also require an understanding of how to get the attention of this customer in an Internet-dominated distribution channel through which hotel rooms, airline and restaurant seats, etc., will be bought and sold. Cyber marketing will replace the idea of strategic marketing.

It suggests that the future of marketing will be shaped by the use of knowledge-based robots to make offers on behalf of the commercial enterprise to the personal software agents of customer. The customer will enter into the transaction only after these agents have agreed to the terms of the offer, thus saving valuable time. At recent think tanks on marketing convened by the International Hotel and Restaurant Association (IH&RA), industry experts identified six major challenges (see Exhibit 3) for tomorrow's marketing manager (Dev & Olsen, 1999).

These reflect the growing complexity of the marketing function as it increasingly embraces emerging technologies. They underline the importance of managing distribution costs as well as choosing channels of distribution, analyzing customers and optimising their lifetime value. Current business models for marketing and managing need to be rethought if firms are to successfully compete in the future.

An additional challenge will be to manage the growing interdependencies in the connected environment and, when possible, control the use of technology in a way that is favorable to the firm. It will require a greater level of skill and knowledge to determine which media opportunities to invest in, in addition to the ability to scan the environment for the "next big thing" that will be decisive in capturing the customer of tomorrow.

EXHIBIT 3. The Six Challenges Facing Tomorrow's Hospitality Marketing Professional

Manage distribution costs
Analyze customers
Rethink the business model
Keep control of technology
Evaluate new media opportunities
The next big thing

The knowledge revolution will redefine not only how the foodservice industry interacts with customers and suppliers, it will also render the business transparent. As customers become aware of quality levels in both front and back of the house, increasing emphasis will be placed on the intangible elements of the transaction.

Customers will be valued for their lifetime worth to the firm and their preferences stored in databases in order to customize products and services. These represent major challenges for an industry built upon the simple idea of providing good food and service.

THE FUTURE OF LABOR AND THE NEW LABOR FORCE ENTRANTS

The knowledge revolution will impact the laws of labor supply and demand. It will change the nature of work and the workplace. As a labor-intensive industry, foodservice outlets have traditionally battled to secure enough labor to produce and deliver quality products and services. Over the next 10 years this challenge will intensify in four ways (see Exhibit 4). First, the demographic changes of the last 20 years will result in a decline in the number of young workers, on whom the industry has relied on in the past. Secondly, young workers will find more challenging, rewarding jobs with better conditions and higher pay in knowledge-related work, and thus have higher expectations of the workplace. Thirdly, alternative sources of labor, such as immigrants, will be restricted owing to the general reluctance of nations to open their borders to new entrants. Finally, other sources of labor such as seniors will put pressure on management to be more socially responsible and provide a more principled, even spiritually-oriented work environment.

These four trends will make it increasingly difficult to obtain quality labor to meet the increased need for higher levels of service and customer interaction. The potential labor shortage will be exacerbated by the growing number of outlets opened as multinationals seek to expand into new overseas markets in response to pressure from the financial community for growth. With labour in demand but supply limited, managers will inevitably have to deal with rising costs. Statistics from a recently published IH&RA report on the casual theme restaurant industry serve to back this up (Olsen & Sharma, 1998).

- According to the U.S. Department of Labor Employment Outlook (Anon, 1994), employment is projected to increase from 127 million to 144.7 million, between 1994-2005. This represents a growth rate of about 14 percent, which is just over one-half that prevailing from 1983 to

1994. The slowing of employment growth reflects the slowing of labor force growth.

- The 1998 edition of the International Labor Organization's Labor Force Statistics Report reveals a similar trend elsewhere. Total Labor Force in OECD countries stood at 41.9 percent (of the total population) compared to 44.5 percent in 1991 (Anon, 1997).
- In the US, the Employment Outlook Document projects that women, Hispanics, Asians, and older workers will significantly increase as a proportion of the labor force (Anon, 1994). A similar trend is likely to be experienced in other parts of the world.
- The ILO Labor Statistics report for 1997 shows that total population in the 15-64 age group has been consistently declining in OECD countries–the current figure stands at 43.1 percent (of the total population) as against 60.5 percent in 1991 (Anon, 1997).
- The same report also identifies the service sector as the maximum driver of this employment growth, accounting for approximately 32.8 percent over the period 1994-2005 (Anon, 1997).
- The report also proposes that Eating and Drinking outlets within the service industry are likely to experience a 14 percent growth over the same period.

Additional pressure on the labor supply will accrue from the service sector in general. As Exhibit 5 indicates, employment in the service sector is growing, while the total employment in the non-service sector is holding steady (Olsen & Sharma, 1998). This trend is more prevalent in the developed world, where most of the demand for eating out exists. Management will have to address these challenges with long term human resource strategies if they want to achieve effective offering and delivery of their products and services.

One way the industry has met this challenge in the past has been to employ newcomers to the labor market, primarily immigrants, who enter with limited knowledge and language skills. In an increasingly nationalist climate, this option may be denied. Current trends indicate that most nations are crafting legis-

EXHIBIT 4. The Future of Labor and the New Labor Force Entrant: Elements of the Force Driving Change

- The growing shortage of qualified employees
- The changing demographics of the workforce
- New leadership skill requirements
- Investing in the human resource

lation to limit immigration (Olsen & Sharma, 1998). As yet, the industry has not proposed any serious means for improving this labor situation.

Labor turnover is an ongoing problem in the foodservice industry, resulting from the types of jobs and the quality of supervision and management that are indigenous to the industry. Low pay entry-level jobs requiring minimum skills offer little to those in search of career opportunity. Physical working conditions also tend to foster the attitude among entrants that they will accept a foodservice job only until something better comes along.

Genuinely investing in people–as opposed to paying lip service to the idea–will have to be embraced and, as opportunities in knowledge-related work continue to expand, the enlightened foodservice firm will have to develop improved leadership capabilities.

Traditional "my way or the highway" styles of leadership will do little to induce loyalty and motivation. Even at higher levels of the business, such as assistant manager and above, the same problem exists. The implication is clear: tomorrow's labor force will expect more of the workplace environment, including leadership.

Management will have to develop resources to meet the expectations of workers who will be seeking much more than a job. Whereas current prospects are asked what they can do for the enterprise, in the future the new entrant will want to know, "what can you do for me?"

The challenge for managers will be creating value *through* people, rather than using them as objects to add value. The new entrant will want to participate in creating value, not just be told how to do it. Access to management and non-management employee education and development will be an increasingly important factor in deciding what company to work for. The assumption that investment in training is wasteful because people tend to move on when it is completed will have to go the way of dinosaurs. The new workforce entrant will simply not join the enterprise without this benefit.

Those destined to reach management level will expect to use the competencies they obtain in order to fulfill job expectations, and to be compensated for achieving them whether they are immediately used in the job or not. Here again, a new leadership and management approach will be required.

A list of needs of tomorrow's hospitality manager have been identified through a series of IH&RA Visioning the Future Workshops and Think Thanks on human resources held since 1995. (See Exhibit 6.) It indicates that an entirely different set of conceptual skills to those currently in vogue will be required, validating the observation of the influential economist Lester Thurow that, "The dominant competitive method of the twenty-first century will be the education and skills of the workforce." (Thurow, 1997). This poses a serious challenge to the foodservice industry if it is going to compete for excellence at all employee levels of the firm. Exhibit 5 summarizes the elements of this force driving change.

WELL BEING

Another emerging trend, among both customers and employees, is the desire for spiritual and physical well-being. In part, this results from the quest to find a "safe haven" in an increasingly complex, uncertain and violent world, and to return to values that have been lost.

One of the social issues driving this concern is the growing disparity between "haves" and "have-nots." This gap has been compounded by the free flow of information showing the "have-not's" what they do not have, and prompting growing numbers to simply take from those who have. Media reports of the looting of homes and businesses of wealthy ethnic Chinese in Indonesia, deadly car jackings in South Africa or murder/suicide rampages in high schools in America offer clear examples of social justice being dispensed by "have not" communities with no faith in government institutions to solve their problems, and increasingly taking matters into their own hands.

The above examples suggest that the potential for personal violence to cross the path of many rather than few is increasing. So too is the concern over the threat to personal health from diseases resulting from new strains of bacteria and viruses. Should these find their way into the food chain and drinking water, they will directly impact the foodservice industry. Additionally, there is growing concern about genetically modified (GM) foods. Presently, there are

EXHIBIT 5. Civilian Employment (sector wise) in Select OECD Countries, 1980-96 (figures in thousands)

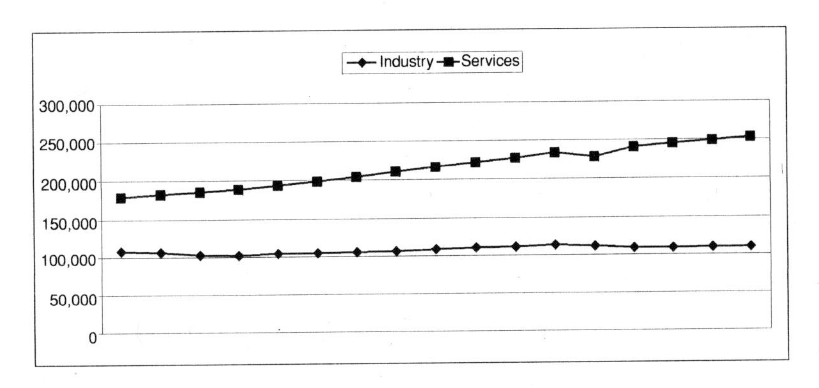

EXHIBIT 6. Leadership Skill Requirements of the Future Hospitality Manager

- A Visionary, employing value adding strategies
- Using and managing knowledge and technology for competitive advantage
- Spanning boundaries of cultures, business environments and management know-how
- A synthesizer and blender of skills and knowledge in a fast changing environment
- A leader in a dynamic and complex, diverse setting
- Strong business skills

conflicting reports in scientific literature regarding the potentially ill effects that these products will have on health. Until evidence is more conclusive, many seek a moratorium on the future sale of such products. This can be expected to add to the impact of this important element of the overall well being force driving change.

As health and security preoccupations grow, the question "Will I be safe?" becomes foremost on the minds of a growing number of travellers. While the issue may not make headline news on a daily basis until a disease outbreak or terrorist attack is dramatized by the media, it will become one of the major issues that the travel and tourism industry will have to face as the world grows more hostile and aging travelers become more cautious. Already, major tour operators such as the UK-based Thomson Travel, Plc, who move millions of tourists out of England and Germany annually are dictating to providers in the destinations they serve the safety and security standards that must be met if business is to be continued.

The oft-predicted travel boom foreseen in the next century is based on the assumption that as the baby boomers in the West retire they will want to travel and they will have the disposable income to do so. While this is intuitively logical, there are some problems with this thinking. The first has to do with the aging process which inevitably reduces physical mobility and heightens susceptibility to disease. The mobility problem will affect the total numbers who can frequent the nearest restaurant, let alone trek through jungles and climb pyramids. The need will be for greater ease of access and less challenging adventures.

Likewise, the aging process will reduce the ability of the human body to handle the invasion of foodborne-related illness. Recent evidence provided by the World Health Organization suggesting that foodborne disease is becoming a major problem worldwide which could negatively affect the older customer (not to mention the very young). A WHO factsheet recognizes the following as the most probable issues that will affect future occurrences of foodborne illnesses (Anon, 1996):

- *The globalization of the food supply*. The value of agriculture and food products as part of total world trade was estimated at US$ 381 billion in 1993 (Anon, 1996). Such trade contributes to the occurrence of foodborne illnesses. For example, a large outbreak of Shigella sonnei infections occurred in Great Britain, Norway and Sweden in 1994 due to contaminated lettuce imported from Southern Europe.
- *Travelers*, refugees and immigrants are exposed to unfamiliar foodborne hazards, while abroad. It is estimated that 90% of all Salmonellosis cases in Sweden are imported (Anon, 1996).
- *Mutations in microorganisms* can lead to the development of more virulent strains of old pathogens. These new strains are likely to have increased resistance to antibiotic remedies, than their predecessors.
- *Changes in demographics* are another reason cited for the emergence of new pathogens. There is an increasing population of "susceptible" persons-as defined by age and/or medical condition.

An additional obstacle to the robust forecasts for growth in the travel industry comes from the cost of health care for the aging population. One of the most coveted possessions for the aging individual is insurance that will provide not simply for their health care needs but prevent financial disaster in case of serious illness.

Therefore it can be expected that the aging consumer will make travel and eating out decisions based upon the risk estimated to their health and also to their insurance coverage. Given that charges associated with medical care for the aging tend to be high and prolonged, and that insurance companies use every means possible to keep costs down, it can be anticipated that more of the responsibility for health care, including prevention, will fall on the individual. For example, smokers can be expected to pay more for insurance coverage since they knowingly are putting their health at risk. It is quite conceivable then that the senior consumer would be expected to pay a higher premium if they travel to parts of the world that put their health and well being at risk.

What can be anticipated in this context will be persistent efforts to set global standards to protect the consumer. Currently, processes designed to insure protection throughout the food chain, such as the Hazard Analysis Critical Control Points (HACCP), have been adopted by private sector firms and/or mandated by government. Given the growing concern for well-being, it can be anticipated that the consumer will demand clear-cut standards and their enforcement. An industry that prides itself on keeping government at bay, may find that, unless it addresses this need, it will have little say in how it is accomplished.

Crime, along with terrorism, poses a growing threat to personal safety and is becoming a major factor in the consumer's decision about where to travel or go out to eat. Hospitality providers are going to have to address this concern, especially in sensitive locations.

As indicated, the consumer's choice will be facilitated by the personal smart agents developing on the Internet. With regard to travel, the "transparency" of destinations is already increasing with the publication of travel advisories by the US Centers for Disease Control and US Department of State, now available through a growing number of infomediaries. For example, the UK-based Tourcheck firm, developed by National Britannia, has a website that monitors such aspects as food hygiene, fire safety, pest control, water quality, etc. Tour operators seek to use this information in order to lower their insurance rates and demonstrate due diligence in the event of customer lawsuits arising from safety and health problems (Olsen, 1999).

It is not difficult to envision a dining out decision influenced by a whole series of filters such as Tourcheck. Reputable guides such as Michelin may likewise integrate measures of well-being as they seek to remain competitive. What is evident is that concern for well-being among consumers–whether young or old–will translate into the intangible elements of a firm's portfolio of competitive method and that firms will advertise their capabilities in this regard as they seek to drive quality standards to new heights.

As a recent IH&RA Think Tank on Safety and Security pointed out, crime, terrorism, food safety and the migration of infectious diseases along with the need to improve the overall quality of information for all stakeholders in the travel chain, will challenge the manager of tomorrow (Olsen & Pizam, 1998). It is essential that safety and health move up the ladder of concern for everyone involved in the travel industry. Exhibit 7 summarizes the elements of this force driving change.

THREATENED NATURAL RESOURCES

According to the Washington, DC based Population Reference Bureau, the world's population will grow to 6.1 billion by 2000, up from its 1996 level of 5.8 billion (Anon, 1997a). Of the 1996 total, 4.7 billion were in developing countries. These accounted for 98% of the growth in the world's population.

As this trend continues, it will put serious strain on natural resources. The most important issue likely to emerge will be the available supply of potable water. Fresh water scarcity will impact every region of the globe with Africa, Asia and South America being the hardest hit. Early signs already exist. In

EXHIBIT 7. Well Being: Elements of the Force Driving Change

- The concern for health and food safety
- The globalization of the food chain
- Concern for personal safety and fear of violence
- The spread of disease
- Aging and personal well being
- Engineering in the food chain
- Consumer and government leadership in improving well being

1994, an estimated 220 million people lacked a source of potable water near their homes (Anon, 1997). Most nations will experience significant reductions in availability of potable water. This shortage will impact the foodservice industry, a large consumer of water.

Exhibit 8 presents a comparison of various types of businesses and their average and peak use of water per day. Although it illustrates differing measures of water usage for each type of establishment, it suggests that the industry could easily become a target of criticism because of its consumption level. For example, let us assume that the average size of a typical restaurant in North America is 6,000 square feet, or 558 square meters. In the minimum-use category, that would approximate to 12 liters/day/square meter or 36 liters/ day/ square meter in the peak-use category. When compared with the other establishments in Exhibit 9, such as hospitals, car washes or service stations, the industry clearly consumes significantly more water.

This level of usage is likely to raise the cost of doing business. As water becomes more scarce, governments will do whatever is necessary to preserve it, for example by increasing the unit cost to restrict use for all but critical needs. As eating out is unlikely to be considered in that category, foodservice operators will have to pay much higher prices-assuming water is available. If not, they may have to invest in alternative sources such as desalinized seawater to meet their needs.

Air quality is also increasingly in danger. The global warming effect of industrialization, with greenhouse gases emitted from fossil fuels and the proliferation of automobiles and the accompanying traffic jams, continues to erode air quality. As it is today, many cities post air quality warnings so that those with respiratory problems do not venture outside. If this situation continues unchecked, and as the numbers of those in the aging categories increase, it will likely impact the total demand curve for the foodservice industry.

Already, in some parts of the world where major cities have high concentrations of pollutants, innovative restaurateurs are offering customers the oppor-

EXHIBIT 8. Per Capita Water Availability Today and in 2025, Selected Countries

Country	Water availability, 1990 (cubic meter per person per year)	Projected Capita Water availability, 2025 (cubic meter per person per year)	Projected Capita Percentage reduction in Water availability, 2025 (cubic meter per person per year)
Africa			
Algeria	750	380	49.33%
Burundi	660	280	57.58%
Cape Verde	500	220	56.00%
Comoros	2,040	790	61.27%
Djibouti	750	270	64.00%
Egypt	1,070	620	42.06%
Ethiopia	2,360	980	58.47%
Kenya	590	190	67.80%
Lesotho	2,220	930	58.11%
Libya	160	60	62.50%
Morocco	1,200	680	43.33%
Nigeria	2,660	1,000	62.41%
Rwanda	880	350	60.23%
Somalia	1,510	610	59.60%
South Africa	1,420	790	44.37%
Tanzania	2,780	900	67.63%
Tunisia	530	330	37.74%
North and Central America			
Barbados	170	170	0.00%
Haiti	1,690	960	43.20%
South America			
Peru	1,790	980	45.25%
Asia/Middle East			
Cyprus	1,290	1,000	22.48%
Iran	2,080	960	53.85%
Israel	470	310	34.04%
Jordan	260	80	69.23%
Kuwait	10	10	0.00%
Lebanon	1,600	960	40.00%
Oman	1,330	470	64.66%
Qatar	50	20	60.00%
Saudi Arabia	160	50	68.75%
Singapore	220	190	13.64%
United Arab Emirates	190	110	42.11%
Yemen	240	80	66.67%
Europe			
Malta	80	80	0.00%

Source: *Water in Crises: A Guide to the World's Fresh Water Resources* (1993), pp. 106

EXHIBIT 9. Water Requirements for Select Municipal Establishments

Type	*Unit*	*Average Use*	*Peak use*
Hotels	Liters/day/square meter	10.4	17.6
Motels	Liters/day/square meter	9.1	63.1
Hospitals	Liters/day/bed	1,310	3,450
Laundry	Liters/day/square meter	10.3	63.9
Car washes	Liters/day/inside square meter	194.7	1,280
Golf-swim clubs	Liters/day/member	117	84
Service stations	Liters/day/inside square meter	10.2	1,280
Apartments	Liters/day/occupied unit	821	1,640
Restaurants	Liters/day/seat	91.6	632
Fast Food Restaurant	Liters/day/establishment	6,780	20,300

Source: *Water in Crisis: A guide to the World's Fresh Water Resources* (1993), pp. 411.

tunity to pay for breathing pure oxygen. A new competitive method perhaps, but in many ways a sad scenario. Exhibit 10 illustrates the rise in atmospheric concentrations of greenhouse and ozone-depleting gases from 1970-94. This is a disturbing trend that will impact negatively on the industry unless efforts are made to correct the problem.

As governments become more concerned with their ability to provide this precious resource, those who contribute to its degradation can be expected to bear the cost burden associated with its preservation. Therefore, foodservices that contribute to pollution can be expected to invest in equipment to limit this pollution or pay taxes or other assessments if they do not reduce their emissions.

Land is another resource that is becoming short in supply. Green space for development is limited in the West as all types of firms expand to meet the needs of the population sprawl. In the United States and Western Europe, there are very few good prime locations for restaurants or hotels, a situation that will not change. Finding enough landfill space to dispose of the enormous amount of waste generated by developed societies is a further challenge.

The developing world has its own land problems. With much of its land arid and uninhabitable under present conditions, and with the current population of 4 billion growing rapidly in these regions, there will also be serious land shortages.

The net impact of this challenge is cost-related land values will continue to rise as firms compete for good locations to conveniently serve the customer. If existing businesses are not earning appropriate returns on the valuable land they are using, others who can will eventually be in a position to acquire this productive resource. This will set growth limits on those foodservice firms seeking to rapidly expand worldwide in order to satisfy hungry investors. Environmentalists will also step in to have their say about preserving green space for future generations compounding the set of interrelated forces driving

EXHIBIT 10. Atmospheric Concentrations of Greenhouse and Ozone-Depleting Gases, 1970-94

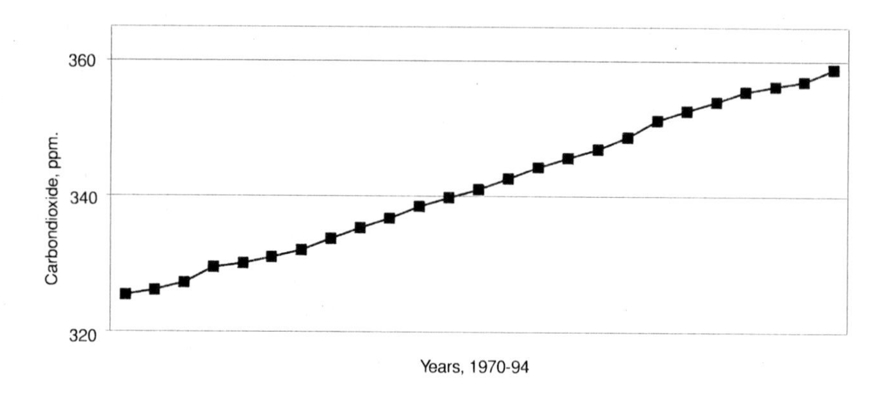

change in the foodservice industry. The implications are clear: management will have to develop strategies based upon the need to conserve threatened resources. Strategies that will have to go further than present programs and become more aggressive if the foodservice firm expects to attract tomorrow's socially-conscious consumer.

TOMORROW'S CHALLENGES

Globalization and free market systems, a knowledge-based environment, labor challenges, a growing concern for health and well being and threatened natural resources are emerging as the drivers of change in the 21st century. Cutting across all major forces is the consumer. The consumer will utilize every element of the knowledge-based environment to achieve the objectives they seek. This environment will make the exchange process between the consumer and the firm fully transparent. For firms to compete in this environment, they will have to concentrate on developing more than just foodservice products of the traditional sense. The intangible elements associated with the forces will become more important in the consumer's eyes over time, especially the provision of wholesome food and safe surroundings. The challenge will be for every foodservice operator to balance tangible and intangible offerings and deliver high quality in both areas.

At the beginning of this section, it was suggested that the manager of tomorrow would have to learn to anticipate change in order to invest in products and services that would capitalize on the opportunities arising from such anticipation. To assess how well the foodservice industry has anticipated the forces driving change to date, a study of the competitive methods utilized by multinational foodservice firms was conducted. These are identified and described in the following section.

COMPETITIVE METHODS OF MULTINATIONAL FOODSERVICE COMPANIES

The multinational foodservice industry has invested in competitive methods in response to the opportunities and threats associated with the forces driving change. Exhibit 11 identifies these methods on the basis of an analysis of primary and secondary sources of information regarding the leading multinational companies listed in Exhibits 12 and 13.

International expansion has emerged as one of the most important competitive methods of multinational foodservice companies confronted by mature domestic markets in their home countries and growing demand in emerging economies. However, this strategy began as early as 1956, when KFC (then Kentucky Fried Chicken) pioneered expansion into Canada. Many have since followed suit and have gradually become multinational foodservice firms. Exhibit 12 illustrates the size and scope of this development. Similarly, foodservice contract firms are consolidating and seeking a global position as illustrated in Exhibit 13. This process can be expected to continue and will accelerate as the globalization of the food supply continues.

Technology emerged as an important investment in competitiveness. Although it has been directed primarily at internal operations, some evidence emerged that it was also being employed in marketing and other outward-oriented functions. Investment in new products and services surfaced as an important competitive method and will no doubt continue to essential for firms to maintain market share.

Internally, multinationals have sought to improve efficiency and competence in response to mature environments and a limited labor supply. They also

EXHIBIT 11. Identified Competitive Methods of Multinational Foodservice Companies

- Strategic expansion decisions into the international marketplace
- Investment in technological development
- Internal competency development
- New product/service development
- Effective communication to the target market
- Competitive pricing strategies

EXHIBIT 12. Estimated Sales and Units of the Top 15 U.S. Based Multinational Chains in 1997

Chain	Overseas Sales ($000)	% of Total Sales	Overseas Units	% of Total Units	# of Countries operating in
McDonald's	16,513,000	49.09	10,752	46.48	115
KFC	4,330,000	51.98	5,117	49.99	73
Pizza Hut	2,525,000	34.95	3,836	30.60	83
Burger King	2,042,410	20.62	2,060	21.46	55
Tim Hortons	749,901	97.14	1,499	94.99	na
Domino's Pizza	680,000	21.51	1,521	25.55	60
Wendy's	625,000	11.95	632	12.14	30
Baskin-Robbins	500,000	46.70	1,927	41.85	na
Subway	500,000	14.71	1,851	14.22	64
Hard Rock	425,000	61.15	55	64.71	30
Dairy Queen	370,000	12.11	733	12.63	25
CoCo's	333,000	53.28	298	60.45	na
Dunkin' Donuts	286,136	13.97	1,386	28.74	37
YogenFrz/Paraduse/Java	281,000	98.25	1,300	99.24	na
Planet Hollywood	267,000	47.51	33	50.77	na

Source: Restaurant Business, 11/1998 and the companies' Web sites, 4/1999

EXHIBIT 13. Three Largest Multinational Foodservice Contract Companies

Company	Headquarters	Number of countries operating in	Estimated sales in 1997
Compass Group Plc	United Kingdom	50	$6.5 billion
ARAMARK	USA	15	$6.4 billion
Sodexho Alliance	France	66	$5.5 billion

stressed the ever-growing reliance on marketing to communicate to a customer who is increasingly wooed by more aggressive competition. As a result, a more competitive pricing strategy is implemented by competitors both at home or abroad.

The challenge is not simply to continue current competitive methods but to develop new ones. This requires more strategic thinkers who can add value in a demanding environment and meet the expectations of global investors.

STRATEGIC EXPANSION DECISIONS INTO THE INTERNATIONAL MARKETPLACE

The commonly used global expansion methods are summarized in Exhibit 14 and include: franchising, management contracts, mergers and acquisitions, and strategic alliances/partnerships/co-branding. Overseas expansion commonly requires a heavy financial commitment, exposing the value of the firm to risk. Many multi-nationals have adopted a master franchise agreements or a management contract as the preferred modes of entry in order to reduce the financial risk, especially in countries where economic and/or political uncertainties exist. For example, in 1998, Wendy's signed a franchise agreement with Wencol, a restaurant company in Colombia (South America), to build 24 Wendy's outlets in three years.

In 1988, Domino's made commitments to have 300 franchise outlets in international markets. While this has been a preferred mode of entry in the past, in recent years host countries have increasingly demanded financial commitment from multinationals. "Pure" franchise or management contract agreements without some investment required have become rare. McDonald's, for example, signed a 50:50 equal partner agreement with a Pakistani company to build the burger stores in Pakistan.

Joint ventures have become an effective entry mode into a new international market. In 1998, Compass Group Plc. formed a 50:50 joint venture with the Brazilian GR SA, Brazil's largest foodservice provider. It also signed an agreement with T.G.I. Friday's to develop restaurants and Friday's American Bar units across Europe and the Asia Pacific region, as well as in Canada and South Africa (T.G.I. Friday's, Friday's American Bar, Friday's Sports Grill and Italianni's are part of Carlson Restaurants Worldwide Inc. with over 500 outlets in more than 380 cities and 45 countries). In 1996, the Philippino Ayala Corporation signed a joint venture agreement to develop 60 Burger King outlets in the Philippines over 5 years.

Strategic alliances/partnerships/co-branding have increasingly emerged as viable ways to capture niche and/or non-traditional market share. McDonald's

EXHIBIT 14. Strategic Expansion into the International Marketplace as a Competitive Method: Definitions

- **Franchise/Master franchise:** A franchiser grants a licensed privilege to a franchisee to use its brand name, standard production, operation and management systems to do business. A master franchise is a licensed franchisee who has the right to do business in the whole country or an entire region.
- **Management contract:** A management team signs a contract with the owner of the facility and provides its expertise for a management fee. It is a commonly used vehicle in institutional, leisure and sport food management outlets.
- **Strategic alliance/Joint venture/Partnership/Co-branding:** These terms are often used interchangeably. Related or non-related businesses may form a business partnership. Such an agreement can require either financial or non-financial involvement. Co-branding, another form of partnership, links foodservice companies with related or non-related brand names in a joint customer service effort.
- **Merger and Acquisition:** To merge and acquire related or non-related businesses to capture market share or reach financial objectives.

and Nippon Oil formed a strategic alliance to serve McDonald's burgers in gas stations in Japan in 1994 and similarly with Mitsubishi Oil Co., in 1998 to pair gas stations with 20 outlets by 2000. In the US, McDonalds has joined with Chevron and Amoco. Baskin-Robbins and Dunkin' Donuts have also explored the idea of joining with gasoline and convenience stores to establish freestanding islands. Subway has started to serve its sandwiches on Ansett Australia airlines, while McDonald's burgers fly on some United Airlines' flights.

Other joint marketing efforts designed to promote related or non-related businesses include the global alliance established in 1998 between Hard Rock Café and the NBA (National Basketball Association of the USA) to open 10 NBA restaurants around the world in three years. In particular, multinationals have courted the entertainment business. Since the signing in 1997 of an advertising deal between McDonald's and Disney World, the fast-food chain has promoted new Disney film releases through its "Happy Meal" product. Subway has teamed up with the Cartoon Network and KFC joined with Warner Brothers. Pizza Hut has promoted the Playstation product of Sony Computer Entertainment America, Taco Bell has offered Chihuahua toys and promoted the movie "Paulie," and all three Tricon Global brands, Pizza Hut, KFC and Taco Bell, united to promote the Star Wars prequel movie, and Burger King has promoted "The Rugrats Movie."

The multinationals have also entered into non-traditional niche markets, such as supermarkets and convenience stores. McDonald's is now in Wal-Mart Megastores and Little Caesar's is in Kmart. Others have signed co-branding agreements with known hotel companies. For example, Pizza Hut signed agreements with Marriott and Best Western hotels. Compass Group Plc., U.K., has a joint venture agreement with Accor, France, to provide catering operations in Australia, New Zealand and Papua New Guinea.

In entering saturated markets in developed countries, mergers and acquisitions become the most effective competitive method for multinational foodservice firms to gain market share in that country. After earlier acquisitions of Dunkin' Donuts and Baskin-Robbins, Allied Domecq Plc. merged with Togo's Sandwich chain in 1997. In 1998, Bain Capital Inc. took over Domino's Pizza and became the pizza chain's parent company. In 1995, Wendy's acquired the Canadian gourmet coffee and baked goods chain Tim Hortons for $400 million. British Compass Group Plc., acquired Canteen, Flik International Corp., Professional FoodService Management Inc., a part of Service America Corp., Daka Restaurant division and SHRM in North America between 1994-97. These activities helped the company to establish its new division, Compass Group USA, which has become one of the largest foodservice management contractors serving North America.

The Sodexho Alliance acquired the British firm Gardner Merchant and Sweden's Partena in 1995, to become the world's leading provider of food and management services. It merged with the food service and facilities management division which was spun off by Marriott International, Inc. to form Sodexho Marriott Services in 1997, giving the French company has immediate access to 4,800 client relationships. It now dominates the North American food management contract market. In 1998, the company took over US Foodservice Concession Company. Mergers and acquisitions can be expected to continue in mature market settings as firms seek to add value through economies of scale.

INVESTMENT IN TECHNOLOGICAL DEVELOPMENT

Applications of new technology have played an important role in achieving competitive advantage for many multinationals. They have focused on computer-based technology such as the Internet to communicate with customers and intranet communication within individual organizations. Management information systems and point of sale (POS) systems to enhance management and operational capabilities are increasingly used. They have also invested in state-of-the-art production equipment to increase quality and production. Exhibit 15 summarizes these technology related competitive methods.

Not surprisingly, almost all the multinational foodservice companies have invested in customer communications through the Internet. Companies use their web-sites to make product/menu announcements, enhance their image, post job openings, provide distribution/location information, put out news releases, recruit franchisees and seek customer feedback. Domino's Pizza has used the Internet to deliver the company's coupons. KFC, Taco Bell and Outback restaurants have offered e-mail postcards, Breugger's Bagel, KFC and Chi-Chi's have introduced on-line ordering for takeout and delivery. ARAMARK has planned to invest $100 million to standardize its information technology infrastructure and set up Extranets in order to speed the order-placing process for customers.

To enhance management effectiveness and efficiency multinational foodservice companies have continued to invest in the computerization of management systems. Tricon Global has embraced the use of Microsoft's ActiveStore architecture for its Taco Bell, Pizza Hut and KFC stores and has created CD-ROM software containing multimedia-marketing materials for its franchisee and regional managers.

Multinationals have also shifted their attention to strategic solutions for global connectivity and shared data capabilities to improve management infor-

EXHIBIT 15. Investment in Technological Development as a Competitive Method

- **Internet communication with the target market:** Through the use of the World Wide Web (WWW), increase communication with customers about the products, services, job offers, etc.
- **Management information systems:** Installation of management information systems to integrate operational and management functions.
- **Production and service-oriented technology:** Utilization of newly designed production equipment and service software to improve the quality of production and service systems.
- **Training and development systems:** Utilization of computer based, interactive, Web-based training and development programs.

mation systems. Allied-Domecq Retailing USA has developed a cohesive MIS infrastructure to integrate its three brands: Dunkin' Donuts, Baskin-Robbins and Togo's Sandwich Shops. Burger King has implemented integration technology for communication, business applications and analyses. Domino's Pizza Inc. has started to utilize its Geographic Information System (GIS) as a database marketing tool to analyze its service areas, competitors and customers. ARAMARK has spent $15 to $20 million to correct its Y2K computer "bug." McDonald's signed a multi-year contract with Progressive Software to develop a restaurant back office system, which will integrate all the management and operational functions, such as time and attendance, payroll, inventory management, recipe management, labor scheduling, forecasting, purchasing/invoicing, financial reporting and Point of Sale systems. Burger King has invested $3.8 million to develop an exclusive "ComPoSer" point of sale system.

Technology has also been used for improving product and service quality. Domino's Pizza Inc. has created a new "hot-bag" technology to keep pizza hot and crisp during delivery. In 1997, McDonald's initiated the company-wide High-Tech Kitchen–"Made For You" system installation to increases the quality of its products and offer "custom made" sandwiches to its patrons. Burger King has installed order confirmation units to improve order accuracy at its drive-thrus. Both Pizza Hut and Domino's Pizza have continuously used air-impingement technology for their pizza ovens, to reduce cooking time by 50 percent, improve product consistency and trim operating costs. Domino's Pizza has also incorporated caller-identification technology to improve service. Burger King is testing a reusable "Smart Card" with a microchip that allows customers to purchase a preloaded $10 or $20 card and deduct the amount of the food s/he purchases via a scanner and can be re-loaded at restaurants or ATM machines.

Training and development is another exciting dimension opened up of technology. Domino's Pizza has developed interactive, learner-centered programs that make use of CD-ROM technology to guide the teenage employee through the steps in making a pizza. It can be expected that computer-based coaches will soon be guiding employees through all parts of the customer exchange process offering hints on how to better serve the customer. As the labor supply situation continues to challenge operators, they will find an increasing need to invest in technology to handle their training needs.

INTERNAL COMPETENCY DEVELOPMENT

Enhancing internal competencies is another area where multinational companies have invested heavily to achieve competitive advantage. With a continued focus on service quality, many firms have established quality control and employee training programs. They have also stressed the importance of management and operational efficiency along with cost control and increased management capacity. Several companies have restructured to keep pace with the changing business environment. Exhibit 16 summarizes the competitive methods under this heading.

With speed an important factor in customer satisfaction, Baskin-Robbins has trained employees to accelerate service. Domino's puts more focus on improving current products and services rather than on developing new products. Subway encourages consistent friendliness and attention to customer service.

To improve overall quality management and customer service, many companies have established customer service centers where customers can file complaints directly if not satisfied with a store's service. Lone Star, McDonald's, Subway, Burger King, Wendy's and Pizza Hut all have representatives to hear customer complaints. The marketing firm, J.D Powers and Associates, ranked McDonalds No. 1 quick-service restaurant company for speed and accuracy, courtesy, appearance of menu, speaker clarity, design and readability of the drive-through menu and consistency in making correct change.

Employee retention has become a major issue for the multinationals in view of the worldwide shortage of labor. In Europe, Tricon Global has established a reward program, CHAMPS (Cleanliness, Hospitality, Accuracy, Maintenance, Product quality, and Speed), whereby employees win points for good performance that can be redeemed for catalogue merchandise. In Puerto Rico, top-performing workers get cash on the spot. In 1995, McDonald's introduced a broad-based stock ownership program to improve morale and productivity.

In Great Britain, both McDonald's and Burger King have increased their minimum wage level and tried to improve the image of employment in the

EXHIBIT 16. Internal Competency Development as a Competitive Method

- **Quality management:** Procedure and measurement to manage product and service quality.
- **Employee training and retention:** Enhance employees' production and service skill through training programs and retain employees through various incentive programs.
- **Organization restructuring:** Through organization restructuring firms seek to become more efficient.

quick service industry. Both provide a bonus program to their restaurant staff. Pizza Hut has solicited employee suggestions on improving working conditions.

Several companies have restructured their organizations and hired value-adding corporate executives in order to keep growing and remain competitive. McDonald's restructured its U.S. operations and created five divisions in 1997. The company's reorganization effort included maximizing productivity through corporate staff support to its field restaurant operations, cost cutting, hiring a new CEO and also creating a vice president of marketing for global alliances.

Tricon Global Restaurants, formed in 1997 when PepsiCo spun off Pizza Hut, KFC and Taco Bell, has worked hard on inter-brand collaboration by developing a standard "operations platform" for the three brands, crossing-franchising, combining media purchases and consolidating purchasing programs. In 1997, new presidents were hired for all the three brands in order to re-engineer the three companies.

NEW PRODUCT/SERVICE DEVELOPMENT

Multinational foodservice chains have devoted time, money, and research to develop new product/services in order to meet the demands of their customers and the challenges of the competition. These include adaptation to local tastes, for example, mutton burgers at McDonald's in India, rice burgers in Japan, spaghetti on KFC's menu in the Philippines.

Between 1993-98, the introduction of new products, new menu concepts, and themes was the primary competitive method focused upon to boost sales and profits. For example, in 1993, KFC toned down its "fried food" image with new "health-conscious" products like popcorn chicken. In 1995, Pizza Hut, Domino's, and Little Caesar's Pizza all added Buffalo Wings to their menus. KFC introduced its "Tender Roast" and offered a "home meal replacement" concept to appeal to more customers. Taco Bell has launched "After 5," a home-meal replacement concept. In 1996, Burger King added

Croissan'wich for breakfast. In 1997, McDonald's experimented with a branded coffee bar concept. It also upgraded the quality of its menu, added new spice ingredients to its products and a new menu line targeted at the adult market. Pizza Hut added pasta to its menu to attract dinner business and introduced a new "Tripledecker Pizza." Burger King introduced new French fries.

In 1998, Hard Rock Café and Worthington Foods Inc. jointly introduced a new item-"All Natural Veggie Burger." Burger King has introduced three new products with the focus on superior taste of the products. Wendy's launched a children's meal promotion with give-away hockey toys. Sodexho/ Marriott has introduced eight new Crossroads Cuisine's. Compass Group has launched a "Profiles In Good Taste" campaign to meet and exceed its customers' expectations. Subway has started to serve breakfast in Ontario, Canada.

The multinational foodservice companies have stepped up efforts to improve on foodservice safety and sanitation. ARAMARK and McDonalds are participating in the U.S. National Restaurant Association's program called AWARE to train employees in customer and employee safety and food sanitation. Sodexho Marriott has established a 24-point quality and food safety checklist based on Hazard Analysis Critical Control Points (HACCP), a system that identifies, evaluates and controls hazards in the food chain.

Many companies have also looked into the re-design, renovation and re-modeling of the eating environment. In 1995, Pizza Hut, KFC and Taco Bell planned to upgrade and shorten the stores' remodeling cycle from 7 to 5 years, while McDonald's focused on updating the store design. They have positioned and repositioned their brand names to obtain market dominance. In each newly entered country, they compete heavily for the best locations in order to reach their desired target market. Exhibit 17 summarizes the competitive methods employed to improve internal competency.

EFFECTIVE COMMUNICATION WITH THE TARGET MARKET

Multi-million dollar media advertising budgets to communicate with target markets have been spent by multinationals to win competitive advantage over local and other international players (see Exhibit 18), via radio, newspapers, magazines and television. In 1994, KFC planned to spend $80 million in 20 countries on advertising, and selected one company to handle its advertising campaigns for all 75 countries in which it operated. In 1997, McDonald's

EXHIBIT 17. New Product/Service Development as a Competitive Method

- **Modifying the menu to adapt to local needs:** Add local appealing products and ingredients, and eliminate products that do not fit the needs and demands of local patron.
- **New product/concept/theme development:** Develop new food products, concepts or themes to tailor the needs and demands of the customers.
- **Safety and cleanliness:** Stress the importance of safety and sanitation in food preparation and services. Establish procedures and measurement for food sanitation and safety training and practice.
- **Chain and brand name domination:** Maintain the chain and brand name image, capture market share and dominate the marketplace.
- **Facility renovation:** Renovate facility to create an appealing eating environment.

spent $578 million on all measured media, while its arch-rival Burger King spent $423 million on promotions. Wendy's and Polgram Filmed Entertainment spent more than $100 million to promote their products on Fox Family Channel. McDonald's used basketball star Michael Jordan, while Tim Horton's used a Canadian hockey star, to promote their products. Subway invested $140 million in its low-fat sandwich promotions in 1998.

Advertising strategies differ from country to country. For example, in Germany, McDonald's promotes "variety" wheras in Australia it emphasizes reinforcing the American brand image with the use of US film stars. In Europe, Burger King marketing focuses on service and taste. Burger King Restaurants of Canada Inc. ran a specific TV campaign at Chinese Canadians in 1998.

Companies have invested in adverting their products on the World Wide Web (as discussed earlier in the technology section) in order to reach the so-called Generation X.

Database marketing has started to gain momentum among the multinational restaurant firms. In 1994, McDonald's first tested the concept by tracking coupon users. In 1998, Pizza Hut used database technology for its coupon marketing. It cut 40% off the cost and increased coupon redemption by 50%, while increasing profitability from coupons by 21%. Since 1997, McDonald's, Dunkin' Donuts, Long John Silvers and Domino's Pizza have all used the CoolSaving site to place their on-line coupons on the Internet for targeted promotions.

To gain a more favorable brand image via programs oriented to their social responsibilities, multinationals have actively been involved in local community services and charities, sponsored local events and participated in local environmental protection programs. In 1998, the newly merged Sodexho Marriott donated more than $40,000 to the U.S. National Fish and Wildlife Foundation for its education efforts and contributed $40,000 to a homeless shelter in Boston, Massachusetts. It also initiated the "Feeding Our Future"

EXHIBIT 18. Competitive Methods of Target Marketing

- **Heavy advertisement:** Invest millions of dollars on TV commercials and other media.
- **Internet advertising and promotion:** Advertise menu items and service through WWW and promote a chain and brand images.
- **Database marketing:** Computerized database program assists a company's target marketing.
- **Sponsorship, community service, and charity:** Through sponsorships, community services and charity activities, to create a positive public image for the company.
- **Environmental awareness:** Initiate plans or programs that involve environmental protection activities, such as, waste reduction, energy efficiency programs or use of recycled products. Such responsible activities gain very positive publicity.

program to donate labor and food to relieve hunger among children in several American cities.

In 1997, McDonald's launched a nationwide tour to support children's literacy donating 25,000 books to needy children. Pizza Hut sponsored for its 13th year a national reading incentive program called "BOOK IT" in which 22 million students from 895 classrooms (70% of the elementary schools in the U.S.) participated. Subway donated free food to students who participated in arts programs in 1998. ARAMARK provided meals to "Habitat for Humanity" volunteers who built housing for the needy in Houston, Texas. It also provided gifts to abused children during the Christmas holiday season in 1998. McDonald's teamed up with cable TV's Nickelodeon Kids to sponsor a "Big Help" volunteer program.

Many foodservice companies have sponsored their national, regional or local sports teams. McDonald's sponsored the U.S. Olympic team during the 1996 Atlanta Olympic Games. In 1998, Wendy's sponsored the U.S. National Hockey League, while Burger King sponsored the ESPN's Sports-Century programs.

To tackle environmental protection issues, McDonald's initiated a waste reduction plan in 1993, which would reduce 80% of the waste generated by its operations. The company spent $600 million on recycled materials between 1991 to 1993. It has also looked into a new multiple energy efficient technology in its new restaurants. In 1994, KFC, Boston Chicken and Kenny Rogers Roasters all worked on environmental friendly packaging for their roasted chicken.

COMPETITIVE PRICING STRATEGIES

Pricing is a common competitive tactic for the multinational foodservice chain. In the developing world, it is difficult to compete with the pricing strategy of local competitors, partly because local players rarely import ingredients or equipment and their labor costs are often lower than their foreign rivals. Multinationals try to reduce their prices to the point where they can compete with local pricing, but this puts considerable stress on profit margins. It often leads to price wars, even in developed countries. In 1997, Burger King promoted its 99-cent Whopper, while McDonald's invested a sizeable portion of its $600 million marketing budget on promoting its 55-cent burger, though it did not obtain the results it wanted.

Purchase coupons, discounting and give-away food items and toys are common practices among the multinationals. For example, Pizza Hut gives away pizza samples, McDonald's offers free ice cream for its Chinese Combo buyers in China. When Burger King opened its 10,000th store in Sydney in late 1998, it gave away cheeseburgers for 10,000 seconds. It has also run a promotional campaign called "Free FryDay." KFC promoted the Mega Meal as a big value meal in 1994. Pricing will become more difficult as the market continues to mature, and consumers increasingly think of such items as burgers as commodities that should not receive any price premium. Exhibit 19 summarizes the pricing tactics of the multinationals.

CONCLUSIONS

According to strategic management theory, multinationals base their competitive methods on the consistent monitoring and evaluation of the forces driving change in the business domain. There should be a co-alignment of these methods and the complex and dynamic environment for the companies to achieve corporate objectives and goals. Performance, in terms of value added, can only be achieved if firms do invest in competitive methods which take advantage of the opportunities in the environment. Exhibit 20 presents an abbreviated look at the forces which have been identified as driving change in this report. The arrows represent each force that has been identified and ex-

EXHIBIT 19. Pricing Strategies as a Competitive Method

- **Price/value relationship:** Through price discounting achieve market share and perceived added value.
- **Discounting war:** Competition based increasingly on discounting prices to attract and maintain market share.
- **Coupons:** Increasing in use as a discounting tactic.

plained in the first section of this report. The ovals represent the competitive methods that have been identified through an analysis of primary and secondary information of the multinational foodservice companies identified in Exhibits 12 and 13.

In comparing the competitive methods with the forces driving change, several observations become evident. First, the multinationals have concentrated on tried and true methods. Improving efficiency, new product development, marketing and pricing strategies, have been characteristic of the industry for decades. And they are more important today as pressure is on to generate more returns in a stringent capital market setting.

Technology has received some attention, but tends to have been focused internally to achieve greater efficiency and quality. It appears that the industry is slow to address the number of variables developing in a knowledge-based environment. In particular, new methods of pricing, the lifetime value of the customer, the advent of the knowledge worker, the role of inter and info-mediaries appears to be missing from the thinking of executives at present.

The changing labor market conditions pose a significant concern at the moment. As was pointed out in the first half of the report, the growing opportunities to work in the knowledge industries is resulting in many vacancies in the foodservice industry. Yet, there is scant evidence of substantive changes in how labor is recruited and managed. Beyond cosmetic attempts such as producing films or videos to improve the image of the industry, little else has been done. And, it is quite probable that the workforce entrant of today and tomorrow will not buy into marketing efforts such as these. Serious, believable change will be required in the management of human resources if this problem is to be remedied.

Therefore, it can be said that in some ways a match is occurring. The international expansion, new products and services development, efficiency and improving competency programs, better marketing and pricing strategies no doubt reflect the pressure of the capital markets for growth in value. It can be anticipated that this will continue. To a certain extent, it appears that the industry is also responding to the technological advances that are taking place but at a delayed rate compared to other industries. The real concerns may lie in the lack of identifiable and measurable competitive methods addressing the well being and human resource forces. For the aggressive firm seeking to achieve long-term success in the future, investments in competitive methods directed at these areas will be essential. The opportunities are extensive.

EXHIBIT 20. A Matching of the Forces Driving Change and the Industry's Competitive Methods, the Challenge Will Be Tomorrow

Forces driving change

Globalization and economic change

- A global capital market system is demanding more
- Valuing the enterprise based upon future earning, not asset value
- An increasing number of investor choices in a global market is increasing competition for available capital
- Demands for improving the response time to competitive threats
- Strategic financial skills are growing in importance

A knowledge based environment

- The accelerating rate of change
- A new model of the customer exchange process
- The need for more knowledge about the customers' individual needs
- Customers gaining more power
- The transparency of the foodservice enterprise
- Market pricing controlled by the customer, not the supplier
- The rise of infomediaries
- Adding value through intangibles
- A new industry and enterprise structure
- Your work is never complete
- The lifetime value of the customer
- Cyber marketing

The future of labor and the new labor force entrant

- The growing shortage of qualified employees
- The changing demographics of the workforce
- New leadership skill requirements
- Investing in the human resource

Well being

- The concern for health and food safety
- The globalization of the food chain
- Concern for personal safety and fear of violence
- The spread of disease
- Aging and personal well being
- Engineering in the food chain
- Consumer and government leadership in improving well being

Exhibit 20 (continued)

Competitive methods of multinationals

Strategic expansion into the global marketplace

*Franchise * Management contract
* Strategic alliances * Merger and acquisition

Investment in technological development as a competitive method

- Internet communication with the target market
- Management information systems
- Production and service oriented technology
- Training and development systems

Internal competency development as a competitive method

- Quality management
- Employee training and retention
- Organization restructuring

New product/service development as a competitive method

- Modifying the menu to adapt to local needs
- New product/concept/theme development
- Safety and cleanliness
- Chain and brand name domination
- Facility renovation

Effective communication to the target market

- Heavy advertisement in TV commercials and other media
- Internet advertising and promotion
- Database marketing
- Sponsorship, community service and charity

Pricing strategies as a competitive method

- Price/value relationship through price discounting
- Discounting war
- Coupons

REFERENCES

Anon. (1994). *Employment Outlook: 1994-2005 Job Quality and other aspects of Projected Employment Growth*, U.S.Department of Labor, Washington, D.C.

Anon. (1997). International Labor Organization, *Labor Force Statistics.*

Anon. (1997a). Singapore Straits Times, May.

Anon. (1996). WHO, Fact Sheet No. 124 (Revised); November.

Anon. (1999). The end of privacy. *The Economist*, May 1, pp. 23.

Connolly, D. J., Olsen, M. & Moore, R. G. (1998, August). The Internet as a distribution channel. *Cornell Hotel and Restaurant Administration Quarterly*, 39 (4), 42-54.

Davis, S. & Meyer, C. (1998). *Blur, the speed of change in the connected economy.* Addison Wesley, Reading, MA.

Dev, C. & Olsen, M.D. (1999). *Executive summary on the IH&RA Think Tanks on Marketing.* International Hotel and Restaurant Association, Paris, France.

Gates, Bill (with Myhrvold, Nathan and Rinearson, Peter). (1995). *The road ahead.* New York: Viking Press.

Linden, E. (1998). *The Future in Plain Sight.* Simon and Schuster, New York, NY.

Olsen, M.D., & Pizam, A. (1998). *Executive summary of the IH&RA Think Tank on Safety and Security*, International Hotel and Restaurant Association, Paris, France.

Olsen, M.D. & Sharma, A. (1998). *Forces driving change in the casual theme restaurant industry.* International Hotel and Restaurant Association. Paris, France.

Olsen, M. & Connolly, D. (1999). *Executive summary of the IH&RA Think Tanks on Technology*, International Hotel and Restaurant Association, Paris, France.

Olsen, M.D. (1999). Macro forces driving change into the new millennium–major challenges for the hospitality manage. *International Journal of Hospitality Management.* Special Issue.

Parkes, C. & Thal, P. (1999). The box in the living room comes to life. *Financial Times*, Weekend Edition, May 8-9, pp.7.

Price, C. & Kehoe, L. (1999). AOL to provide interactive television service. *Financial Times*, May 12, pp.18.

Schein, E. (1996). Leadership and Organizational Culture in *The Leader of the Future*, (eds. Hesselbein F., Goldsmith, M & Beckhard, R.) Jossey Bass Inc. New York, NY.

Schiesel, S. (1999). AT&T is thinking (and spending) big on interactive TV. *The International Herald Tribune*, May 8-9, pp. 11.

Soros, G. (1998). *The Crisis of Global Capitalism.* PublicAffairs, Perseus Books Group, New York, NY.

Tapscott, D. (1996). *The digital economy: Promise and peril in the age of networked intelligence.* New York: McGraw-Hill.

Thurow, L. (1997). Changing the nature of capitalism, in *Rethinking the future*, Gibson, R. ed. Nicholas Brealey Publishing.

Food Safety: A Public Crime

Kim Harris

SUMMARY. Food safety in the foodservice industry is an issue that is a priority at the federal level. In 1999 the National Food Safety System formed the Coordinating Committee to combat the problem. Since the formation of this group, outreach programs have extended globally to assist other countries with this issue. According to Susan Alpert, national leader on the Food Safety Initiative, many states are experiencing the same food safety challenges. Programs and standards have been established for each state; however, the problems of state implementation of a high quality food safety program is becoming a growing issue. This study investigated the importance of food safety training for food handlers and compares the current training and inspections for those who manage foodservice establishments. Data is presented from one high tourism state serving as an example of the violations to and the training given on food safety. The study presents the food safety training program required, inspection form used, and categories of violations most often experienced. Issues related to combating food safety are presented as well as suggestions for improving food safety training and inspection process. *[Article copies available for a fee from The Haworth Document Delivery Service: 1-800-HAWORTH. E-mail address: <getinfo@haworthpressinc.com> Website: <http://www.HaworthPress.com> © 2001 by The Haworth Press, Inc. All rights reserved.]*

Kim Harris, PhD, is Associate Professor, Department of Hospitality Management at Florida State University, Tallahassee, FL.

[Haworth co-indexing entry note]: "Food Safety: A Public Crime." Harris, Kim. Co-published simultaneously in *Journal of Restaurant & Foodservice Marketing* (The Haworth Hospitality Press, an imprint of The Haworth Press, Inc.) Vol. 4, No. 3, 2001, pp. 35-64; and: *Quick Service Restaurants, Franchising and Multi-Unit Chain Management* (ed: H. G. Parsa and Francis A. Kwansa) The Haworth Hospitality Press, an imprint of The Haworth Press, Inc., 2001, pp. 35-64. Single or multiple copies of this article are available for a fee from The Haworth Document Delivery Service [1-800-HAWORTH, 9:00 a.m. - 5:00 p.m. (EST). E-mail address: getinfo@haworthpressinc.com].

KEYWORDS. Sanitation, food safety, food safety inspections, training

INTRODUCTION

According to the National Institute of Allergy and Infectious Diseases (2000), foodborne illnesses cost the nation $6 billion dollars annually. Food safety is considered a public responsibility of the foodservice industry and the government at large. The federal government established the Food Safety Initiative in 1999 to support the industry by providing funds for education, research, production and handling guidelines, and a variety of support services to assist the industry and consumers with safe food handling.

According to Dennis Baker, Associate Commissioner for Regulatory Affairs of the FDA, it is the mission of the United States government to supply information and support programs for those seeking to provide safe food products to consumers. These agencies include, but are not limited to, the United States Department of Agriculture (USDA), Food and Drug Administration (FDA), and Department of Health at the national and state levels. These agencies provide one or several services such as food safety inspections, education and training for state officials, support staff, and outreach programs for the public at large. Each agency's quest is the same; to provide food safety information and expert referrals to those required to implement food safety guidelines or to these simply interested in the topic of food safety.

The agencies mentioned are further supported by organizations and other experts serving as independent consultants, food production companies, education and training organizations, or professional hospitality organizations such as the National Restaurant Association's Educational Foundation (NRA-EF), the American Hotel and Motel Association (AHMA), and a number of state organizations representing the hospitality industry. There are numerous entities available from which to gain the history as well as cutting edge information on food safety. The FDA's Office of Regulatory Affairs (ORA) is responsible for providing regulated food products from the point of "farm to table" (U.S. FDA, 1999, http://www.vm.cfsan.fda.gov/~dms/fsirp995.html), and according to Marjorie David, education team leader, "Not only must we provide people with information, we must do it in a manner that results in changing unsafe food handling behaviors to safe food handling behaviors . . . constant reinforcement of educational messages is important to sustaining behavior change."

Food safety programs are targeted at producers, retailers, and government entities providing food products. The food service industry and inspection programs are custom-designed to address the particular products, service, equipment, and personnel within these categories. This paper focuses on food service operations and the food handlers who work in them.

Programs for Foodservice Facility Managers

Each state is required to have a food safety training course offered to food service facility managers (US Food and Drug Administration, 2000). Hourly wage employees serving as food handlers may obtain training in their place of employment or formal training and certification on a voluntary basis. In many states, foodservice managers are expected to train all staff who may come into contact with food, food equipment, or food contact surfaces in safe food handling practices. To date, only one state, California, requires food handlers to obtain state mandated training and certification in the area of food handling and safety. For managers responsible for training foodhandlers in states where external training and certification for foodhandlers is not required, major challenges are ever present. Time to deliver adequate training, access to high quality training programs, adequate analysis of trainee learning levels, programs offered in a variety of languages, high turnover and costs associated with training a transient workforce, specific information related to a food facility's menu, and variety of appropriate training equipment, are some of the factors confronting many food service-related training efforts (Chung & Hoffman, 1998; Deery & Shaw, 1999; Enz & Potter, 1998; Harris & Bonn, 2000).

Food Safety Requirements

Requirements for food safety training vary by state. Some states require the food service facility operator to be trained and certified by state food safety experts. These experts can be employed by federal, state, or private agencies such as the Department of Business and Professional Regulations (DBPR), U.S. Department of Agriculture (County Extension Service), Environmental Protection Agency (EPA), Centers for Disease Control (CDC), state colleges and universities, trainers representing the National Restaurant Association (NRA), or trainers from independent small businesses. The state regulatory agency enforces the food safety codes and regulations as developed by the FDA and other federal agencies, and does this by providing training program and inspection services.

Federal and state agencies have informative web sites that provide training materials, food safety initiatives and regulations, and resources. For example, *http://www.hospitality.org* represents the state of Florida's governmental web site and a variety of food safety programs, experts, and support agencies.

The food safety program in the state of Florida was selected for this study as an example of a state program. This program offers food safety training at no charge for a half-day seminar and printed materials if the trainee is a licensed food facility operator. For those seeking training and do not possess a food facility operators license, a $6 fee is charged. The fee of $30 is charged for walk-ins to the certification exam, which is administered by a state university. Food safety certification is not required of food handlers who are not considered management. However, training in food safety, which can be delivered by the foodservice facility manager, is required (Florida DBPR, 2000); yet, testing for specific competency levels on the topic of food safety are not required. Certification is only required for one manager at the particular site and this manager does not have to be in the facility at all times. As long as one manager has been trained and certified in food safety, the licensee of the food facility meets state food safety certification requirements.

Food Safety Training Programs

Food safety training programs are in abundance from a variety of suppliers. These include, but are not limited to, the National Restaurant Association, state restaurant associations, private businesses specializing in food and hospitality-related training, private consultants, state agencies and the federal government. Most programs provide training in some or all core areas that may include biological, chemical, and physical hazards to food safety, personnel uniforms and hygiene, facilities and equipment, foodborne illnesses, microorganisms of most concern, hazard analysis and critical control points (HACCP), pest control, safety, signage and regulations, developing a food safety program, and time and temperature controls. As food codes change, organizations and agencies update programs accordingly, which on an average, has been approximately every 2-5 years depending on the emerging pathogen or other findings (FDA Food Code 1997, 1999). Training categories were developed to properly and completely inform food handlers of the criticality of food safety from farm to table and how to manage food safety in a commercial environment (Progress and Perspective, Food Safety Initiative, Annual Report, April 2000).

While prevention of foodborne illnesses is of primary concern, significant emphasis should be given to controlling habits and conditions that contribute to contamination and spoilage of food products and the cleanliness and sanitation of facilities and equipment. This is because food handlers and consumers are the number one cause of all foodborne-illnesses (Loken, 1995).

Two programs compared in this study are those offered by the NRA's Educational Foundation (EF) and a program most widely used for managers in

Florida. According to Charlotte Eoff (Administrative Assistant, Florida's Hospitality Education Program), the food safety program offered by the NRA-EF is now reciprocally accepted in Florida; however, the state's food safety training program is often the program of choice. This preference may be due to several reasons. The state program meets the guidelines required by the Department of Business and Professional Regulations (DBPR) of the state of Florida, and it is free to trainees. A $6 fee is charged on each licensee to cover the costs of the program for all trainees. As for duration of training, the session is three hours in length and testing for certification follows at a later date. Certification requires trainees to be tested by another entity. Testing for certification varies in cost depending on whether the trainee is a licensed foodservice facility manager/operator and whether the training course was taken. The maximum certification exam cost per person is $30.

In comparison, the NRA-EF ServSafe training program is at least eight hours long, and in some cases, is expanded to a two-day session. This time also includes testing for certification. Managers certified in the state program must renew their certification every five years, and those certified through *ServSafe* program have no renewal specifications. However, the EF suggests that those certified must follow the certification mandates of the state in which they are managing a food service facility. The EF also offers retraining, retesting, and free information updates upon request or at their website at http://www.edfound.org.

With regard to cost, the EF certification program, when offered by a contracted trainer, varies in cost ranging from $500 to $1,500 for a group rate, which includes the cost of testing and certification. Those trainees who achieve a score of 90% or higher (trainer status) on the certification exam can purchase training materials for under $500, depending on the training package purchased, and administer training in their facility. Trainer status can greatly reduce the cost of certification training for an organization employing certified managers who have achieved the appropriate competency level. An in-house trainer certified to deliver the *ServSafe* course can customize the food safety program for his or her company and deliver the program as many times as needed to employees, thereby eliminating the costs of a consultant.

The state of Florida's program materials are printed by the state in the form of a front and back copied, stapled handout, entitled *Food Manager's Workshop Manual*. There are significant differences between the EF's and Florida's programs in the amount of content, time given to subject areas, quality of information presented, and materials used in presentations and those given to students.

Goal of Florida's Food Safety Program

The goal of the state program is to train managers in food safety and provide information to foodservice providers on safe handling of food, safety in the facility, and inspection regulations. The training topics correlate with the inspection form categories, enabling managers to maintain food facilities to meet state mandated sanitation and safety compliance expectations. Inspectors are also trained accordingly and, in addition, check for posted regulatory signage and information guidelines, sanitation certifications of managers, and readily available state-mandated educational information.

The state agency responsible for the Food Safety program employs 216 inspectors for seven districts and they are responsible for inspecting 66,338 restaurant and lodging properties. This agency is also responsible for elevator safety, licensure, compliance, program policy, and hospitality education. In 1998/99 it operated on a budget of $17.6 million and employed 331 full-time employees. A portion of this budget and personnel is dedicated to food safety training, licensing food service facilities, ensuring compliance with state rules and regulations, and conducting inspections to monitor sanitation and safety.

Time to Train, and Quality and Quantity of Training Materials

Currently, the training program used by Florida comprises a 20-item pretest and a 50-item practice test. The training content, mostly double-spaced and a variety of graphics and tables, is 10 pages in length (5 pages, 10 pages of content on front and back). Of this content, one-third of a page is devoted to Hazard Analysis and Critical Control Points (HACCP) and is the only information appearing on the page. The remaining nine pages present information on each of the following topics: (1) types of contaminants, (2) common bacterial illnesses, potentially hazardous foods, (3) managing employee health and prohibiting bare hand contact with ready-to-eat foods, (4) time and temperature controls, ware washing and sanitizing, (5) food code requirements, and (6) required signs, posting, and plans.

The last page is a list of the seven steps to avoid foodborne illness, and again, is the only information appearing on the page. There is little or no information on food contact surfaces, utensils, and equipment; personnel; how to properly thaw or reheat foods (although temperatures and time is given), how to conduct a HACCP program, receiving and storing foods, how to use a variety of cleaning and sanitizing chemicals, safety, or the proper way to calibrate or use a thermometer. Training as to how to wash hands, restrict hair, care for uniforms, or personal hygiene is not included. The program is written on an approximately 6th to 8th grade reading level, which is common to training programs targeted to managerial staff (Noe, Hollenbeck, Gerhart, & Wright, 2000).

Florida's Hospitality Education Program (HEP) has three trainers. Each uses a short manual, shares operational examples and presents illustrations during training. Discussion and event sharing among trainees during the three-hour training session is also prevalent. After the training session, trainees take a practice test before taking the certification exam. The University of Southern Florida administers the certification exam, which is required of at least one on-duty manager in a designated foodservice facility. The passing rate on the exam for those trained by the state using this program is not known; however, the current rate of certification is 2,007 per month and growing. The number of food managers certified using the state program totaled 15,000 in 1997; 18,000 in 1998, and 22,000 in 1999 (Charlotte Eoff, HEP).

The *ServSafe* program, offered by NRA's EF, is available in several formats. The most popular is the text and accompanying compact disc titled, *Serving Safe Food: A Practical Approach to Food Safety*. This program is, at a minimum, taught in an eight-hour time frame and, in some instances, requires up to sixteen hours of instruction. The elongated instructional program is an option if the client chooses laboratory exercises, games, or other interactive instruction in addition to the standard program. Materials can include the use of a compact disc slide show, posters, handouts, case studies, role plays, question and answer sessions, puzzles, videos, games, and word-find activities. Videos are an optional tool that can be used as stand-alone training or to supplement the text and CD course content. Small group problem-solving exercises, individual problem-solving assignments, and graphics displaying problem scenes requiring the trainee to pick out all of the problem situations are also available to trainers to use. Practice questions are also available for each topic.

The topics in the NRA-EF program include segments on food safety, food safety hazards, safe food handling, HACCP, adapting HACCP principles for an operation, purchasing and receiving food, storing food safely, keeping food safe during preparation and service, sanitary facilities and equipment, cleaning and sanitizing, developing an integrated pest management program, and regulatory agencies and inspections.

Also included are appendices, a glossary, list of terms, answer keys for all practice questions, more information for further reading, a directory of food safety resources and an index. Certification exams are proctored by NRA approved training officials. Exams are professionally written, measured, and evaluated. According to the Director of Certification Programs for the NRA's EF, the passing rate on the certification exam for those using *ServSafe* is 85-89% depending on the test administered. This passing rate serves as a national average. The number of foodservice managers holding ServSafe certification is not known.

RESEARCH METHODOLOGY

This study investigated the number and nature of violations given to food service facilities in hotels, restaurants, and snack bars in Florida, one of the most tourist-visited states in the United States. The information was then compared to the information received in the training program designed to reduce and eliminate such violations. Food facility inspection data obtained from the state's Divisions of Hotel and Restaurants, from July 1994 to July of 1999 were analyzed. Data was then transferred onto a compact disc (CD) to be used on a personal computer (PC). The statistical package SPSS Professional Package, Version 9 for IBM compatibles, was used to analyze the data.

Frequencies, descriptive statistics, and Analysis of Variance (ANOVA) tests were used to present the results. The number of cases analyzed total 466,295 ($N = 466,295$). The accepted level of significance was based on an alpha level of $p < .05$. The ANOVA test was applied to study the differences in violations among districts and a year by year basis. These differences provided a basis for studying the geographic areas receiving the most violations as well as the years to investigate the regions with the most food safety problems. Also investigated were the training improvements made in response to violations, food or safety code changes, and support offered to reduce problems. Managers trained and certified in food safety by the state program, on the average, achieved a 99% passing score.

Testing categories of the exam were compared to the food safety violations to study the quality and amount of information given as related to the types of violations incurred. This was done to investigate training as compared to the areas of inspection (and thus required food safety knowledge) according to the state guidelines. Frequencies were recorded to calculate the number of standard violations and critical violations as a total, as well as to determine the categories in which the violations occurred to determine the areas considered problematic. The focus of this study is on the non-critical violations. Further studies would include critical violations. The following hypotheses were developed for this study:

H1: There are no significant differences in the number of violations occurring among inspection categories on the inspection form used by state inspection officials.

H2: There is no significant difference in the number of food inspection violations among the seven districts in Florida.

H3: There is no difference in the type and number of violations when comparing one year to another.

H4: The food safety training program *ServSafe* does not compare equally to the food safety program offered by the State of Florida (in amount of content, training activities, passing rate, and testing procedures).

RESULTS AND DISCUSSION

The results of the data are first presented as frequencies of violations, both critical and standard. The inspection form used for inspections is recreated in the table to reflect the categories established by the state statutes. Total violations for each line item are given. Note that multiple violations can occur in the category, as each category has multiple areas of inspection.

Table 1 identifies the number of violations that are considered "non-critical" or standard as well as a column for critical violations. A non-critical violation includes those that receive a non-acceptable check on the inspection form and suggestions by the inspector as to what should be done to correct the problem before the next inspection. Critical violations may result in a shorter time given to correct the problem with re-inspection to follow. Fines, temporary or permanent closure can also result from critical violations, depending on state reprimands for violations. Inspections are unannounced and they occur on a regular basis several times per year.

Year of Inspection and Violations Incurred

Analysis of Variance (ANOVA) results for differences in districts showed that the two largest cities in the state have the most variety and number of food facilities, suppliers, and number of employees, and thus differing levels of learning and language differences. These districts also receive the highest number of inspection violations. Since this finding is not profound, it does not appear as a table.

Tables 2 through 12 identify the significant differences reflected in an ANOVA test using Year of Inspection as the factor (independent variable). The years reflect those in which the most violations incurred as well as type of violation (category). While ANOVA tests also produced a large number of critical violation differences among the years, due to the length limitations of

TABLE 1. Food Inspection Violations in One High Tourist-Visited State–1994-1999

Category on Inspection Form	Violations	Critical Violations
Source		
1a. Source, Approved		
1b. Source, Wholesome, sound condition		
Total Violations for 1 a-b	0	8,578
2. Source, Original container, properly labeled	25,311	0
Total Violations for Source Category	**25,311**	**8,578**
Temperature		
3a. Temp., Cold food		
3b. Temp., Hot food		
3c. Temp., Foods properly cooked/reheated		
3d. Temp., Foods properly cooled		
Total Violations for 3 a-d	0	78,878
4. Temp., Facilities to maintain proper temp.	11	9,304
5. Temp., Thermometers provided and conspicuously placed	88,537	0
6. Temp., Potentially hazardous food properly thawed	14,767	0
7. Temp., Unwrapped or potentially hazardous food not reserved	0	824
8a. Temp., Food protection during storage, prep., display, service, transportation		
8b. Temp., Cross-contamination, equipment, personnel, storage		
8c. Temp., Potential for cross-contamination, storage practices, damaged food segregated		
Total Violations for 8 a-c	126,856	0
9. Temp., Foods handled with minimum contact	23,833	0
10. Temp., In use food dispensing utensils properly stored	57,485	0
Total Violations for Temperature Category	**311,489**	**89,006**
Personnel		
11. Personnel, Personnel with infections restricted	0	455
12a. Personnel, Hands washed and clean, good hygienic practices		

TABLE 1 (continued)

Category on Inspection Form	Violations	Critical Violations
12b. Personnel, Proper hygienic practices, eating, drinking, smoking		
Total Violations for 12a-b	0	62,207
13. Personnel, Clean clothes, hair restraints	10,360	0
Total Violations for Personnel Category	**10,360**	**62,662**
Food Equipment and Utensils		
14. FEU, *Food* contact surfaces designed, constructed, maintained, installed, located	88,203	0
15. FEU, *Non-food* contact surfaces designed, constructed, maintained, installed, located	91,407	0
16. FEU, Dishwashing facilities designed, constructed, operated	32,656	0
17. FEU, Thermometers, gauges, test kits provided	61,037	0
18. FEU, Pre-flushed, scraped, soaked	3,585	0
19. FEU, Wash, rinse, water clean, proper temperature	3,225	0
20a. FEU, Sanitizer concentration		
20b. FEU, Sanitizing temperature		
Total Violations for 20 a-b	0	36,742
21. FEU, Wiping cloths clean, used properly, stored	64,439	0
22. FEU, Food contact surfaces of equipment and utensils clean	178,463	0
23. FEU, Non-food contact surfaces clean	133,645	0
24. FEU. Storage/handling of clean equipment, utensils	62,803	0
Total Violations for Food, Equipment, and Utensils Category	**719,463**	**36,742**
Single Service Articles		
25. SSA, Single service items properly stored, handled, dispensed	45,573	0
26. SSA, Single service articles not reused	2,977	0
Total Violations for Single Service Articles Category	**48,550**	0
Water and Sewerage/Plumbing		
27. WSP, Water source safe, hot and cold under pressure	0	23,995
28. WSP, Sewage and waste water disposed properly	0	14,935

TABLE 1 (continued)

Category on Inspection Form	Violations	Critical Violations
29. WSP, Plumbing installed and maintained	55,921	0
30. WSP, Cross-connection, back siphonage, backflow	0	26,860
Total Violations for Water, Sewerage/Plumbing Category	**55,921**	**38,930**
Toilet and Handwashing Facilities		
31. THF, Toilet and handwashing facilities, number, convenient, accessible, designed, installed	0	46,066
32. THF, Restrooms with self closing doors, fixtures operate properly, facility clean, supplied with hand soap, disposable towels or hand drying devices, tissue, covered waste receptacles	141,659	0
Total Violations for Toilet and Handwashing Facilities Category	**141,659**	**46,066**
Garbage and Refuse Disposal		
33. GRD, Containers covered, adequate number, insect and rodent proof, emptied at proper intervals, clean	64,409	0
34. GRD, Outside storage area clean, enclosure properly constructed	27,983	0
Total Violations for Garbage and Refuse Category	**92,392**	**0**
Insect and Rodent Control		
35a. IRC, Presence of insects/rodents, animals prohibited		
35b. IRC, Outer openings protected from insects, rodent proof		
Total Violations for 35 a-b	0	79,337
Total Violations for Insect and Rodent Control Category	**0**	**79,337**
Floors, Walls, and Ceilings		
36. FWC, Floors properly constructed, clean, drained, covered	97,294	0
37. FWC, Walls, ceiling, and attached equipment, constructed, clean	133,098	0
38. FWC, Lighting provided as required, fixtures shielded	76,004	1
39. FWC, Rooms and equipment vented as required	8,587	0

TABLE 1 (continued)

Category on Inspection Form	Violations	Critical Violations
Total Violations for Floors, Walls, and Ceilings Category	**314,983**	**1**
Other Areas		
40. OA, Employee lockers provided and used, clean	10,623	0
41a. OA, Toxic items properly stored		
41b. OA, Toxic items labeled and used properly		
Total Violations for 41a-b	0	65,697
42. OA, Premises maintained, free of litter, unnecessary articles, cleaning and maintenance equipment properly stored, kitchen restricted to authorized personnel	47,464	0
43. OA, Complete separation from living/sleeping area, laundry	739	0
44. OA, Clean and soiled linen segregated and properly stored	5,534	0
Total Violations for Other Areas Category	**64,360**	**0**
Safety		
45. Safety, Fire extinguishers proper and sufficient	0	104,065
46. Safety, Exiting system adequate, good repair	0	63,852
47. Safety, Electrical wiring adequate, good repair	0	47,073
48. Safety, Gas appliances properly installed, maintained	0	5,793
49. Safety, Flammable/combustible materials properly stored	0	5,306
Total Violations for Safety Category	**0**	**226,089**
General		
50. General, Current license properly displayed	0	23,125
51. General, Other conditions sanitary and safe operation	69,845	0
52. General, False/misleading statement published or advertised relating to food/beverage	452	0
53a. General, food management certification valid		
53b. General, employee training valid		

TABLE 1 (continued)

Category on Inspection Form	Violations	Critical Violations
Total Violations for 53 a-b	0	76,741
54. General, State Clean Indoor Air Act	11,305	0
55. General, Automatic gratuity notice	190	0
56. General, Copy of Chapter 509, State Statutes, available	10,974	0
Total Violations for General Category	**92,766**	**99,866**
Information		
57. Information, Hospitality Education Program information provided	8,001	0
58. Information, yes, smoke free	5,023	0
Total Violations for Information Category	**13,024**	**0**

TABLE 2. Analysis of Variance for Violation 2 on the Inspection Form

Source: Original Container: Properly Labeled

Source	df	Sum of Squares	Mean Squares	F Ratio	Prob.
Between Groups	6	24.0622	4.0104	90.0787	.0000
Within Groups	541,402	24,103.64	.0445		
Total	541,408	241,127.70			

TABLE 3. Analysis of Variance for Violation 4 on the Inspection Form

PHF Temperature Control: Facilities to Maintain Product Temperatures

Source	df	Sum of Squares	Mean Squares	F Ratio	Prob.
Between Groups	6	.0006	.0001	3.8469	.0008
Within Groups	541,402	12.9991	.0000		
Total	541,408	12.9997			

TABLE 4. Analysis of Variance for Violation 13 on the Inspection Form

Personnel: Clean Clothes, Hair Restraints

Source	df	Sum of Squares	Mean Squares	F Ratio	Prob.
Between Groups	6	159.1067	26.5178	646.8528	.0000
Within Groups	541,402	22,194.8143	.0410		
Total	541,408	22,353.9210			

TABLE 5. Analysis of Variance for Violation 16 on the Inspection Form

Food Equipment and Utensils: Dishwashing Facilities Designed, Constructed, Operated (Washing)

Source	df	Sum of Squares	Mean Squares	F Ratio	Prob.
Between Groups	6	107.1067	17.9172	317.2260	.0000
Within Groups	541,402	30,578.7951	.0565		
Total	541,408	30,686.2981			

TABLE 6. Analysis of Variance for Violation 25 on the Inspection Form

Single Service Articles: Single Service Items Properly Stored, Handled, Dispensed

Source	df	Sum of Squares	Mean Squares	F Ratio	Prob.
Between Groups	6	110.5674	18.4279	317.5920	.0000
Within Groups	541,402	45,851.4064	.0847		
Total	541,408	45,961.9738			

TABLE 7. Analysis of Variance for Violation 30 on the Inspection Form

Single Service Articles: Single Service Items Properly Stored, Handled, Dispensed

Source	df	Sum of Squares	Mean Squares	F Ratio	Prob.
Between Groups	6	6.4431	1.0739	78.6550	.0000
Within Groups	541,402	7,391.6059	.0137		
Total	541,408	7,398.0490			

TABLE 8. Analysis of Variance for Violation 31 on the Inspection Form

Toilet and Handwashing Facilities: Toilet and Handwashing Facilities, Number, Convenient, Accessible, Designed, Installed

Source	df	Sum of Squares	Mean Squares	F Ratio	Prob.
Between Groups	6	5.8570	.9762	59.5524	.0000
Within Groups	541,402	8,874.5010	.0164		
Total	541,408	8,880.3580			

TABLE 9. Analysis of Variance for Violation 33 on the Inspection Form

Garbage and Refuse Disposal: Containers Covered, Adequate Number, Insect and Rodent Proof, Emptied at Proper Interrvals, Clean

Source	df	Sum of Squares	Mean Squares	F Ratio	Prob.
Between Groups	6	132.8341	22.1390	211.7173	.0000
Within Groups	541,402	56,613.7176	.1046		
Total	541,408	56,746.5502			

TABLE 10. Analysis of Variance for Violation 34 on the Inspection Form

Garbage and Refuse Disposal: Containers Covered, Adequate Number, Insect and Rodent Proof, Emptied at Proper Intrervals, Clean

Source	df	Sum of Squares	Mean Squares	F Ratio	Prob.
Between Groups	6	.0045	.0007	4.9606	.0000
Within Groups	541,402	80.9834	.0001		
Total	541,408	80.9879			

TABLE 11. Analysis of Variance for Violation 36 on the Inspection Form

Floors, Walls, and Ceilings: Floors Properly Constructed, Clean, Drained, Covered

Source	df	Sum of Squares	Mean Squares	F Ratio	Prob.
Between Groups	6	429.2749	71.5458	463.1708	.0000
Within Groups	541,402	83,630.1643	.1545		
Total	541,408	84,059.4392			

TABLE 12. Analysis of Variance for Violation 41 on the Inspection Form

Insect and Rodent Control: Presence of Insects/Rodents, Animals Prohibited

Source	df	Sum of Squares	Mean Squares	F Ratio	F Prob.
Between Groups	6	.3350	.0558	37.1639	.0000
Within Groups	541,402	813.4381	.0015		
Total	541,408	813.7732			

this paper, these were not included. There are violations in each category except categories 9 and 12. A random selection of non-critical violations from the remaining 12 categories is presented.

Each table represents an ANOVA result considered significant from each of the main categories on the inspection form (N = 14); however, the list of differences in this study represents only sampling of the differences on the factor *Year of Inspection* for non-critical violations. Table 1 presents a review of the critical violations.

The number of violations relates to the number of food facilities such as those in restaurants, hotels, motels, and bed and breakfasts over a six-year period (N = 466,295). The post-hoc test, Modified LSD Bonferroni, was used to identify differences based on means (Howell, 1987), using a significance level of .05.

Identifying districts is most helpful in disseminating the results, as it is to be assumed that low population and low tourist areas would have significantly different results in food safety violations compared to highly populated and high tourist-traffic areas. As large numbers of customers patronize foodservice outlets more facilities are opened and a larger number of distributors supply a wider variety of products. This increased traffic and availability results in a higher probability of food violations due to the increased handling, storage requirements, service demands, and training required to properly handle the variations of product.

Over the period of study, 1994 and 1998 experienced significant changes in food safety training, inspection procedures, and codes in Florida. Due to the number of E-Coli and Salmonella outbreaks in this state in 1994 prompting new and improved training, new food codes in 1997 and 1999 and compliance guidelines were required (FDA, 1999). This change in code prompted updated training programs, which should have had an impact on inspections and thus a positive change in the number of violations.

Table 2 reflects the significant differences among the violations for category 1, *Source*. With an alpha level of 0.05, the effect of *Year of Inspection* was statistically significant, F (5; 466,278) = 134.011, p = .000. Bonferroni's post hoc test shows a significant difference at the $p < 0.05$ level between the following years:

Differences When Base Year Data Was Compared to Data in Following Years

Base Year	Compared to Year	Compared to Year	Compared to Year	Compared to Year	Compared to Year
1994	1998	1999			
1995	1998	1999			
1996	1997	1998	1999		
1997	1996	1998	1999		
1998	1994	1995	1996	1997	
1999	1994	1995	1996	1997	1998

Table 3 reflects the significant differences among the violations for category 2, *Potentially Hazardous Food Temperature Control*. With an alpha level of 0.05, the effect of year of inspection was statistically significant, F (5; 466,278) = 29.568, p = .000. Bonferroni's post hoc test shows a significant difference at the $p < 0.05$ level between the following years:

Differences When Base Year Data Was Compared to Data in Following Years

Base Year	Compared to Year	Compared to Year	Compared to Year	Compared to Year	Compared to Year
1994	1996	1999			
1995	1997	1998	1999		
1996	1994	1997	1998	1999	
1997	1995	1996	1999		
1998	1995	1996	1999		
1999	1994	1995	1996	1997	1998

Table 4 reflects the significant differences among the violations for category 3, *Personnel*. With an alpha level of 0.05, the effect of year of inspection was statistically significant, F (5; 466,278) = 246.512, p = .000. Bonferroni's post hoc test shows a significant difference at the $p < 0.05$ level between the following years:

Difference When Base Year Data Was Compared to Data in Following Years

Base Year	Compared to Year	Compared to Year	Compared to Year	Compared to Year	Compared to Year
1994	1995	1996	1997	1998	1999
1995	1994	1997	1998	1999	
1996	1994	1997	1998		
1997	1994	1995	1996	1998	1999
1998	1994	1995	1996	1997	1998
1999	1994	1995	1996	1997	1998

Table 5 reflects the significant differences among the violations for category 4, *Food Equipment and Utensils*. With an alpha level of 0.05, the effect of year of inspection was statistically significant, F (5; 466,278) = 135.344, p = .000. Bonferroni's post hoc test shows a significant difference at the $p < 0.05$ level between the following years:

Difference When Base Year Data Was Compared to Data in Following Years

Base Year	Compared to Year	Compared to Year	Compared to Year	Compared to Year	Compared to Year
1994	1996	1997	1998	1999	
1995	1996	1997	1998	1999	
1996	1994	1995	1996	1997	1998
1997	1994	1995	1996	1998	1999
1998	1994	1995	1996	1997	1999
1999	1994	1995	1996	1997	1998

Table 6 reflects the significant differences among the violations for category 5, *Single Service Articles*. With an alpha level of 0.05, the effect of year of inspection was statistically significant, F (5; 466,278) = 112.574, p = .000. Bonferroni's post hoc test shows a significant difference at the $p < 0.05$ level between the following years:

Difference When Base Year Data Was Compared to Data in Following Years

Base Year	Compared to Year	Compared to Year	Compared to Year	Compared to Year	Compared to Year
1994	1995	1996	1997	1999	
1995	1994	1998	1999		

Base Year	Compared to Year	Compared to Year	Compared to Year	Compared to Year	Compared to Year
1996	1994	1998	1999	1997	
1997	1994	1998	1999		
1998	1995	1996	1997	1999	
1999	1994	1995	1996	1997	1998

Table 7 reflects the significant differences among the violations for category 6, *Water and Sewerage/Plumbing*. With an alpha level of 0.05, the effect of year of inspection was statistically significant, F (5; 466,278) = 80.787, p = .000. Bonferroni's post hoc test shows a significant difference at the $p < 0.05$ level between the following years:

Difference When Base Year Data Was Compared to Data in Following Years

Base Year	Compared to Year	Compared to Year	Compared to Year	Compared to Year	Compared to Year
1994	1996	1997			
1995	1996	1997			
1996	1994	1995	1998	1999	
1997	1994	1995	1998	1999	
1998	1996	1997	1999		
1999	1994	1995	1996	1997	1998

Table 8 reflects the significant differences among the violations for category 7, *Toilet and Handwashing Facilities*. With an alpha level of 0.05, the effect of year of inspection was statistically significant, F (5; 466,278) = 277.656, p = .000. Bonferroni's post hoc test shows a significant difference at the $p < 0.05$ level between the following years:

Difference When Base Year Data Was Compared to Data in Following Years

Base Year	Compared to Year	Compared to Year	Compared to Year	Compared to Year	Compared to Year
1994	1997	1998	1999		
1995	1997	1998	1999		
1996	1997	1998	1999		
1997	1994	1995	1996	1998	1999
1998	1994	1995	1996	1997	1999
1999	1994	1995	1996	1997	1998

Table 9 reflects the significant differences among the violations for category 8, *Garbage and Refuse Disposal*. With an alpha level of 0.05, the effect of year of inspection was statistically significant, F (5; 466,278) = 165.460, p = .000. Bonferroni's post hoc test shows a significant difference at the $p < 0.05$ level between the following years:

Difference When Base Year Data Was Compared to Data in Following Years

Base Year	Compared to Year	Compared to Year	Compared to Year	Compared to Year	Compared to Year
1994	1996	1997	1998	1999	
1995	1996	1997	1998	1999	
1996	1994	1995	1998	1999	
1997	1994	1995	1996	1998	1999
1998	1994	1995	1996	1997	1999
1999	1994	1995	1996	1997	1998

Table 10 reflects the significant differences among the violations for category 10, *Floors, Wall, and Ceilings*. With an alpha level of 0.05, the effect of year of inspection was statistically significant, F (5; 466,278) = 344.572, p = .000. Bonferroni's post hoc test shows a significant difference at the $p < 0.05$ level between the following years:

Difference When Base Year Data Was Compared to Data in Following Years

Base Year	Compared to Year	Compared to Year	Compared to Year	Compared to Year	Compared to Year
1994	1995	1996	1997	1998	1999
1995	1994	1996	1997	1998	1999
1996	1994	1995	1997	1998	1999
1997	1994	1995	1996	1999	
1998	1994	1995	1996	1999	
1999	1994	1995	1996	1997	1998

Table 11 reflects the significant differences among the violations for category 11, *Other Areas*. With an alpha level of 0.05, the effect of year of inspection was statistically significant, F (5; 466,278) = 152.121, p = .000. Bonferroni's post hoc test shows a significant difference at the $p < 0.05$ level between the following years:

Difference When Base Year Data Was Compared to Data in Following Years

Base Year	Compared to Year	Compared to Year	Compared to Year	Compared to Year	Compared to Year
1994	1997	1998	1999		
1995	1997	1998	1999		
1996	1997	1998	1999		
1997	1994	1995	1996	1998	1999
1998	1994	1995	1996	1997	1999
1999	1994	1995	1996	1997	1998

Table 12 reflects the significant differences among the violations for category 13, *General*. With an alpha level of 0.05, the effect of year of inspection was statistically significant, F (5; 466,278) = 313.601, p = .000. Bonferroni's post hoc test shows a significant difference at the $p < 0.05$ level between the following years:

Difference When Base Year Data Was Compared to Data in Following Years

Base Year	Compared to Year	Compared to Year	Compared to Year	Compared to Year	Compared to Year
1994	1995	1996	1997	1998	1999
1995	1994	1996	1997	1998	1999
1996	1994	1995	1998	1999	
1997	1994	1995	1998	1999	
1998	1994	1995	1996	1997	1999
1999	1994	1995	1996	1997	1998

Training Program Compared to Violations Incurred

Comparing sections of information contained in the training materials to the non-critical violations (N = 466,295), the findings were informative. This comparison was done to investigate the amount and quality of information as compared to the number of violations occurring in the topics presented.

The sections (topics), which contained categories of specific interest (number of categories appear in parentheses), appearing on the inspection form include the following: Source (1); PHF Temperature Control (2); Personnel (3); Food Equipment and Utensils (4); Single Service Articles (5); Water and Sewerage/Plumbing (6); Toilet and Handwashing Facilities (7); Garbage and Refuse

Disposal (8); Insect and Rodent Control (9); Floors, Walls, and Ceilings (10); Other Areas (11); Safety (12); General (13); Information (14)

Sections of the inspection were compared by coding pages in the training handout according to the topics of information presented. The categories provided detailed areas of concern for inspectors to check or to use for specific coding. For example, under the topic of Temperature Control, there are thirteen specific details that must be checked during an inspection, such as *hot food at proper temperature* and *foods properly cooked/reheated.*

Sentences were also counted and converted to page length encompassed to give an indication of "amount," if the sentence addressed a particular segment of the inspection but appeared under a title that did not apply. In the training materials, margins were at .5 on the outside, 1" on the inside or binding margins, .75 on top, and 1" to 2.25 on the bottom depending on whether text or graphics were used.

The final page, page 10, ended with a short list of seven items and a final bottom margin of 6 inches. Text was double-spaced and for a full page, considering the format, 28 lines of text could be included. It should be noted that out of 466,295 non-critical inspections, a multitude of violations can occur under a specific section. Therefore, the number of non-critical violations indicated below represented the total incurred during the 466,295 inspections that were conducted.

Section 1, *Source,* incurred a total of 25,311 non-critical violations. The amount of material covered in the most widely used training program in the state included zero information. Critical violations totaled 8,578.

Section 2, *PHF* (Potentially Hazardous Food) *Temperature Control,* incurred a total 311,489 non-critical violations. The amount of material covered included 1 page on temperatures, which appeared in graph form. This information presented temperatures for holding; danger zone range; cook temperatures; chill temperatures; reheat temperatures; discarding time and temperature rule; and date marking. Critical violations totaled 89,006.

Section 3, *Personnel,* incurred a total of 10,360 non-critical violations. The amount of material covered included approximately two and a half pages of information (graphs included) on the transfer of viruses, bacteria, and parasites; prohibiting bare hand contact with ready-to-eat food, and managing employee health. Critical violations were greater in this category, totaling 62,662.

Section 4, *Food Equipment and Utensils,* incurred a total of 719,463 non-critical violations. The amount of material covered 0.75 of one page including a graph of a three-compartment sink set-up for the manual dishwashing procedure. Critical violations totaled 36,742. Section 5, *Single Service Articles,* incurred a total of 48,550 non-critical violations. The amount of material covered included one sentence. Critical violations totaled 0.

Section 6, *Water and Sewerage/Plumbing,* incurred a total of 55,921 non-critical violations. The amount of material covered was zero. Critical violations totaled 38,930. Section 7, *Toilet and Handwashing Facilities,* incurred a total of non-critical 141,659 violations. There was no information covered in this section. While information was presented on the importance of employees using handwashing facilities and the illnesses caused by failure to adequately practice hygienic procedures, nothing was presented on the importance of the facilities, equipment, or supplies that must be present to pass the inspection. Critical violations totaled 46,066.

Section 8, *Garbage and Refuse Disposal,* incurred a total of 92,392 non-critical violations. No amount of material was covered in this section. Again, a sentence appeared that indicated that employees should wash their hands after "engaging in other activities that contaminate the hands" (p. 12, Food Manager's Workshop Manual), but nothing was mentioned of garbage and refusal disposal and the proper management of refuse, containers, rodent control, garbage storage. Critical violations totaled 0.

Section 9, *Insect and Rodent Control,* incurred a total of 0 non-critical violations (all were considered critical). Zero information was covered in this section. This section requires managers to eradicate rodents and insects, secure potential entry areas for pests, and prohibit animals from entering the production facility. No information was mentioned in the training materials of pest or rodent control management. Critical violations totaled 79,337.

Section 10, *Floors, Walls, and Ceilings,* incurred a total of 314,983 non-critical violations. No amount of material was covered. Critical violations totaled 1. Section 11, *Other Areas,* included the cleanliness of employee lockers, toxic items properly stored, toxic items properly labeled, unnecessary articles stored properly, separation of living and sleeping areas, and clean linen properly stored. This segment incurred a total of 64,360 non-critical violations. The amount of material covered in the most widely used training program in the state included zero information. Critical violations totaled 0.

Section 12, *Safety,* included fire extinguishers "proper and sufficient," exiting system adequate, electrical wiring adequate, gas appliance properly installed, and flammable/combustible materials properly stored. This segment incurred a total of non-critical 0 violations. Critical violations totaled 226,089. No information is provided in this section.

Section 13, *General,* included current licenses displayed, food management certification current, False/Misleading Statements published, employee training validation, Clean Air Act displayed, Gratuity Notice, and State Statutes Available. This segment incurred a total of 92,766 non-critical violations. The amount of material covered in the most widely used training program in the state included a half page on required signs, posting, and plans. Critical violations totaled 99,866.

Section 14, the final section, *Information,* included Education Program information provided, and Smoke Free. This segment incurred a total of 13,024 non-critical violations. The amount of material covered in the most widely used training program in the state included zero information. Critical violations totaled 0.

DISCUSSION

Food safety is an issue that is receiving global attention. In the United States, the topic has become an important concern as the federal government has increased research, education, and training materials to better equip the food industry of the risks associated with food safety. It is the goal of each state to reduce the number of violations incurred by food service facilities and food production sites. However, the number of biological, physical, and chemical hazards is large, as well as the codes for preventing, reducing, and eliminating each change in environment, element, or emerging pathogen. The result is a food safety program that is undergoing constant change and requiring more trained human resources to meet the food safety training needs in each state.

This study investigated the number of violations experienced by one high tourist-visited state, Florida, and then compared the training programs offered to food service managers. The purpose of this comparison was to study the amount and quality of information given to trainees compared to the food safety inspections to which they are subjected on a monthly basis. Food safety inspections are conducted by state inspection agents using approved forms developed according to current food code and state and federal food safety regulations. All food safety training programs are designed with the intent of adequately preparing food service managers to comply with the code and to operate facilities and manage employees according to food safety guidelines. To date, only one state requires hourly food handlers (non-managerial staff) to be formally trained and certified in food safety. The state used as an example in this study does not require food handlers to be certified or necessarily trained in safe food handling practices, although training on this topic is highly encouraged by food safety trainers when preparing managers for food safety certification.

The results of this study indicate that the training program offered by Florida is significantly different in amount of information, time to train, and materials given to students compared to the NRA-EF ServSafe program. While this state program produces performance scores that are significantly high, the number of violations incurred in facilities where managers have been trained continue to be high. There are three trainers delivering the Florida food safety and sanitation program. These trainers are responsible for seven districts, which is further divided into 67 counties.

The training program offered provides significantly less information than the *ServSafe* national program. The state training program reports a 99% passing rate on the exam for the current year, and 22,000 managers were trained in the 1999 year. For the past three years, the number of managers trained has steadily climbed by approximately 2,007, and last year, this number increased by approximately 4,000. The number of foodservice managers trained under the ServSafe program is low compared to those trained and certified by the state's program; however, the exact number of ServSafe certified managers is not known.

The ANOVA results indicate a number of significant differences producing F values of .0000. This is in part due to the large sample size, which results in greatly reduced F values. Creating subsets of the data was done, reducing the sample size; however, the concluding results were no different. Therefore, the sample size, while quite large, was used in its entirety.

According to the results of this study, Florida's food training program must undergo frequent and aggressive updates, and state program officials should institute follow-up training programs for those currently certified. In addition, increasing the number of trainers as well as the number of inspectors would be recommended. Developing and supporting the implementation of food safety training in grade schools and high schools would also aid in improving the accomplishment of the food safety initiative as well as help form properly food handling habits at home as well as in the workplace.

Hypotheses Addressed

Hypotheses one, two and three were rejected. The fourth and final hypothesis was accepted. With regard to hypothesis one, the number of non-critical violations vary in each category of the inspection form. In some categories such as *Temperature, Personnel, Food Equipment and Utensils, Insect and Rodent Control, Safety, Other Areas* (storage of toxic chemicals), and *General* (posting of required paperwork or displays), the total number of non-critical violations was zero, as all of the violations were deemed *critical* for every inspection in the last six years. The various tables indicate that the number of non-critical violations varied significantly from year to year between the categories.

Hypothesis two was rejected due to the significant difference in the number of non-critical violations occurring from district to district. Frequencies indicate that Districts 2, 3 and 4 yielded 55% of the total non-critical violations. Large cities in these districts to help identify the districts include Palm Beach, Tampa, and Orlando respectively. It is interesting to note that District 1, comprised of two counties and containing the city of Miami, yielded 15% of the non-critical violations. Non-critical violations for the remaining districts were significantly lower. It should be noted that district size is based on per capita and number of facilities, not geographical size.

Hypothesis three was rejected due to the significant differences found in the non-critical violations cited in the six years. For those violations of non-critical nature, thirty-eight yielded significant differences at the $p < .01$ level based on the year cited. This indicates that training needs to be targeted in these areas. It should also be noted that turnover in inspection staff may also have an effect in the type of violation and the frequency with which violations are cited.

Lastly, hypothesis four was rejected after comparing Florida's training manuals and support materials provided during training. They contained much less information and the materials were presented in a less visually rich manner. Also, time spent with trainees was relatively less both between trainee and trainer and trainee interaction with other trainees.

Factors Affecting the Quality of Food Safety Training and Inspections

A number of variables impact an inspection as well as training program quality. Emerging pathogens, revised food codes, change in inspectors (each may have own areas of inspection interest and thus concentrate on one or several areas of inspection than all areas), suppliers, delivery methods, facility and employee management, employee turnover, language barriers (complicates training), and skills, knowledge, and abilities of food handlers are but a few of the possible reasons for significant differences in violations incurred.

Most areas on the inspection form evaluated during an inspection are those processes, tasks, and habits required of foodhandlers, not managers. Therefore, the results suggest that food safety and inspection training, communication, consistency of inspection processes and emphasis on items on the inspection form, frequency of inspections, and possibly consequences of poor scores needs improvement. More specifically, categories highlighted in inspection form need special attention and can guide a training program for both the inspector and foodhandler.

Not included as ANOVA results were the total number of district differences in violations; only a selection of differences from each inspection category were presented. Critical violations were reported as raw number violations, and in reviewing this table, it should be noted that although non-critical violations may not have been noted, in some cases, the violations were so severe that they were considered "critical" in a particular category. It should also be noted that in several instances, violations and critical violations occurred in categories in which there is no training material given in the training manual used by the state.

In many areas where violations of both categories occurred, the information given in the training manual is significantly limited. In comparison to the program offered by NRA's EF, the amount of training material is less. The NRA's EF program, ServSafe, is more comprehensive on every topic and offers presentations, activities and practice session in significantly more variety.

Of critical concern is the review of the Florida's state food safety training program used as compared to what trainees must know about food safety. Lacking in the training materials is a high percentage of the information appearing as important on the state inspection forms. The inspection form that communicates this information is pertinent for managing a safe and sanitary food facility according to federal and state regulations.

Finally, while both training certification programs used in this study recommend follow-up certification and training on the topic of food safety, enforcing this expectation has not been implemented as intended. Since only one manager per facility must have a current certification in the state in question, managers of food facilities can be considerably behind in training but have a current, certified manager (within 5 years) on staff and still meet state regulations. Follow-up training and testing for those with current certificates does not occur although changes in food codes and new, emerging pathogens surface on an average, in less than five years.

Limitations

Data used in this study represents one state; therefore, the results are not to be generalized to all states or food safety training programs. The conclusions are also to be considered when evaluating the status and quality of training programs used in each state, understanding that states may use a program of their choice as long as it has been approved by the agency responsible for food safety and inspections. Reasons why the years reflect such significantly different results needs further investigation. Also, several inspection forms were used over the years, somewhat complicating the interpretation of violations.

As the forms changed, new categories were added and others removed. Also, some categories were not tracked.

As this is a study considered to be in the pilot stages, research into the specific variables is planned as a follow-up investigation. Research also needs to be conducted on the competency and training programs of inspectors.

REFERENCES

Administrative Rules, Public Lodging and Food Service Establishment, *Department of Business and Professional Regulations, Division of Hotels and Restaurants*, July 2, 1998.

Bopry, J. (1999). The warrant of constructivist practice within educational technology. *Educational Technology Research and Development*, 47(4), 5-26.

Chung, B., & Hoffman, K. D. (1998). Critical incidents. *Cornell Hotel and Restaurant Administration Quarterly*, 39(3), 66-71.

Deery, M. A., & Shaw, R. N. (1999). An investigation of the relationship between employee turnover and organizational culture. *Journal of Hospitality & Tourism Research*, 23(4), 387-400).

Enz, C. & Potter, G. (1998). The impacts of variety on the costs and profits of a hotel chain's properties. *Journal of Hospitality & Tourism Research*, 22(2), 142-157.

Florida Division of Business and Professional Regulations, Tallahassee, FL. Website: http://www.state.fl.us/dbpr/hr.

Food and Drug Administration, 1999. United States FDA, http://vm.cfsan.fda.gov/~dms/fsirp995.html, *Progress and Perspective, Food Safety Initiative, FY '99 Annual Report*, Education.

Gagne, R. M., Briggs, L. J., & Wager, W. W. (1988). *Principles of Instructional Design*, (3rd ed.). New York: Holt, Rinehart and Winston, Inc.

Harris, K. J., & Bonn, M. A., (2000). Training Techniques and Tools: Evidence from the Food Service Industry. *Journal of Hospitality & Tourism Research*.

Howell, D. C. (1987). *Statistical Methods of Psychology* (2nd ed.). Boston: PWS Publishers.

Loken, J. K. (1995). The HACCP food safety manual. John Wiley & Sons: New York.

National Institute of Allergy and Infectious Diseases (2000). Foodborne diseases. *http://www.niaid.nih.gov/factsheets/foodbornedis.htm.*

Progress and Perspective, Food Safety Initiative, Annual Report, Fiscal Year 1999. Website: http://vm.cfsan.fda.gov/~dms/fsirp994.html.

Noe, R. A., Hollenbeck, J. R., Gerhart, B., & Wright, P. M. (2000). *Human Resource Management* (3rd ed.). Boston: Irwin McGraw-Hill.

Changing Food Consumption Patterns and Their Impact on the Quick Service Restaurant Industry

Sung-Pil Hahm
Mahmood A. Khan

SUMMARY. During the past two decades there have been significant changes in U.S. food consumption patterns. Consumers are purchasing smaller quantities of milk, eggs, pork, and beef. They are consuming more poultry, fish, fruits and vegetables. Such trends in food demand are important since they may require corresponding changes in marketing and production strategies for the foodservice industry. This study investigates the driving forces that influence consumer food preferences and the trends in food consumption patterns, and consequently their influence on the Quick Service Restaurant industry. *[Article copies available for a fee from The Haworth Document Delivery Service: 1-800-HAWORTH. E-mail address: <getinfo@haworthpressinc.com> Website: <http://www.HaworthPress.com> © 2001 by The Haworth Press, Inc. All rights reserved.]*

KEYWORDS. Economic change, demographic change, quick service restaurants, change in eating habits

---

Sung-Pil Hahm, PhD, RD, is Assistant Professor, Kyung Sung University.
Mahmood A. Khan, PhD, RD, is Professor and Director, Northern Virginia Graduate Center, Department of Hospitality and Tourism Management, Virginia Tech.

[Haworth co-indexing entry note]: "Changing Food Consumption Patterns and Their Impact on the Quick Service Restaurant Industy." Hahm, Sung-Pil, and Mahmood A. Khan. Co-published simultaneously in *Journal of Restaurant & Foodservice Marketing* (The Haworth Hospitality Press, an imprint of The Haworth Press, Inc.) Vol. 4, No. 3, 2001, pp. 65-79; and: *Quick Service Restaurants, Franchising and Multi-Unit Chain Management* (ed: H. G. Parsa and Francis A. Kwansa) The Haworth Hospitality Press, an imprint of The Haworth Press, Inc., 2001, pp. 65-79. Single or multiple copies of this article are available for a fee from The Haworth Document Delivery Service [1-800-HAWORTH, 9:00 a.m. - 5:00 p.m. (EST). E-mail address: getinfo@haworthpressinc.com].

INTRODUCTION

During the past two decades, there have been significant changes in U.S. food consumption patterns. The per capita consumption of some foods has increased, while the consumption of others has declined. Consumers are purchasing less beverage milks, eggs, pork, and beef, while consuming larger quantities of poultry and fish, fruits, and vegetables. The most significant trend in food consumption behavior has been the shift away from animal products. In the particular case of meats, per capita consumption of red meats has declined substantially from 1980 to 1996. American consumers are increasingly demanding healthier and more convenient products. Consumers are increasingly consuming more fruits, fresh vegetables, cereals and other crop products (Table 1).

These changing consumption patterns might be driven by many factors. First, the purchasing decisions of consumers are the end result of complex interactions among economic, sociocultural, and psychological factors. For example, some of the previous studies view the change in food demand as a response to variations in the demographic and socioeconomic composition (Pollack & Wales, 1981; Capps & Havlicek, 1984; Smallwood, Blaylock, & Blisard, 1992). In these studies, researchers focused on demographic factors such as sex, education, age, and income level as important indicators. In addition, some researchers used not only the income and demographic effects, but also the price effects on patterns of consumer expenditures (Kokoski, 1986; Lee, 1990; Falconi, 1991).

Second, the change of food preference can be affected by different factors. In this regard, lifestyle and noneconomic factors may play an important role in the consumption patterns of many food products. One of the important factors is the consumer health concern over blood cholesterol level and saturated fat intake. According to Chavas (1983), consumer concern regarding fat and cholesterol might have produced an important shift in meat preferences. Also Wohlgenant (1986) attempted to explain the cause of the consumption change explicitly. He found that one-third of the unexplained decrease in the demand for beef and pork and one-half of the unexplained increase in the demand for poultry are due to quality changes and perceived health benefits in the respective products.

Potential changes in food demand is an important issue because such changes may require corresponding changes in marketing and production strategies for the foodservice industry. Therefore, it is essential to distinguish between changes in consumption due to variations in economic and sociodemographic factors (i.e., prices and income variation) and the variations associated with changes in consumer preferences. The former require strategy

TABLE 1. Per Capita Food Consumption Changes

BY SEGMENT		DIFFERENCE BETWEEN 1980 AND 1996			BY INDIVIDUAL PRODUCT		DIFFERENCE BETWEEN 1980 AND 1996		
	Unit #	1980	1996	Percent		Unit #	1980	1996	Percent
Decrease					**Decrease**				
Coffee	*Gallons*	27	22	−18.5%	Whole milk	*Gallons*	16.5	8.4	−49.1%
Eggs	*Number*	271	236	−12.9	Shell eggs	*Number*	236	174	−26.3
Beverage milk[2]	*Gallons*	27.6	24.3	−12.0	Table spreads	*Pounds*	15.8	13.4	−15.2
					Pork	"	52.1	46.0	−11.7
					Beef	"	72.1	64.2	−11.0
					Potatoes	"	51	49	−3.9
Increase					**Increase**				
Bottled water	*Gallons*	2.4	12.4	416.7	Fat-free milk (skim)	*Gallons*	1.3	3.9	200.0
Cheese[3]	*Pounds*	17.5	27.7	58.3	Diet drink	*Gallons*	5.1	11.7	129.4
Carbonated soft drinks	*Gallons*	35.1	51.9	47.9	Rice	*Pounds*	9	19	111.1
Flour and cereals[7]	*Pounds*	145	198	36.6	Turkey	*Pounds*	8.1	14.6	80.2
Vegetables[5]	*Pounds*	336	412	22.6	Processed eggs	*Number*	35	62	77.1
Fruit juice	*Gallons*	7.2	8.7	20.8	Corn products	*Pounds*	13	23	76.9
Added fats and oils[4]	*Pounds*	57.2	64.9	13.5	Frozen Potatoes	"	35	60	71.4
Fruits[5,6]	"	258	283	9.7	Chicken	"	32.7	49.8	52.3
Total meat[1]	"	179.6	191	6.3	Light milk (1% & 0.5%)	*Gallons*	1.8	2.6	44.4
					Regular drink (nondiet)	*Gallons*	29.9	40.2	34.4
					Wheat flour	*Pounds*	117	148	26.5
					Reduced fat milk (2%)	*Gallons*	6.3	8.0	27.0
					Salad and cooking oils	*Pounds*	21.2	26.0	22.6
					Shortening	"	18.2	22.2	22.0
					Lard and beef tallow	"	3.7	4.5	21.6
					Fish	"	12.4	14.7	18.5
					Tomatoes for canning	"	64	74	15.6

Notes: Data are per person per year. [1]Boneless weight. Includes lamb, mutton, and veal. [2]Includes flavored milk and buttermilk. [3]Excludes full-skim American, cottage, pot, and baker's cheese. [4]Total fat content. Individual items shown on a product-weight basis. [5]Farm weight. [6]Includes fruit juice. [7]Includes oat, barley, and rye products. Source: Clauson (1999)

related to the economic side, while the latter require development of new product innovations.

In this study, an investigation is made regarding the forces driving change in consumer food preferences and the emerging trends in food consumption patterns, and finally how these changes have influenced the Quick Service Restaurant (QSR) industry. Furthermore, this study will focus only on the U.S.

market, as was done by researchers such as Alston and Chalfant (1988, 1991, and 1992), Swofford and Whitney (1986, 1988), Lee (1990), and Jensen and Bevins (1991).

ECONOMIC CHANGE

Disposable Income and Food Expenditure

One of the factors influencing consumer demand is the increase in consumer income. Income has been shown to be an important indicator in studies of consumption patterns. For example, low-income individuals usually exhibit less healthy dietary patterns such as lower fruit and vegetable consumption (Kushi, 1988).

During the past twenty years, personal disposable income has greatly increased, mainly as a result of sociocultural changes, which include smaller households, an increase in the dual-income family, and more single parent households. These types of households probably spend less time preparing meals than do traditional, single-earner families. In other words, today's consumers spend less time in the kitchen and are increasingly shopping for conveniently prepared food products that fit faster-paced lifestyles (Lin & Frazao, 1997; Price, 1998).

Disposable personal income has increased to 4,978 billion dollars in 1998 from 1,640 billion dollars in 1980, which represents an increase of over 200% over 18 years (Table 2). According to a report by the USDA's Economic Research Service (ERS), a total of 15.4 percent of disposable personal income was spent on food in 1980, including 10.4 percent for food at home, and 4.9

TABLE 2. Food Expenditures by Families and Individuals as a Share of Disposable Personal Money Income

Year	Disposable Income	Expenditures for food					
	Billion dollars	At home		Away from home		Total	
1980	1,639.8	10.4%	$ 170.8	4.9%	$ 81.1	15.4%	$ 251.9
1998	4,978.4	7.6%	$ 375.0	5.2%	$ 259.0	12.8%	$ 635.0

	Food expenditures at 1988 prices		
	Food at home	Food away from home	Total
	Million 1988 dollars		
1980	$ 242,511	$ 175,461	$ 417,972
1998	$ 283,994	$ 267,942	$ 551,937

Source: Clauson (1999)

percent for food away from home. In 1998, a total of 12.8 percent of disposable personal income was spent on food, including 7.6 percent for food at home, and 5.2 percent for food away from home.

The total percentage expenditure for food has decreased. However, in real dollar amounts, food expenditures have increased from 417,972 million dollars in 1980 to 551,937 million dollars when converted to 1988 prices. The increase in food expenditures results mainly from food away from home, and roughly half of the dollars spent for food was used for food consumed away from home.

The fastest growing segment in terms of total expenditures on food away from home is recreational places–463 percent–followed by eating and drinking places, at 314 percent. The proportion of eating and drinking places account for 67.3 percent in 1998 compared to 63.1 percent in 1980. The lowest increase–224 percent–was in the segment including schools and colleges. However, the proportion of schools and colleges decreased from 9.2 percent to 7 percent between 1980 and 1998. On the other hand, the proportion of the recreational places segment grew from 2.5 percent to 4 percent between 1980 and 1998 (see Table 3).

The demand for food away from home has grown and likely will increase faster than the demand for traditional food cooked in the home since Americans lead faster-paced lifestyles and no longer have a lot of time for preparing meals. This trend is also caused by growth in household disposable incomes and an increase in dual-income households. Consequently, consumers are increasingly spending a higher proportion of their total food budgets at restaurants, and other places away from home. This increase in food expenditures away from home is simultaneously associated with an increase in restaurant selection options (Table 4).

TABLE 3. Total Expenditures on Food Away from Home

Year	1980	1998	Percentage Difference
Segments	Expenditures in Million		
Eating and drinking places	75,883 (63.1%)	238,483 (67.3%)	314%
Hotels and motels	5,906 (4.9%)	17,364 (4.9%)	294%
Retail stores, direct selling	8,158 (6.8%)	25,319 (7.1%)	310%
Recreational places	3,040 (2.5%)	14,085 (4%)	463%
Schools and colleges	11,115 (9.2%)	24,891 (7%)	224%
All other	16,194 (13.5%)	34,212 (9.7%)	211%
Total	120,296	354,354	308%

Source: Clauson (1999)

TABLE 4. Top-100 Chain Restaurant Units and Sales Trends of Change (sales in thousand dollars)

	1985				1995				CHANGE FROM 1985 TO 1995	
	Sales		Units		Sales		Units		Sales	Units
Quick service	$36,864	70.9 %	62,157	81.7 %	$70,832	71 %	101,973	85 %	192 %	164 %
Mid-scale Restaurants	$10,204	19.6 %	11,300	14.8 %	$15,019	15.1 %	11,552	9.6 %	147 %	102 %
Upscale Restaurants	$4,982	9.6 %	2,645	3.5 %	$13,862	13.9 %	6,254	5.2 %	278 %	236 %
Top-100 Restaurants	$52,020	100 %	76,102	100 %	$99,714	100 %	119,779	100 %	192 %	157 %
Restaurant Industry	$113,400		287,000		$205,928		369,377		253 %	129 %
Top-100 share	45.9%		26.5%		48.4%		32.4%			

Source: Nation's Restaurant News Research (1998)

Demographic Changes

From 1980 to 1998, the U.S. resident population increased from 227.2 million to 270.3 million, which amounts to a 19 percent growth rate. Along with the population change, per capital food expenditure has increased from $1,357 to $2,798 at the current prices, which is almost a double-fold increase. Calculated at 1988 prices, it can be seen that per capita food expenditures at home decreased slightly compared to a dramatic increase in away from home expenditures (see Table 5). In fact, the increase in per capita food expenditures was caused mainly by food away from home, indicating that more people consume food products away from home. Additionally, other demographic changes such as population growth may influence these food consumption statistics.

Several previous studies pointed out that demographic characteristics such as sex, age, household size and education, share some responsibility in explaining the pattern of household food consumption (Huang & Raunikar, 1978; Salathe, 1979). Additionally, family composition, lifestyles, and other non-economic factors may have affected food consumption patterns. For example, the average size of the household unit has declined from 2.76 members per unit in 1980, to 2.62 members per unit in 1998. This indicates an increase in the overall number of households.

In addition, the participation of women in the labor force has increased substantially, and the average number of children under 18 years of age in the family unit has declined from 1.05 to 0.99 (U.S. Bureau of Census, 1980, 1998). Along with population growth, the effects of increased longevity and a de-

TABLE 5. Per Capita Food Expenditures

	U.S. resident population, July 1 (in millions)	Current prices			1988 prices		
		At home	Away from home	Total	At home	Away from home	Total
1980	227.2	$ 828	$ 529	$ 1,357	$ 1,067	$ 772	$ 1,839
1998	270.3	$ 1,487	$ 1,311	$ 2,798	$ 1,051	$ 991	$ 2,042

Source: Clauson (1999)

crease in the number of younger people (under 18) are linked with growth in the mature population by numbers as well as by percentage. The increase in the proportion of mature Americans has contributed to a greater demand for healthier food products.

The Mature Consumer

The appearance of a large number of mature consumers on the demographic horizon was first noted in the professional literature in the late 1950s. In 1958, Robert D. Dodge predicted in the *Journal of Retailing* that the mature market would become an important segment. "The numbers in this age group," stated Dodge, "will continue to increase, and problems will lessen if marketers understand the nature and characteristics of the market" (p. 75).

The mature market refers to individuals age 55 and older (Moschis, 1992). Using this definition, the size of this market was estimated at 58 million in 1999, according to the U.S. Bureau of the Census (1999). The great majority of this group consists of people between the ages of 55 and 64 (24 million or 41%), with another 18 million (31%) between the ages of 65 and 74. Thus, this segment of the population is skewed to the younger ages in the segment. Moreover, the market's age distribution has changed as the 76 million baby boomers begin to swell into the ranks of the mature consumer market. In approximately fifty years, there will be an even age distribution of older Americans across the three main age brackets (55-64, 65-74, 75+), and its total size is expected to double (Moschis, 1992).

The mature market is a viable and growing market. It differs from markets comprised of younger age groups because of a number of experiential, biophysical, and psychosocial factors. These and other factors that offset consumer behavior are also responsible for the wide heterogeneity in the older population. Thus, it becomes necessary for marketers to carefully examine and analyze the mature market before developing market strategies.

Empirical research on the mature market has been done by Morgan and Levy (1996), which described the characteristics of the over 50 age group.

They separated mature customers into three categories based on food preferences, as the "Nutrition Concerned" (46 percent), the "Fast and Healthy" (38 percent), and the "Traditional Couponers" (16 percent). Those in the "Nutrition Concerned" group cook most of their meals and pay attention to food labels, while the "Fast and Healthy" group are not interested in reading food labels. The "Nutrition Concerned" group also consider that fast food restaurants do not provide healthy food for mature customers, while the "Fast and Healthy" group favor fast food restaurants. They found that all three groups are more willing to eat fresh vegetables and fruits. The aging population already has partially influenced food consumption patterns in general. For example, the consumption of high cholesterol and fat-containing foods, such as eggs and red meats, have decreased steadily.

As the mature market grows, marketers for foodservice chains must consider how they will cope and prosper with the changing food needs. These changes force companies to compete to provide the senior market with the types of food they prefer, in order to increase or at least maintain a market share. This will most likely force many foodservice operations to think differently about their target markets. In the quick service restaurant segment, the focus is likely to shift from the younger generation to a greater emphasis on the older consumer. For instance, McDonald's restaurant is now serving a bagel sandwich on its breakfast menu and a convenient salad to appease the health conscious consumer.

Multiculturalism

According to the U.S. Bureau of the Census (1999), the U.S. population is growing at a diminishing rate along with an increasing diversity in U.S. culture. The most significant trend in population change is that the growth rate of racial ethnic groups is increasing while the Caucasian group growth rate is decreasing. The ethnic groups can be classified into 5 categories: Caucasian (White); African-American (and Black); Hispanic; Native American, Eskimo, and Aleut; Asian, and Pacific Islander. The statistical data in 1999 indicates that Whites (196,113 in thousands) constitute the largest segment, followed by Blacks (33,125 in thousands), Hispanics (31,365 in thousands), Asian and Pacific Islanders (10,251 in thousands), and American Indian, Eskimo, and Aleut (2,024 in thousands), respectively. As indicated in Table 5, American culture is becoming more diversified. Among the ethnic groups, the Asians are the fastest growing, followed by the Hispanic group. The White population has increased only 4 percent from 1980 to 1999, while the non-white population has increased by 26 percent. This can be an important factor that Quick Service Restaurants should monitor.

Driven by the development of the multicultural mix in the U.S., and the consumer desire for change, fast food ethnic chains flourished and achieved an early success with minimal or no entry barriers. The leading companies representing these segments are Taco Bell, serving Mexican fast food, and Manchu Wok and Panda Express, serving Chinese fast food. The ethnic food category has not formed a new power or a threat for the existing quick service restaurant chains at this time. However, as long as the diversified population continues to grow, the market share of ethnic restaurant chains will grow, and competition will increase.

Due to the rapid increase in the number of mature customers, combined with the trend toward a diversified population, America faces a significant change in the food consumption patterns as seen in the last twenty years. The impacts of demographic changes should be continuously monitored, and applied to a readjusting marketing strategy for the quick service segment of the foodservice industry.

CHANGING FOOD HABITS

Food Consumption Changes Between 1980 and 1996

Food consumption patterns in the United States have changed dramatically during the last two decades. Empirical results showed that food preference changes have been a significant factor in explaining both declining beef and pork consumption, and increasing poultry consumption per capita (Chavas, 1983; Wohlgenant, 1986, Alston & Chalfant, 1988, 1991, 1992; Putnam & Allshouse, 1998).

These consumption changes are attributed to consumer health concerns over fat and cholesterol intake, and also reflect a demand for convenience food. It is estimated that consumer per capita consumption of beef has decreased by twenty-four percent over the last three decades, and that of poultry has increased by sixty-five percent. This trend is expected to continue into the near future (Putnam & Allshouse, 1998; U.S. Bureau of Census, 1999).

Table 1 shows the annual per capita consumption patterns for 1980 and 1996. The consumption patterns of meat products (beef, pork and poultry) have changed steadily over the last seventeen years. Within those meat products, the important trend is the growing consumer preference for poultry rather than red meat. Chicken and turkey consumption has increased rapidly; total consumption has increased 52 percent and 80 percent respectively from 1980 to 1996, compared to the total consumption of meat products, which has increased slightly from 179.6 pounds per capita to 191, representing a 6.3 percent increase. On the other hand, the consumption of beef and pork has decreased by 11 percent, along with a 19 percent increase in fish consumption.

Among dairy products, consumption of cheese has increased from 17.5 pounds to nearly 27.7 pounds, which is a 53.8 percent increase, while beverage milk consumption has decreased from 27.6 to 24.3 gallons. In the subgroups of beverage milk, reduced fat, light, and skim milk consumption has increased by 27, 44.4, and 200 percent, respectively, while whole milk consumption declined by 49.1 percent. Now consumers buy more low fat and skim milk than whole milk. It is clear that consumers have been substituting one product for another in the case of subgroups of meat and dairy products. Other drinking products, such as coffee, decreased from 27 to 22 gallons, but carbonated soft drinks consumption increased from 35.1 to 51.9 gallons, and the biggest increase being for the bottled water from 2.4 to 12.4 gallons per capita, which is a 416.7 percent increase.

Since the 1980s, the annual per capita use of fresh potatoes decreased from 51 to less than 49 pounds in 1998, which is almost a 4 percent decrease, while per capita consumption of frozen potatoes (mostly as French fries) increased from 35 pounds to over 60 pounds. The increase in consumption of frozen potatoes may be linked with the increase in the consumption of added fats and oils. Consumption of added fats and oils increased from 57.2 pounds to 64.9, which is a 13.5% increase. It seems that most of the fats and oils' consumption increase is caused by food away from home, based on frequent restaurant visits, and also is caused by customer menu selection habits and a fat and oil-added cooking style. Other than frozen potatoes, the consumption of fruits and fresh vegetables also increased by 10 percent and 20 percent, respectively.

The process of changing consumer preferences should be seen as a continuous process, since consumers are adjusting their preferences gradually over time rather than at a particular time period. Since the estimated preference changes for several commodities were non-zeros and show a continuous direction (negative for beef, beverage milk, and eggs; and positive for poultry, cheese, and diet products), it might be said that a shift of preferences away from high fat and cholesterol and in favor of a demand for healthy food has occurred.

The United States Surgeon General issued a report in 1988 providing strong evidence that diets high in total fat and low in fiber are positively related to high rates of major chronic diseases (*The Surgeon General's Report*, 1988). Given the scientific evidence, it was recommended that a major priority for dietary change in the U.S. population should be a reduction in total fat intake and an increase in the consumption of fruits, vegetables and whole grain products. Table 1 shows positive consumption patterns based on recommendations in the Surgeon General's report.

Taste elasticity is increasing, showing that consumers are more responsive to taste changes. This is because health information is more readily available, consumers are more health conscious, and the demand for convenience increases as more married women are working.

Reaction from Quick Service Restaurants

With today's hectic lifestyles, timesaving convenience products are increasingly in demand. Perhaps one of the most obvious examples is the quick service restaurant segment. The sales-growth of the total restaurant industry was 253 percent, and unit growth was 129 percent from 1985 to 1995, while the top-100 chain restaurants increased their sales by 192 percent and their units by 157 percent from 1985 to 1995. In addition, the market share of the top-100 chain restaurants increased from 45.9 percent to 48.4 percent in sales, and from 26.5 percent to 32.4 percent in number of units (see Table 6).

The rate of growth in consumer expenditures on fast food has led most of the other food-away-from home segment. Since 1985, the amount consumers spent at quick service restaurants outlets grew 192 percent (through 1995), compared with 147 percent growth in mid-scale restaurant expenditures among the top-100 chain restaurants. In addition, units of quick service restaurants increased 164 percent between 1980 and 1998, while mid-scale and up-scale restaurants units increased 102 percent and 236 percent, respectively. The increase in restaurant units and sales indicate that the restaurant industry is growing, and is related to consumer eating and food preference patterns.

The quick service restaurants tend to be responsive to consumer concerns pertaining to nutrient intake and overall public health. The switch to vegetable oils for deep-frying by the largest fast food hamburger chains (i.e., McDonald's, Wendy's, and Burger King) was an act in response to consumers' concern over the health effects of saturated fat intake. Moreover, Kentucky Fried

TABLE 6. Population Change by Ethnic Group

	1990	1999	Percent Change from 1990 to 1999
	As of July 1 (in thousands)		
White	188,581	196,113	4%
Non-white	60,858	76,765	26.1%
Black	29,397	33,125	12.7%
Hispanic	22,575	31,365	38.9%
Asian	7,084	10,251	44.7%
Native	1,802	2,024	12.3%

Source: Yax (1999)

Chicken changed its name to KFC in order to decrease the stigma placed on it from the word "fried." In addition, KFC added new non-fried items to their menu offering.

However, attempts to capture the consumer demand for healthier meal options have not always been successful. In particular, the hamburger segment has tried to clean up its image concerning high fat and unhealthy food. In 1991, McDonald's introduced the McLean Deluxe sandwich, which used a 91 percent fat-free beef patty, and Taco Bell also introduced "Border Lights" low-fat menu items in 1994. Both these products failed to capture the customers' attention. However, the grilled chicken sandwich has proven to be more successful and remains on the menus of most major hamburger chains.

Along with increasing consumption of chicken, Wendy's has introduced "Chicken Burritos" and the "Spicy Chicken Sandwich." And Burger King has introduced the "Chick'n Crisp," "Italian Chick'n Crisp Sandwich" and the "Chicken Club Sandwich." Whether these chicken products have showed great success or not, we think the introduction of such items will be necessary in the future, even though they may not be "Star" menu items immediately after their introduction. But consumers perceive these new healthy products to be a response to the changing consumer consumption patterns.

Recently McDonald's expanded its business segment through acquiring other restaurant companies, including Chipotle Mexican Grill, Donatos Pizza, and Boston Market. This may be the result of high competition in the hamburger segment as well as a response to demographic and consumer food preference changes that include increasing demand for chicken products, Mexican foods and pizza. McDonald's strategy of expansion could be a significant threat to their competitors, whether they belong to quick service restaurant segment or not.

Although the consumption of chicken products has increased, some of the chicken chains suffered setbacks. This could be explained by the fact that consumers are bypassing quick service restaurants and opting instead to pick up fully prepared dinners at local supermarkets. Some grocery store deli-departments have expanded their menus to include fully prepared entrees and side dishes for take home purposes as home meal replacements. This changing consumer behavior could hurt the upscale quick-service chains like Boston Market and Kenny Rogers Roasters.

In addition, several sandwich chains appeared in the market as a response to consumers' need for healthy food and fresh food. From a marketing point of view these chains have shown dramatic success because of their timely market positioning. Some of the deli chains such as Subway, Blimpie Subs and Salads, and Schlotzsky's Deli are trying to take advantage of the emerging trend of healthy menu offerings. These chains seem to be in the mature life cycle stage in terms of unit sales growth and market share. This proves that even though consumers are concerned about healthy issues about food, they look for tasty food and have a desire for what they habitually eat at restaurants.

CONCLUSIONS

There can be little doubt that the level of competition in the industry in the 1990s was rising to heights never before experienced by restaurant owners and operators. To survive in the new millennium, restaurant companies need to know what customers want, need and are willing to pay. Understanding the changes in customer preferences is one of the most important things for restaurant companies to consider. This understanding could lead to launching new menu items (i.e., chicken products) as well as new cooking systems (i.e., "made for you" by McDonald's), and changing menu boards to a large font size for mature customers, even though these changes may not guarantee immediate success. But failure to adapt to the changes could lead to early demise.

The direction of changes in taste is clear. First, consumers are continuing to shift preferences in favor those products considered as healthy in a diet (grains, fish, poultry, diet drinks) and are against other products perceived as unhealthy such as beef, pork, eggs, and desserts due to their high content of fat, calories and cholesterol. Second, most people just do not have extra time to spare for food preparation and prefer to spend money on food away from home. These changes are mainly caused by socio-economic-demographic changes and increase in need for more quality leisure time. Third, various multicultural demographic categories offer opportunities to establish new restaurant segments, and increase competition within the saturated quick service segment in the foodservice industry.

The foodservice industry, and in particular the quick service segment, can be characterized as a mature operating environment, and a volatile, uncertain and complex environment (Denoble & Olsen, 1984; Parsa and Khan 1992). Foodservice marketing in general is facing severe challenges in today's intensely competitive environment. These challenges include deregulation, new and innovative methods for purchasing and paying, the proliferation of new products, the multiplication of distribution channels, the explosion of couponing, and the slippage of network advertising efficiency.

Given the uncertain, volatile, and complex environmental conditions facing the foodservice industry, it is appropriate to consider the impact of these factors on the functional area of marketing. While the trends discussed in the present paper are by no means exhaustive, it is believed that they represent the most important and pressing issues facing the food service industry. Quick service restaurant firms need to understand the relationship between changing food patterns and their impact on QSR menus.

REFERENCES

Alston, J. M., & Chalfant, J. A. (1988). Accounting for changes in tastes. *Journal of Political Economy*, 96, 391-410.

Alston, J. M., & Chalfant, J. A. (1991). Can we take the con out of meat demand studies? *Western Journal of Agricultural Economics*, 16, 36-47.

Alston, J. M., & Chalfant, J. A. (1992). Consumer demand analysis according to GARP. *Northeastern Journal of Agricultural and Resource Economics*, 21, 125-139.

Capps, O., Jr., & Havlicek, J., Jr. (1984). National and regional household demand for meat, poultry and seafood: A complete system approach. *Canadian Journal of Agricultural Economics*, 32, 93-108.

Chavas, J. P. (1983). Structural change in the demand for meat. *American Journal of Agricultural Economies*, 65, 148-153.

Clauson, A. (1999, November). *Food market indicators briefing room* [On-line]. Available Internet: ers.usda.govhttp:// briefing/foodmark/ expend/Data/history/ history.htm

DeNoble, A. F., & Olsen, M. D. (1986). The food service industry environment: market volatility analysis. *F.I.U. Hospitality Review*, 4(2), 89-100.

Dodge, R. D. (1958). Selling the older consumer. *Journal of Retailing*, Summer, 73-81 & 100.

Falconi, C. A. (1991). *An estimation of an Almost Ideal Demand System for U.S. food with household and aggregate data*. Unpublished doctoral dissertation. University of Minnesota, St. Paul.

Jensen, K., & Bevins, S. (1991). The demand for butter, margarine and oils: A nonparametric test for evidence of structural change. *Southern Journal of Agricultural Economics*, 23, 59-63.

Huang, Chung-Liang, & Raunikar, R. (1978). Estimating the effect of household age-sex composition on food expenditures. *Southern Journal of Agricultural Economics*, 10, 151-158.

Kokoski, M. F. (1986). An empirical analysis of intertemporal and demographic variations in consumer preferences. *American Journal of Agricultural Economics*, 66, 894-907.

Kushi, L H. (1988). Educational attainment and nutrient consumption patterns: the Minnesota heart survey. *Journal of the American Dietetic Association*, 88, 1230-1236.

Lee, Hwang-Jaw (1990). *Nonparametric and parametric analysis of food demand in the United States*. Unpublished doctoral dissertation. The Ohio State University, Columbus.

Lin, Biing-Hwan, & Frazao, E. (1997). Nutritional quality of foods at and away from home. *FoodReview*, 20(2), 33-40.

Morgan, M. C., & Levy, D. J. (1996). *Segmenting the mature market*. New York: American Demographics Books.

Moschis, G. P. (1992). *Marketing to older consumers*. Westport, CT: Quorum.

Nation's Restaurant News. (1998). Top-100 chain restaurant units and sales trends of change. *Nation's Restaurant News Research*.

Parsa, H.G. and M.A. Khan (1992). Menu trends in quick service restaurant industry during the various stages of industry life cycle 1919-1988. *Hospitality Research Journal*, 15(1) 93-107.

Pollack, R., & Wales, T. J. (1981). Demographic variables in demand analysis. *Econometrica,* 49, 1533-1551.

Price, C. C. (1998). Sales of meals and snacks away from home continue to increase. *FoodReview*, 21(3), 28-30.

Putnam, J. J., & Allshouse, J. E. (1998). U.S. per capital food supply trends. *Food-Review,* 21(3), 2-11.

Salathe, L. E. (1979). The effects of changes in population characteristics on U.S. consumption of selected foods. *American Journal of Agricultural Economics,* 61, 1036-1045.

Smallwood, D. M., Blaylock, J. R., & Blisard, N. W. (1992). How did household characteristics affect food spending. 1980-1988. United States Department of Agricultural Research Service. *Agricultural Information Bulletin,* 643.

Swofford, J. L., & Whitney, G. A. (1986). Nonparametric tests of utility maximization and weak separability for consumption, leisure and money. *Review of Economics and Statistics,* 69, 458-464.

Swofford, J. L., & Whitney, G. A. (1988). A comparison of nonparametric tests of weak separability for annual and quarterly data on consumption, leisure and money. *Journal of Business and Economics Statistics,* 6, 241-246.

United States, Department of Commerce. Bureau of the Census. (1980). *Resident population estimates of the United States by sex, race, and Hispanic origin.* Population Estimates Program, Population Division.

United States, Department of Commerce. Bureau of the Census. (1999). *Resident population estimates of the United States by sex, race, and Hispanic origin.* Population Estimates Program, Population Division.

United States, Department of Health and Human Service. (1988). *The Surgeon General's report on nutrition and health* (DHHS (HRSA) Publication No. 88-50210). Washington, DC: U.S. Government Printing Office.

Wohlgenant, M. K. (1986). Discussion: assessing structural change in the demand for food commodities. *Southern Journal of Agricultural Economies,* 15, 1-6.

Yax, L. K. (1999). *Population estimates* [On-line]. Available Internet: census.gov/population/www/estimates/nation/ intfile3-1.txt

Understanding Mature Customers in the Restaurant Business: Inferences from a Nationwide Survey

WonSeok Seo
Vivienne J. Wildes
Fred J. DeMicco

SUMMARY. The study reported here examined the differences between intangible service experiences of mature and younger customers in quick-service, casual, and fine-dining restaurants. The authors used ten unpleasant service experiences to identify the significant experiences that can impede satisfaction in three types of restaurants. No unpleasant service experiences were significant for either mature or young customers. However, in the casual dining restaurants, the (1) lack of product knowledge and (2) inattentive servers created significantly unpleasant service experiences for mature customers. In the quick service restaurant, the (1) lack of product knowledge, (2) inattentive servers, (3) rudeness, and (4) lack of cleanliness produced significantly unpleasant service experiences which in turn impeded the satisfaction for the mature customer. The reasons for these findings are discussed, and the authors suggest that the restaurant industry pay more attention to the mature population in order to

WonSeok Seo and Vivienne J. Wildes are Doctoral Candidates at Pennsylvania State University.

Fred J. DeMicco, PhD, RD, is Professor and ARAMARK Chair, Hotel, Restaurant and Institutional Management, University of Delaware.

[Haworth co-indexing entry note]: "Understanding Mature Customers in the Restaurant Business: Inferences from a Nationwide Survey." Seo, WonSeok, Vivienne J. Wildes, and Fred J. DeMicco. Co-published simultaneously in *Journal of Restaurant & Foodservice Marketing* (The Haworth Hospitality Press, an imprint of The Haworth Press, Inc.) Vol. 4, No. 3, 2001, pp. 81-98; and: *Quick Service Restaurants, Franchising and Multi-Unit Chain Management* (ed: H. G. Parsa and Francis A. Kwansa) The Haworth Hospitality Press, an imprint of The Haworth Press, Inc., 2001, pp. 81-98. Single or multiple copies of this article are available for a fee from The Haworth Document Delivery Service [1-800-HAWORTH, 9:00 a.m. - 5:00 p.m. (EST). E-mail address: getinfo@haworthpressinc.com].

increase both customer satisfaction levels and profits. A literature review precedes a discussion of the characteristics and service requirements of mature restaurant customers.

KEYWORDS. Mature customers, restaurant, service, newspapers

INTRODUCTION

It is only recently that the restaurant industry began diligently to count "mature market," people 55 years and older. In part, this recent attention reflects the mature market's increasing size. According to the U.S. Bureau of the Census (1998), the population of United States will grow 22 percent from 2000 to 2025, while the mature population is projected to increase 72 percent during the same period. Additionally, the mature population promises to comprise nearly 30 percent of the United States' population by the year 2025 (U.S. Bureau of the Census, 1998). Two social factors explain most of this increase in the mature market. First, advancing medical science helps people live longer. Second, the so-called "baby-boomers" have already reached middle age and will join the mature population soon.

In addition to the growing numbers, their comparatively high discretionary buying power is another reason that the mature market is so attractive to the restaurant industry. In fact, as Certron, DeMicco and Williams (1996) noted, nearly half of all the discretionary income in the United States belongs to those aged 55 or older. Fully aware of this proportionately high amount of disposable income, restaurants find it a far from simple task to respond to the special needs of the mature population. Williams, DeMicco and Kotschevar (1997) suggested that the needs of the mature consumer could be quite different from other age groups, and the restaurant industry must recognize this group as distinctly different and independent if they want to attract a greater share of the mature market.

Service Quality

Meanwhile, the term "service quality" generates numerous definitions because it is intangible, heterogeneous, and impossible to separate production and consumption (Parasuraman, Zeithaml & Berry, 1985). Many researchers attempted to define and explain the dimensions or determinants of service quality, but their results have so far generated disagreement, regarding not only those dimensions and determinants, but also the methodologies used.

For example, Groonos (1983) described the service quality as both technical and functional; " technical quality" referring to "what is delivered," and "functional quality" referring to "how it is delivered." Groonos (1983) considered such functional quality as food server's behavior to be critical to forming perceptions of service quality. Shank and Nahhas (1994) singled out Groonros's description as important in restaurant marketing since one should use both functional and technical quality to evaluate service quality in restaurants.

Lehtinen (1983) concluded that service quality consists of "process quality and output quality." In the restaurant, process quality refers to such mealtime service amenities as the physical appearance of one's food server, while output quality refers to the customer's impression of the service after the meal.

While all researchers have helped the hospitality industry define and measure the service quality, perhaps the most comprehensive study in service quality was conducted by Parasurman, Zeithaml and Berry in 1985. The study posited five distinct dimensions that explain service quality:

1. Tangibles–physical facilities, equipment and appearance of the personnel.
2. Reliability–ability to perform the promised service dependably and accurately.
3. Responsiveness–willingness to help customers and provide prompt service.
4. Assurance–knowledge and courtesy of employees and their ability to inspire trust and confidence.
5. Empathy–caring, individualized attention the firm provides its customers.

Although these five dimensions appear frequently in the literature, this particular approach has some weaknesses. To start, some studies of restaurant management have failed to support the five dimensions approach. Richard, Sundaram, and Allaway (1994), for example, failed to confirm the five dimensions of service quality on choice behavior of the home-delivery pizza market. They found a sixth dimension of "outcomes," to add to the five dimensions of service quality. This additional "outcome" measured whether the restaurant had made the pizza exactly the way the patron liked. Bojanic and Rosen (1994) also tested five dimensions of service quality in a chain restaurant, but failed to support them. They found that empathy was divided into two dimensions: knowing the customer and access to service. Additionally, Fu (1998) measured service quality in the mature diners market, but found that only tangibles remained in her model; reliability-responsiveness merged as a second dimension, and assurance-empathy was grouped as a third dimension.

Another weakness in the five-dimension approach lies in a broadness that makes it difficult to provide specific cues to explain service quality. For example, Bojanic and Rosen (1994) suggested that reliability was one of most significant dimensions in overall restaurant quality, but one cannot tell which specific service attributes among the several attributes contributed most to overall restaurant quality.

While the service quality theory presents conceptual background for the restaurant marketers, they should also understand the characteristics of their target markets. The following section describes the market of interest to the present study.

The Characteristics of Mature Customers

Much research has been conducted to identify the characteristics of mature customers. These characteristics can be grouped into four categories. First, mature customers (over 55 years of age) are more likely than younger customers (under 55) to eat at family and mid-scale restaurants (Shank and Nahhas, 1994). According to a 1988 National Restaurant Association survey, mid-scale restaurants are more popular for mature customers among all restaurant types. People age 55 to 64 said they typically visit mid-scale restaurants almost once for every three times they eat out.

Second, mature customers are more likely to eat out on weekdays than younger customers. In fact, during weekdays, almost 63 percent of all restaurant customers are older than 65 (Pederson and DeMicco, 1993). Additionally, mature consumers differ from younger customers regarding the time of day they like to dine out. The National Restaurant Association's 1988 study found that as people age, they are more likely to eat out for breakfast and lunch, while dinner remains more popular with younger customers.

Third, mature customers consider convenience and companionship to be the two major reasons for eating out (Knutson and Patton, 1993). With regard to convenience, many older people find cooking for themselves more costly than eating out, since many of them reside in only one or two-person households. Moreover, physical limitations can interfere with cooking at home. In regard to companionship, Pederson and DeMicco (1993) found that older people are particularly interested in the social aspects of eating out.

Fourth, mature customers are more conscious of nutrition than younger customers, and in order to follow a healthful lifestyle or to accommodate physical problems, they tend to choose foods that are low in fat, cholesterol, and sodium (Pederson & DeMicco, 1993; Knutson and Patton, 1993). Menu items that mature customers prefer include vegetables and baked or broiled fish, poultry, and potatoes (NPD CREST, 1988). They also appreciate having nutritional information available to them (Fintel, 1990). Williams et al. (1997) suggested that food servers should become knowledgeable about the nutritional value of foods to help mature customers choose healthy menu items. Lago and Poffley (1993) suggested that restaurant operators provide nutrition information and nutritious food to their mature customers to gain more business and their loyalty.

Service Requirement

For a restaurant striving to attract the mature customer, understanding the service requirements is just as important as understanding their characteristics. The following statement summarizes the service requirements for mature customers: "Mature customers want to have a welcoming, pleasant, and comfortable atmosphere when they eat out" (Fu, 1998).

Knutson and Patton (1993) found that about one-fourth of mature customers consider service quality a crucial factor in their restaurant selections. Moreover, they consider friendly service and individualized attention more important than younger customers do, and the promptness of restaurant service is less of a concern for mature customers than for younger customers (Shank and Nahhas, 1994). Shank and Nahhas (1994) concluded that the prompt service is less important to mature customers in selecting a restaurant because they consider eating out a social event and they have more leisure time for dining out than younger people. Shank and Nahhas (1994) also suggested that older people consider a convenient location more important than younger consumers, partly because of physical and transportation limitations.

Knutson and Patton's (1993) survey revealed that a majority of the mature respondents considered themselves occasionally mistreated in restaurants. For example, some of them felt rushed to finish eating so others could have the table. They also felt restaurant employees did not care about them. Another important finding in Knutson and Patton's (1993) survey is that many mature customers believed that restaurants should improve the lighting and design easier-to-read menus. More than a third of seniors said that they had difficulty reading menus because the print was too small or the laminated covers reflected too much glare (Knutson and Patton, 1993). Williams et al. (1997) recommended that the menus use bold instead of fine print and should be printed on paper that contrasts highly with the print. In addition, older consumers require better lighting not only in restaurants, but also in the parking lots for their safety and convenience (Knutson and Patton, 1993).

PURPOSE OF THE STUDY

The literature on mature customers generally recommends ongoing research into the development of the dining experience for mature guests. While many studies have focused on service quality dimensions, empirical studies of specific intangible service experiences provided by restaurant wait staff and their relationship with customer satisfaction has been largely ignored.

Thus, the study reported here addressed four objectives:

1. Examine the difference in the satisfaction level of mature (older than 55) and younger customers (55 years or younger) in three types of restaurants (quick-service, casual and fine-dining) as well as between mature and younger customers.
2. Identify the significant unpleasant service experience that impede satisfaction for mature and younger customers in three types of restaurants.
3. Determine differences between the mature and younger customers in the frequency of unpleasant service experiences.
4. Determine differences in complaint behavior between mature and young customers experiencing unpleasant restaurant service.

METHODOLOGY

The Survey

Data used in the analysis was accumulated in response to *The Penn State-USA Today Restaurant Service Survey*. The survey was developed at Penn State University and ran in the *USA Today* newspaper during October 1996. The 14-question survey asked *USA Today* readers their opinions on restaurant service. The authors realize that the *USA Today* readership is far from a random sample of the US population, and that those who answered the survey were self-selected. Nevertheless, *USA Today* proved an excellent vehicle for distributing a survey to a sample of some 5.2 million people. Moreover, the newspaper appears daily at the doors of several hundred thousand hotel rooms around the world, and the affinity business travelers have with the newspaper is well known. One can safely say that the *USA Today* readership is well-traveled and well-read. Table 1 shows the characteristic of the respondents.

The study produced a total of 5,992 responses; 2,372 mailed and 3,620 delivered electronically. The authors used only mailed survey responses for this study, since comparatively few mature customers used the on-line survey. Returned questionnaires that did not contain complete information were discarded. As a result, a total of 2,170 usable survey responses were used for this study.

TABLE 1. Characteristics of the Respondents

Characteristics	n	%
Gender		
Female	900	41.5%
Male	1,270	58.5%
Age		
18-25	108	5%
26-35	325	15%
36-45	542	25%
46-55	501	23%
56-65	412	19%
Over 65	282	13%
Income		
Less than $30,000	290	13.4%
$30,001-$45,000	527	24.3%
$45,001-$60,000	466	21.5%
$60,001-$75,000	313	14.3%
$75,001-$100,000	279	12.9%
Over $100,000	295	13.6%

The questions used in this study involved four types of information: (1) satisfaction levels in quick-service, casual, and fine-dining restaurants; (2) the frequency of unpleasant service experiences; (3) complaint behavior; and (4) demographics

Levels of satisfaction of each type of restaurant were measured on a 4-point scale where 1 is "very satisfied" and 4 is "not very satisfied," and the frequency of ten unpleasant service experiences were also measured on a 4-point scale where 1 meant "not at all" and 4 meant "very frequently." The ten unpleasant service experiences were developed by food service expert panels of quality researchers at Pennsylvania State University and the University of Nevada, Las Vegas. The ten unpleasant service experiences includes:

- Rude servers
- Inattentive servers
- Forgetful servers
- Servers who do not know the product
- Uncleanliness
- Intrusive service
- Rudeness
- Unwillingness to correct problems
- Incorrect billing

These 10 service experiences were representative of the five service dimensions of service quality. Uncleanliness and rudeness represents tangible dimensions of service quality. Slow service and intrusive service represents reliable dimensions of service quality. Forgetful servers, unwillingness to meet needs, and incorrect billing represents the responsiveness dimension of service quality. Finally, servers who do not know the product and inattentive servers respectively represent the assurance and empathy dimension of service quality.

The frequency of three complaint behaviors–verbal, written in letter form, and those who did not return–were evaluated on 4-point scale with 1 as " all the time" and 4 as "never." The last section of the survey consisted of demographic information (e.g., age, gender, and income level).

RESULTS AND DISCUSSION

Satisfaction in Three Different Types of Restaurants

Overall, the fine dining restaurant was the most satisfactory, and quick service restaurants were the least satisfied restaurant experiences by both young and mature customers. Table 2 shows that 93.5 percent of the respondents called themselves "satisfied" or "somewhat satisfied" in fine-dining restaurants, and 87.1 percent of them agreed that they were "satisfied" or "somewhat satisfied" in casual dining. However, only 69.6 percent of the respondents were "satisfied" or "somewhat satisfied" in quick-service restaurants. This percentage is quite low compared to casual and fine dining, and this result suggested that quick-service restaurants need major improvements in their service to satisfy customers.

TABLE 2. Percentage of Respondents Who Are Satisfied or Somewhat Satisfied in Different Types of Restaurants

	Overall	Younger customers (Under 55 years)	Mature customers (55 years or greater)	P- value
Quick service/fast food (no table service)	69.6%	71.3%	65.2%	.000*
Casual dining (table service)	87.1%	86.2%	91.3%	.001*
Fine dining	93.5%	94.1%	92.7%	.128

* = Significance at 0.05

Chi-square analysis tested the differences in satisfaction levels between the mature and younger customers for three types of restaurants. There are no significant differences in satisfaction in fine dining restaurants between the mature and younger customers. However, significant differences emerged in satisfaction in casual and quick-service restaurants between mature and younger customers. As Table 2 shows, mature customers (91.3%) are more likely to be satisfied in casual dining restaurants than younger customers (86.2%), and mature customers (65.2%) are less likely to be satisfied in quick-services restaurant than younger customers (71.3%). This result suggested the explanation of Shank and Nahhas' (1994) findings that mature customers are indeed less likely to eat at quick-service restaurants, and mature customers are more likely than younger customers to eat at mid-scale restaurants.

As shown in Table 2, satisfaction levels of the three types of restaurant were varied. The next section contains an investigation of which unpleasant services yielded significant customer satisfaction in three types of restaurants.

Significant Service Experiences for Satisfaction

The regression analyses were used to identify the most significant unpleasant service experiences that impeded customer satisfaction in three types of restaurants. The authors chose not to use confirmatory factor analysis to test service quality dimensions since the dimension approach is too broad to identify specific service experiences for overall customer satisfaction. By using ten unpleasant service experiences as independent variables and satisfaction level of three types of restaurant, the authors hoped to provide specific unpleasant service experiences that impede customer satisfaction in each type of restaurant.

For casual dining and quick service restaurants, it is necessary to use two separate regression analyses to identify specific service experiences for each age group, especially since there were significant differences in satisfaction between younger and mature customers from earlier findings (Table 2). The dependent variable was the satisfaction level of each age group, and the independent variable was the ten unpleasant service experiences. Subsequent analyses showed some differences between the two populations.

In casual dining restaurant, (1) inattentive servers and (2) lack of product knowledge were significant unpleasant service experiences for mature customers, while (1) inattentive servers and (2) slow service inattentive servers were significant unpleasant service experience that impeded younger customers' satisfaction in casual dining restaurants (Table 3).

In quick service restaurants, (1) lack of product knowledge, (2) inattentive servers, (3) rudeness and (4) uncleanliness emerged, respectively, as signifi-

TABLE 3. Significant Unpleasant Service Experiences That Can Negatively Affect Customer Satisfaction in Casual Dining Restaurant

Independent variables (Unpleasant service attributes)	Mature Customers		Younger Customers	
	Coefficient	Sig	Coefficient	Sig
Slow service	−5.1E-02	.183	−0.210	.000*
Inattentive Servers	−0.287	.000*	−0.232	.000*
Forgetful Servers	−4.0E-02	.221	−8.8E-02	.176
Servers who don't know the product	−0.310	.000*	−3.2E-02	.312
Uncleanliness	2.9E-02	.384	.68E-02	.234
Intrusive service	−1.37	.097	−1.21	.087
Rudeness	1.7E-02	.411	−3.1E-02	.407
Unwilling to correct problems	3.1E-02	.169	.23E-02	.471
Incorrect billing	−1.9E-02	.431	.25E-02	.361

Dependent Variable: customer satisfaction in casual dining restaurant
R^2 for mature customers: 19.5%
R^2 for younger customers: 18.7%
* = Significance at 0.05

cant unpleasant experiences impeding mature customer satisfaction (Table 4). For younger customers, (1) slow service, (2) inattentive servers, (3) rudeness and (4) uncleanliness emerged, respectively, as significant unpleasant service experiences (Table 4).

There are some common findings between casual dining and fine dining restaurants. In both casual dining and fine dining restaurants, inattentive servers were proved to yield significant unpleasant service experiences for both age groups. Additionally, lack of product knowledge was significant in unpleasant service experiences only for younger customers, and the lack of product knowledge was also significant for mature customers in casual dining and fine dining restaurants

The result of regression analyses for casual dining and fine dining restaurants suggested that mature customers seem more tolerant or forgiving of slow service than younger customers, since slow service was significant only for younger customers. This result is consistent with the Shank and Nahhas (1994) study that showed older customers are less likely to rate speed of service as an important factor in selecting a restaurant. Shank and Nahhas (1994) found that mature customers are more likely to agree with the statement, "It is okay if the wait staff is too busy to respond to my requests promptly." Previous research

TABLE 4. Significant Unpleasant Service Experiences That Can Negatively Affect Customer Satisfaction in Quick Service Restaurant

Independent variables (Unpleasant service attributes)	Mature Customers		Younger Customers	
	Coefficient	Sig	Coefficient	Sig
Slow service	–5.2E-02	0.101	–0.291	0.000*
Inattentive Servers	–0.137	0.000*	–0.188	0.000*
Forgetful Servers	–5.8E-02	0.073	–7.4E-02	0.199
Servers who don't know the product	–0.277	0.000*	–4.7E-02	0.309
Uncleanliness	–0.198	0.000*	–1.51	0.000*
Intrusive service	2.2E-2	0.387	–5.9E-02	0.376
Rudeness	–0.212	0.000*	–0.287	0.000*
Unwilling to correct problems	2.4E-02	0.327	–2.9E-02	0.511
Incorrect billing	–2.1E-02	0.432	–6.8E-02	0.236

Dependent Variable: customer satisfaction in quick service restaurant
R^2 for mature customers: 25.1%
R^2 for younger customers: 23.3%
* = Significance at 0.05

suggests two reasons mature dinners are less concerned with timeliness. First, they see dining as a social event; second, mature customers tend to have more leisure time than younger customers (Pederson and DeMicco, 1993).

Lack of product knowledge emerged as a significant impediment to satisfaction for mature customers in casual dining and fine dining restaurants. The need mature customers have for more information about the menu items than do the less nutrition-conscious younger customers, may explain this result. Fintel (1990) found that mature customers appreciate the availability of nutritional information. Thus, it seems wise for a restaurant's wait staff to be informed about the nutritional value of the foods they serve so they can guide mature customers.

In fine dining restaurants, since there are no significant differences in satisfaction levels between younger and mature customers (Table 2), a single regression analysis was used for both age groups. However, all ten unpleasant experiences were not significant. In other words, those ten unpleasant experiences did not impede customer satisfaction in fine dining restaurants. This result explains why fine dining restaurants were the most satisfactory restaurants for both age groups.

The R-square of the regression model ranges from 17.7 percent to 25.1 percent, and this range means that the ten unpleasant service experiences in this study explain only 17.7 percent to 25.1 percent of customer satisfaction in three types of restaurants. These results show that there are many other factors than just these ten unpleasant service experiences that explain customer satisfaction in three types of restaurants. However, these R-square results also suggest that those ten unpleasant service experiences could reduce up to 25.1 percent of the satisfaction level. Few restaurants can afford to ignore this high percentage of customer satisfaction. Because of this important factor, the findings are important.

The Frequency of Unpleasant Service Experiences by Age (55 or under versus over 55)

Chi-square analyses were conducted to determine whether significant differences existed between the younger customers (55 years or younger) and mature customers (older than 55 years) in terms of the perceived unpleasant service food servers provide. Significant differences emerged between the mature and younger customers in frequency of experiences in (1) slow service, (2) inattentive servers, (3) forgetful servers, and (4) servers who did not know the products.

Of the items listed in the survey, as shown in Table 5, slow service emerged as the most frequently perceived unpleasant service, and it was the primary

TABLE 5. Percentage of Respondents Who Encounter These Problems Very Frequently or Somewhat Frequently

	Overall	Younger customers (under 55 years)	Mature customers (55 years or older)	P-value
Slow service	80.2%	83.3%	74.1%	.000*
Inattentive Servers	67.6%	70.9%	61.3%	.000*
Forgetful Servers	55.1%	57.5%	50.4%	.000*
Servers who don't know the product	51.2%	49.0%	56.1%	.000*
Uncleanliness	41.6%	42.7%	39.5%	.178
Intrusive service	31.3%	31.9%	30.1%	.297
Rudeness	29.3%	29.5%	28.9%	.312
Unwillingness to meet needs	19.4%	20.4%	17.5%	.182
Incorrect billing	13.5%	12.9%	14.8%	.175

* = Significance at 0.05

problem for both age groups (80.2%). However, there are significant differences between the mature and younger customers in frequency of slow service. Younger customers (83.3%) perceived themselves as receiving slow service more frequently than mature customers (74.1%).

The second and third most frequently perceived unpleasant services for all age groups were "inattentive servers" (67.6%) and "forgetful servers" (55.1%), respectively. Again, there are significant differences between the mature and younger customers in frequency of experiences of inattentive servers and forgetful servers. Younger customers more frequently perceived inattentive and forgettable servers than mature customers (Table 5). One wonders whether servers provide more attention to their mature customers than to younger customers. In any event, more than half of the mature respondents said they encounter inattentive servers "very frequently" or "somewhat frequently." This result would seem unacceptable for the restaurants industry since, as shown earlier in the regression analyses, inattentive servers appear to be one of unpleasant service experiences most liable to decrease mature customers satisfaction in both casual and quick service restaurants.

Lack of product knowledge ranked as the fourth most frequently experienced problem by both age groups (51.2%), and there are significant differences between the mature and younger customers (Table 5). Mature respondents (56.1%) experienced this service deficiency more frequently than younger respondents (49%). This appears to be another service area where the restaurants industry can improve. Although lack of product knowledge emerged as a significant impediment to satisfaction for mature customers both in casual dining and fine dining restaurants, still 56.1% of mature customers regularly experienced the servers who do not know their products. Restaurant servers would do well to become informed about their products so as to guide mature customers.

Complaint Behavior

Again, the authors performed Chi-square tests to examine the differences between the complaint behavior of differences between the mature and younger customer. The most frequently used complaint behavior for both mature customers and younger customers was to not return to the restaurants, and more younger customers (70.1%) than mature customers (66.9%) said they do decline to return to restaurants when they have had an unpleasant experience there. Significant differences emerged between mature and younger customers in terms of complaint behaviors. Table 6 shows mature customers (32.7%) are more likely to complain verbally than younger customers (28.7%). Similarly,

TABLE 6. If I Have Bad Service in a Restaurant, I . . .

	Overall	Younger customers (under 55 years)	Mature customers (55 years or greater)	p-value
Complain verbally (all the time or frequently)	29.3%	27.7%*	32.7%*	.000*
Write a letter (all the time or frequently)	7.9%	6.6%*	10.7%*	.000*
Do not return (all the time or frequently)	68.9%	70.1%	66.9%	.087

* = Significance at 0.05

mature customers (10.7%) are more likely to complain with a letter than younger customers (6.6%).

These results suggest that, overall, mature customers are more likely to complain than younger customers when they have had poor service. This finding contradicts previous behavioral research suggesting that dissatisfied mature adults are generally less likely to "pursue their rights" than younger ones (Bearden, 1983 and Lafarge 1989). Solomon, Supernant, Czepiel, and Gutman (1985) suggest that complaint behavior in restaurants is more acceptable to mature customers because of previous experiences and established service scripts.

CONCLUSION

As pointed out earlier, attracting more mature customers promises to become crucial to the success of the restaurants industry in view of the increasing size of this market. This study revealed that fine dining restaurants were the most satisfactory restaurants, and quick service restaurants were the least satisfactory restaurants by both young and mature customers. Additionally, quick-service restaurants satisfy mature customers less reliably than casual restaurants. What can the restaurant industry do to increase the satisfaction of mature customers? This study provided answers to this question by specifying unpleasant service experiences liable to lessen satisfaction for mature customers.

In fine dining restaurants, almost 93 percent of customers were satisfied, and there were no unpleasant service experiences that were significant for both mature and young customers.

Unlike fine dining restaurants, there is room for improvement in casual dining and quick service restaurants. In casual dining restaurants, (1) lack of product knowledge and (2) inattentive servers provided significant unpleasant service experiences for mature customers. Inattentive servers were significant for both age groups, and lack of product knowledge was significant for mature customers only.

In quick service restaurants, (1) lack of product knowledge, (2) inattentive servers, (3) rudeness, and (4) uncleanliness were significant unpleasant service experiences that impeded the satisfaction of mature customer. Again, lack of product knowledge was significant for mature customers only, as it was in casual dining restaurants. Therefore, this area is in need of definite improvement.

The most frequently used complaint behavior for mature customers was not to return the restaurants. This means that if the restaurants industry does not fix the significant unpleasant service that this research suggests, they will lose their valuable customers. In the next section, suggestions will be made as to how the restaurants industry could improve their service to increase mature customers' satisfaction.

Management Implications

The results of this study have important marketing implications for the restaurant industry. To increase the level of satisfaction among mature customers, restaurant managers should make sure that (1) their employees have good product knowledge; (2) all servers are attentive to their guests; (3) servers are kind to their guests, and (4) the restaurant is always clean. More specifically, restaurant managers can usefully consider these four behaviors associated with the four general principles just enunciated:

Regarding the Employees' Product Knowledge

This research suggested that significant service for both casual dining and quick service restaurants is necessary, and, more importantly, was imperative particularly for mature customers. Therefore, restaurants should make sure that their employees have good product knowledge for their mature customers. Mature customers are nutrition-conscious and want nutritional information about such food components as calories and fat percentage. Also, mature customers may be concerned about preparation methods and ingredients. Thus, educating food servers about nutritional information and other menu related information always helps. Also, suggestive selling of healthful menu items among mature guests can be both helpful and profitable, and is a good strategy.

Regarding Attentive Servers

This research revealed that it was also significant service for both casual dining and quick service restaurants that was important. Therefore, food servers in casual dining restaurants should look for opportunities to provide extra attention to the their mature customers. In quick service restaurants, counter clerks must be patient when mature customers order, since they may need extra time to choose and to phrase their orders. Cleaning staff members should pay attention to their mature customers since they may need some help finding a seat or getting up due to physical problems.

Regarding Kindness

This research found that significant service was required in quick service restaurants for both age groups. However, friendly interaction between employees and mature customers can be very important, regardless of the restaurants type, since eating out is a social event for mature customers. Promoting friendly attitudes toward mature customers can be crucial. A smile is always a good strategy.

Regarding Cleanliness

Significant service in quick service restaurants for both age groups was important, but it should not be ignored in fine and casual dining restaurants. Clean counters, clean tables, and clean employees can be crucial for not only mature customers, but also younger customers. Make sure that door handles, windows, and bathrooms are clean, and it is a good idea to post a cleaning schedule publicly to let customers know that the restaurant is cleaned and checked regularly.

Limitations

There are three important limitations in this study. First, the ten unpleasant service experiences used in this study were not enough indicators of customers' satisfaction since those experiences produced low r-square. Definitive conclusions may require more than ten service experiences. Additionally, although the same ten unpleasant service experiences used to predict satisfaction in different type of restaurants were used, those service experiences could differ, depending on the restaurant types. Finally, this research does lack external validity, since the *USA Today* readership is not representative of the entire population of the United States. Although this study has some limitations, it does provide the restaurant industry with helpful advice about those services likely to increase customer satisfaction.

REFERENCES

Bearden, W. and J. E. Teel (1983), "Selected Determinants of Consumer Satisfaction and Complain Reports," *Journal of Marketing Research*, 20, (February), 21-28.

Bojanic, D. C., and Rosen, L. D. (1994), "Measuring Service Quality in Restaurants: An Application of the SERVQUAL Instrument," *Hospitality Research Journal*, 18(1), 4-14.

Cetron, M., DeMicco, F. J., and Williams, J. A. (1996), "Restaurant Renaissance: Present and Future Trends in Foodservice Management," *The Futurist*, 8-12.

Groonos, C. (1983), "A Service Quality Model and Its Marketing Implication," *European Journal of Marketing*, 18, 36-44.

Fintel, J. (1990), "Gold Among the Silver and Gray: How to Market in an Aging America," *Restaurant USA*, 17-22.

Fu, Y. (1998). *The Influence of Service Quality on Older Diners' Restaurant Loyalty*. Unpublished dissertation, The Pennsylvania State University, PA: State College.

Knutson, B., Stevens, P., Patton, M., and Thompson, C. (1992), "Consumers' Expectations for Service Quality in Economy, Mid-price and Luxury Hotels," *Journal of Hospitality & Leisure Marketing*, 1(2), 27-43.

Lafarge, M. (1989), "Learned Helplessness as an Explanation of Elderly Consumer Complaint," *Journal of Business Ethics*, (May), 359-366.

Lehtinen, J. R. (1983), *Customer Oriented Service System*. Unpublished Working Paper, Helsinki: Service Management Institute.

NPD CREST. (1988), *Marketing to seniors*. Consumer reports on Eating Share Trends [CREST] report, Park Ridge, IL.

Parasuraman, A., Zeithaml, V. A., and Berry, L. L. (1985), "A Conceptual Model of Service Quality and Its Implications for Future Research," *Journal of Marketing*, 49, 460-469.

Pederson, B. & DeMicco, F.J. (1993), "Restaurant Dining Strategies: Attracting Nutrition-Conscious Future Seniors," *Florida International University Hospitality Review*, 11(2), 7-17.

Richard, M. D., Sundaram, D. S., & Allaway, A. W. (1994), "Service Quality and Choice Behavior: An Empirical Investigation," *Journal of Restaurant & Foodservice Marketing*, 1(2), 93-109.

Shank, M. D. and Nahhas F. (1994), "Understanding the Service Requirements of the Mature Market," *Journal of Restaurant & Foodservice Marketing*, 1(2), 23-43.

Solomon, M., C. Supernant, J. Czepiel, and E. Gutman (1985), "A Role Theory Perceive on Dyacdic Interactions: The Service Encounter," *Journal of Marketing*, 49, (Winter), 99-111.

U.S. Bureau of the Census (1998), *Statistical Abstract of the United States*: Washington D.C.

Williams, J. A. and DeMicco, F. J., and Kotschevar, L. (1997), "The Challenges That Face Restaurants in Attracting and Meeting the Needs of Older Customers," *Journal of Restaurant and Foodservice Marketing*, 2(4), 49-63.

Consumer Attitude Toward Adding Quick-Service Foods to In-Flight Meals

Juline Mills
Joan Marie Clay

SUMMARY. This study examines consumer attitude toward the addition of quick-service foods (QSF) to domestic coach class in-flight meals. Multiple regression analysis is used to test two hypotheses. Findings indicate that airline passengers who are satisfied with current airline food do not want QSF added to in-flight meals. However, airline passengers who currently enjoy QSF would like to see the addition of QSF to in-flight meals.

KEYWORDS. Quick-service foods (QSF), frequent flyers, airlines, customer satisfaction, co-branding

Airline food . . . reminds us that we're sticky and cramped, tired and hungry. No one seems to enjoy any of it. But isn't that the way it always is with in-flight meals? (Kessler, 1998)

Juline Mills, MS, is a Doctoral Student at Purdue University.
Joan Marie Clay, PhD, is Associate Professor and Chair, Hospitality Management, University of North Texas.

[Haworth co-indexing entry note]: "Consumer Attitude Toward Adding Quick-Service Foods to In-Flight Meals." Mills, Juline, and Joan Marie Clay. Co-published simultaneously in *Journal of Restaurant & Foodservice Marketing* (The Haworth Hospitality Press, an imprint of The Haworth Press, Inc.) Vol. 4, No. 3, 2001, pp. 99-112; and: *Quick Service Restaurants, Franchising and Multi-Unit Chain Management* (ed: H. G. Parsa and Francis A. Kwansa) The Haworth Hospitality Press, an imprint of The Haworth Press, Inc., 2001, pp. 99-112. Single or multiple copies of this article are available for a fee from The Haworth Document Delivery Service [1-800-HAWORTH, 9:00 a.m. - 5:00 p.m. (EST). E-mail address: getinfo@haworthpressinc.com].

INTRODUCTION

Some type of food and beverage service has been available to passengers since airlines began offering commercial passenger service (Borchgrevink, 1999). On October 11, 1919, the first in-flight meal, a box lunch, was served on a flight from London to Paris (Slippery Enterprises, 1994). Since those early beginnings, in-flight meal service has moved from extensive menus in the 1970s and 1980s to become the subject of jokes and jibes and the brunt of many comedy routines (George, 1998; Spector, 1999).

In the early 1990s, the airlines responded by cutting costs using crude and obvious methods such as cramming seats closer together, removing olives from salads, then removing salads altogether (McCartney, 2000a). Battered by billions of dollars in losses, the airline industry tried to save money any way it could (Maxon, 1997). In 1999, many airlines tried to make a comeback by spending more to treat their passengers better, but the issue of in-flight meals is still a sore point. Industry analysts predict that the five hundred million passengers expected to board United States domestic flights are less likely to get an airline meal than at any time in modern commercial aviation history (Buckley, 1999).

Passengers expecting more from airlines have been vocal about their desire to have more and better foodservice available on all flights (Schwartz, 2000). Faced with this situation, consumers have derived alternative solutions for poor airline foodservice such as taking their own food on flights (Nomani, 1999). Passengers are also seeking more meals on the ground at airport concessionaire stands and quick-service outlets before boarding their flights (Earle, 1998; Moran, 1998). Thus, airlines are losing a possible income source to airport food outlets, upscale hotels and other restaurants that specialize in preparing meals for travel (Blank, 1999).

Generally, airline passenger attitude and expectation of in-flight menu items is low. The image and quality of food offered on United States airline domestic flights lacks consistency and is described by some consumers as inadequate or inferior (Nomani, 1999;Williams, 1995). Research is needed to determine an appropriate course of action or solution for improving in-flight meals. Given this background this study proposes Quick-Service Foods (QSF) as the alternative to current in-flight meals. Thus, the purpose of this study is to determine whether or not a relationship exists between adding quick-service foods to domestic coach class in-flight meals and airline passenger satisfaction.

LITERATURE REVIEW

If an organization is to remain competitive in today's service-oriented economy, it must address the needs of its customers. According to a British Traveler survey of U.S. domestic airlines conducted in 1996, meals ranked eleventh in importance to airline carriers after scheduling, punctuality, safety, speedy check-in, route network, comfort, price, departure lounges, mileage programs and cabin staff. However, airline industry analysts contend that this low ranking might not be as limited in importance as is suggested but could be the deciding factor in differentiating an airline with emphasis on quality service from the rest of the competition. For example, though food may not rank high among the top factors in choosing between airlines, Singapore Airlines states that food can evoke emotional sentiments and it is certainly a key service element for its repeat passengers. The airline contends that if the meal served is not appetizing the flight is also considered bland to passengers (Tu, 1997).

Foodservice is not a major factor in generating brand preferences for airlines in the United States, but there is sufficient evidence that it could become one if appropriate changes were made (Lombardi, 1996). The 1998 National Survey of Frequent Flyers' Attitudes toward Airline Food Service states that 56% of frequent flyers would make an effort to fly an airline whose food was superior, while 43% percent would like to see new and popular food items being introduced on airline menus (*Business Wire*, 1998).

Many airlines are in need of a more "intelligent concept of catering" as they are bogged down by foods that are unsuitable for in-flight conditions (Thorpe, 1998). In recognizing this problem of "plastic meats and pallid fruits," some airlines have stopped serving food while others have announced new food-free short-haul flights (Sturgis, 1998). In addition, the quality of meals is sometimes adjusted based on passenger load: Hot lunches for flights with lots of business travelers, but snack sacks for a plane load of cheap ticket vacationers (McCartney, 2000b). Previous research also indicates that no meals are offered on flights shorter than an average of ninety minutes; on longer flights food is generally served during standard meal times (Williams, 1995).

Co-Branding of In-Flight Meals and Quick Service Foods

There is no doubt that branding wields power and that the quick-service industry has that power. When people recognize a brand, it makes them feel at home. A brand can also add prestige and an innate sense of quality to a dining experience (Grattan, 1998). Along with quality image, brands generate considerable profit. If a customer perceives a message of guaranteed quality in a recognized brand name, it will win regardless of the cost of the item (McDermid,

1997). For example, when pupils in the London Borough of Havering, England were served burgers in McDonalds look-alike wrappers, sales rose dramatically (Food Service Management, 1996).

The idea of the quick-service industry adding menu items to airline in-flight meal service is termed "co-branding." Co-branding is the pairing of two or more recognized brands within one space. The idea behind co-branding is that several brands together can command more power through customer awareness and traffic than a single brand-name operation (Boone, 1997). Co-branding in the airline industry is not a new concept. United Airlines introduced Starbucks Coffee in 1998 on its domestic flights along with Ben & Jerry's Ice Cream and Eli's Cheesecake. United Airlines used these brands to redefine its in-flight dining experience (*Bon Appetite*, 1999). However, the long-term success of this partnership is unknown as customers complained about the quality of the airline version of these products. Consumer comments included the fact that the coffee being offered tasted nothing like the real Starbucks Coffee (Kessler, 1998).

The possibility of quick-service businesses forming an alliance with the airline industry is an untapped field. Airlines could benefit from an affiliation with quick-service restaurants, as they are well-established business phenomena with identifiable brands. The importance of quick-service food items has not gone unrecognized (Binkley & Eales, 1998). Big fortunes in the foodservice business have been made in the quick-service segment which continued in 1998 to dominate the Top Ten Who's Who in Restaurant Sales (Restaurants and Institutions, 1998). Quick-Service Restaurants (QSR) have been grabbing market share through concept evolution, strong marketing, and promotion programs. QSR chains are familiar icons that consumers identify with, and many have solid reputations (Aaker, 1991). Quick-service franchising builds on three well-established business phenomena: product standardization, economies of scale, and mass marketing (Lundberg, Stavenca & Krishnamoorthy, 1995).

This standardization of product coupled with reduced costs could assist the airline industry with addressing the problem of inconsistent products being served on flights. Based on the current food cost per passenger, quick-service menu items also cost less to prepare. The average quick-service meal retails to the customer at less than five dollars for a combination meal of, for example, a hamburger, fries and a drink (Muller, 1997). The average food cost for in-flight meal service ranges from twenty-two cents for just peanuts and beverage service on Southwest Airlines to a high of $7.92 on American Airlines for all classes of service (Coles, 1997). For airlines looking to their foodservice in an effort to increase sales and achieve product differentiation, pairing with the

QSF industry might be the way to go (*Food Management*, 1996a; Barfield, 1997; Woodyard, 1999).

In addition, acceptance of foods that are served on airlines is one of the problems that passengers currently have with in-flight meals. However, customer acceptance and frequency of use are two major drivers of success within the QSF industry. Quick-service food as a comfort item is important especially for teenagers and small children. Many small children are afraid of flying; the placing of quick-service foods on airlines could possibly improve confidence and relaxation because of the familiarity with this food (Cebrzynski, 1999).

Another benefit that the quick-service industry can provide is that it is a familiar food, and most passengers are aware of the ingredients and preparation methods. This is not always the case with in-flight menu items. For example, nuts and foods containing peanut derivatives have been known to cause fatal reactions, resulting in deaths on airline flights (Fitzsimons, 1997). Previous experience with and knowledge of the products would allow consumers to identify immediately the possibilities of allergic reactions.

The airline industry would not be the only segment that stands to gain from this relationship. The quick-service business would also benefit from entering in-flight meal service as a steady growth in the number of restaurants and a less than anticipated increase in the demand for dining services has made the restaurant industry highly competitive (Sundaram, Jurowski & Webster, 1997).

The traditional restaurant markets are saturated; chains seeking to expand are looking elsewhere (*Food Management*, 1996a). Studies indicate that competition is intense in the quick-service business, and quick-service chains are targeting new non-traditional locations in which to invest (Shubart, 1998; Muller, 1997; Pettijohn, Pettijohn & Luke, 1997). Quick-service franchisers are targeting new locations in schools, colleges, hospitals and airports. Many quick-service restaurants are opening non-traditional sites and are planning to continue to grow in these areas in the future. McDonald's in recent years has formed an association with Wal-Mart Inc. giving them the opportunity to provide meals to customers in their larger stores. Shoppers get the opportunity to pause while shopping to enjoy a familiar meal (Shubart, 1998) Given this background, the possibility of implementing quick-service foods on airline menus could achieve similar relationships as with McDonald's and Wal-Mart. The opportunity thus exists for both industries to increase market share, customer satisfaction and ultimately profits (*Business Wire*, 1998).

Customer Satisfaction

The variable of customer satisfaction is instrumental in measuring the possible success of quick-service items being implemented on domestic coach class in-flight meals. Customer satisfaction has traditionally been viewed as the outcome of confirmed or disconfirmed expectations. Confirmed expectations tend to yield satisfaction while disconfirmation of expectations tend to yield dissatisfaction (Rosenberg, 1956; Aronson & Carlsmith, 1967; Fishbein & Icek, 1975; Miller, 1977; Lewis & Chambers, 1989). Customer satisfaction is a complex, relative, and individual subjective evaluation of a life experience and is difficult to quantify and predict. Studies of consumer behavior emphasize customer satisfaction as the core of postpurchase evaluation and an essential aspect of the purchase process (Czepiel and Rosenberg, 1977; Yuskel and Rimmington, 1998). For the purposes of this study customer satisfaction is addressed at the macro level (overall picture) and is defined as the fulfillment or gratification of customers wishes. Customer satisfaction is created by exceeding expectations, delivering quality, and targeting customer preferences (Barsky, 1995; Garman, 1996).

In saturated markets, as with both the airline and the quick-service industry, customer satisfaction is thought to be one of the most valuable assets, serving as an exit barrier, thus helping the firm to retain its customers (Gundersen, Heide & Olson, 1996). According to Sellars (1998) customers tell twice as many people about bad experiences as good ones. Complainers left unhappy can ruin a company's image. A study conducted by TARP Inc. shows that happy customers tell 5-6 people, unhappy customers tell 8-10, and 10% of the unhappy people tell 20 or more people. If customer satisfaction is not met, then companies lose profitability, growth, and customers, as quoted in Love (1998).

The potential benefits of improved in-flight meals include heightened advantage over competitors in product differentiation, increased customer retention, and positive word-of-mouth communication (Heskett, 1994; Yuksel & Rimmington, 1998). In addition, the American Customer Satisfaction Index (ASCI) also showed that when customer satisfaction scores increased or decreased, the stock prices of those companies have followed the moves three months later (Lieber, 1998; Martin, 1998). If customers become satisfied with the product or menu item that is being offered then the opportunity exists for them to repurchase the product. The airline industry has in recent years become interested in developing loyal guests who continually repurchase their product through relationship marketing. The airline industry's approach so far has focused largely on transactional tactics such as frequent-flyer programs. A major weakness of this approach is that any other company can copy these tactics (Bowen & Shoemaker, 1998). Food, on the other hand, is difficult to duplicate and may be a possible solution to garnering airline passenger loyalty.

Thus, the current study is designed to explore the relationship(s) between customer satisfaction and the addition of QSF to domestic coach class in-flight meals as another solution to building loyal passengers. To achieve the objective two hypotheses, stated in null form, were developed.

H_1: Among frequent-flyers, there is no relationship between current satisfaction with airline food and the satisfaction derived by adding quick-service foods to in-flight meals.

H_2: Among frequent-flyers, there is a no relationship between satisfaction with quick-service foods and satisfaction derived by adding quick-service foods to in-flight meals.

Based on these hypotheses two independent and one dependent variable were developed. The independent variables are: (1) Customer Satisfaction with Airline Food, and (2) Customer Satisfaction with Quick-Service Foods. The dependent variable is Customer Satisfaction with adding Quick-Service Foods to in-flight meals.

METHODOLOGY

The population for the research study consists of frequent flyers. A frequent-flyer is any individual who uses the services of an airline at least once per month. The sample consisted of domestic frequent-flyers from the United States. A minimum sample size was determined using the simplest "rule of thumb" ($n > 50 + 8m$) for testing regression (Wetherill, 1986). Where: n = number of participants required and m = the number of independent variables (4). Thus the minimum sample size required for the study was eighty-two.

A listserv of frequent flyers was obtained from a website that is the Internet's premier site for frequent flyer program information and advice. This company produces an award-winning weekly electronic mail (e-mail) newsletter. The newsletter delivers news, updates, trends and opinions with an insider perspective but consumer focus on the airline industry. Through this newsletter the survey was sent to the participants.

The questionnaire consisted of four sections. Sections one and two represented the independent variables while section three represented the dependent variable. Each section contained ten questions and used a five-point Likert-type rating scale to measure the independent and dependent variables. The instrument then concluded with a section on demographics. Figure 1 gives a sample of questions asked for both the independent and dependent variables.

In developing the questionnaire content validity of the instrument was first established through a review of related literature on customer satisfaction.

FIGURE 1. Sample Research Questions

Variables	Questions
Customer Satisfaction with Airline Food ***(Five-point Likert-type scale)***	Listed below is a series of statements asking you to describe your level of satisfaction with current airline foods. Please Indicate your level of agreement with each statement. ***Please check (√) only one box.*** 1. I am satisfied with the quality of meals on airlines. 2. Airline meals offer satisfaction for price paid.
Customer Satisfaction with QSF ***(Five-point Likert-type scale)***	Listed below is a series of statements asking you to describe your level of satisfaction with fast-foods. Please indicate your level of agreement with each statement. ***Please check (√) only one box*** 1. Fast-foods appeal to me. 2. The quality of fast-food items is good.
Customer Satisfaction with Adding QSF to In-Flight Meals ***(Five-point Likert-type scale)***	The statements below refer to the use of fast-food on in-flight meal service. Please indicate your level of agreement with each statement. ***Please check (√) only one box*** 1. I would prefer fast-food on flights because of the familiar taste. 2. I would prefer fast-food on flights because of its quality.

Second, several experts in the field examined the instrument. Third, a pilot study was conducted using thirty frequent flyers from the North Texas region. This group was chosen randomly and did not include any members of the sample population. Respondents were asked to complete the questionnaire and to analyze the instrument's clarity, content and computer interface. Of the twenty-one surveys returned from the pilot study twenty were useable. Cronbach's Alpha was computed and Standardized Item Alpha for all questions ranged from 0.93 to 0.53. Preliminary regression results also revealed that beta values for the independent variables did not equal zero. Based on the results of the pilot study the instrument was refined.

The data collection phase of the survey yielded 156 responses. The website received a total of 420 hits within the two week period it was posted, resulting in a 37% response rate. The returned instruments were reviewed and the data verified. Since the current study is focused on U.S. domestic coach class passengers, international passengers, first class passengers and airline employees were removed from the database. The total sample size of 156 was reduced to 94.

RESULTS AND DISCUSSION

The following demographic information was gathered in the study: age, gender, and number of flights per month. Thirty-eight percent (n = 36) of the respondents were between the ages of twenty-one and thirty-four. Fifty-five percent (n = 52) were male. Seventy-two percent (n = 68) flew one to three times per month with the airline of choice being American Airlines (26.6%), followed by Delta Airlines (19.1%), and United Airlines (17.0%).

Before statistical analyses were performed negatively worded items were reversed and re-coded. "Don't Know" responses from the Likert-type scale as well as items which received no response were coded as missing values. For both the independent and the dependent variables the multiple measurements of each variable were then summed to arrive at a single score for each variable. Cronbach's Alpha = 0.82, indicating a high level of reliability. A multiple regression linear model was used to assess the relative strength of the two independent predictor variables resulting from the hypotheses on the dependent variable. The multiple regression analysis was then performed using the following relationship:

$$FUT\_AIR = A + \beta_1\ CS\_AIR + \beta_2\ CS\_FASTFOOD + E$$

KEY:

Dependent Variable:
FUT_AIR–Customer Satisfaction with adding QSF to in-flight meals.

Independent Variable:
CS_AIR–Customer Satisfaction with the current Airline Food.
CS_FASTFOOD–Customer Satisfaction with Quick Service Foods.

Other Terms:
A = the value of FUT_AIR when all the independent variables are zero,
β = the coefficients assigned to the independent variables, and
E = an error term.

The resulting equation yielded a R^2 value of 0.44 and an adjusted R^2 of 0.43, providing an indication of the goodness of fit of the model. Thus, 44% of the variations in customer satisfaction with quick-service meals on coach class in-flight meals can be explained by customer satisfaction with airline food and customer satisfaction with quick-service foods. The model yielded F statistic of 35.64 and a p value of less than 0.001. This F statistic is significant and therefore the model is significant.

Collinearity diagnostics were conducted to determine the degree to which each independent variable is explained by the other independent variables. Variance Inflation Factor (VIF) scores were 1.02 for both CS_AIR-Customer Satisfaction with Airline Food and CS_FASTFOOD-Customer Satisfaction with QSF. Tolerance scores were 0.98 for both CS_AIR and CS_FASTFOOD. The multicollinearity level is therefore minuscule. Individual regression coefficients were then tested to determine the significance of each independent variable. Table 1 displays the resulting beta values and t-statistics.

Based on these results the two independent variables, CS_AIR–customer satisfaction with current airline food and CS_FASTFOOD–customer satisfaction with quick-service foods contributed significantly to the prediction of customer satisfaction with adding quick-service foods to domestic coach class in-flight meals (FUT_AIR).

The results for CS_AIR ($\beta_2 = -0.422$, $t = -3.09$) suggest that a negative relationship exists between airline passenger satisfaction with current airline food and the satisfaction they would derive from adding QSF to domestic coach class in-flight meals. Thus, if airline passengers are satisfied with current airline food they perceive that they will not be satisfied with QSF being added to coach class in-flight meals.

The results also show that a significant relationship exists between CS_FASTFOOD–customer satisfaction with QSF and the dependent variable FUT_AIR–satisfaction with adding QSF to domestic coach class in-flight

TABLE 1. Customer Satisfaction of Adding Quick-Service Foods to Domestic Coach Class Inflight Meals

Independent Variables	Standardized Coefficients β	Standard Error	t-Statistics	Significance Level
Constant		3.745	2.811	0.006
CS_AIR	−0.245	0.136	−3.099	0.003
CS_FASTFOOD	0.647	0.107	8.177	0.000

Dependent Variable: FUT_AIR; $p < 0.001$, F stat = 35.64; $R^2 = 0.44$, adjusted $R^2 = 0.43$.

KEY:

Dependent Variable:
FUT_AIR - Customer Satisfaction with adding QSF to in-flight meals.

Independent Variable:
CS_AIR - Customer Satisfaction with Airline Food.
CS_FASTFOOD - Customer Satisfaction with QSF.

meals ($\beta_2 = 0.872$, $t = 8.17$). Thus, if airline passengers are satisfied with QSF they perceive that they will be satisfied with the addition of QSF to domestic coach class in-flight meals.

Based on the results H_1 and H_2 are rejected. The results of H1, frequent flyer current satisfaction with airline food is not related to satisfaction of adding QSF to domestic coach class in-flight meals, are in keeping with earlier findings of studies which examine consumer expectation with novel and familiar foods. Pliner (1982) suggests that familiarity can drive food selection. This is evident, as airline passengers already familiar with current airline food and unsure of how QSF would appear, prefer to remain with current in-flight foods. In addition, Cardello, Maller, Masor, Dubose and Edelman (1985) contend that the greater the degree with which a consumer's experience with a product matches his/her expectation, the greater the liking for that product will be. Thus, it is less likely that the consumer will want to try another product.

The most significant finding in this study is found when testing hypothesis two. This reveals that there is a relationship between frequent flyer satisfaction with QSF and satisfaction of adding QSF to in-flight meals. This suggests that airline passengers because of their previous familiarity and satisfaction with QSF would react favorably to its addition to domestic coach class in-flight meals. This finding supports Cardello (1994) earlier conclusions that one's expectation of a food has a profound effect on its acceptability in any setting. Bell and Meiselman (1995) also contend that consumers evaluate the merits of foods and the source from which the food is being selected when making food choices.

This study revealed the presence of two strong relationships. In order to establish causality and ultimate development of the concept, further research is needed. In using the results from this study, both the airline and quick-service industries should conduct in-depth analysis into customer satisfaction and adding QSF to airline domestic coach class in-flight meals. If QSF were to be added to airline in-flight meals, then the brands that are chosen must be viewed from a "boredom-busting" perspective. The biggest issue with branding as it relates to the foodservice business is that customers grow tired of the same thing, day-in and day-out. Foodservice establishments rotate menus and stagger promotions in order to keep customer interest alive (*Food Management*, 1996b). The same would hold true for the airline industry. If the airline industry were to add QSF then a variety instead of a single quick-service food item would have to be considered.

The results of this study indicate that if consumers are happy with the food they currently receive on airlines they would not want this food to change to QSF. On the other hand, if consumers are satisfied with QSF they would want to see QSF added to airline coach class in-flight meals. In order to make this concept viable, other factors in the industry should be considered such as preparation methods and facilities, holding time for foods, and the finances necessary to undertake such a venture. Further research into this topic could include experimenting with QSF on airlines to see which items would be acceptable to airline passengers. Would airline passengers, given a choice of QSF, prefer McDonald's, Burger King, or Taco Bell?

REFERENCES

Aaker, D. (1991). *Managing brand equity: Capitalizing on the value of a brand name.* New York: The Free Press.

Aronson, E. & Carlsmith, J. (1967). Performance expectancy as a determinant of actual performance. *Journal of Abnormal and Social Psychology*, 65(3), 178-182.

Barfield, M. (1997). Domestic food spending for top US air carriers drops again. Spending for international destinations up. *Onboard Services*, 29(6), 1.

Barsky, J. (1995). *World-class customer satisfaction.* New York, N.Y.: Irwin Professional Publishing.

Bell, R. & Meiselman, H. (1995). The role of eating environments in determining food choice. In D.W. Marshall (ed.), *Food choice and the consumer.* London, England: Blackie Academic & Professional, 292-310.

Binkley, J. & Eales, J. (1998). Demand for fast food across metropolitan areas. *Journal of Restaurant and Foodservice Marketing*, 3(1), 37-50.

Blank, D. (1999). Hotels, "in-flight" meals take wing as airlines serve less food. *Nation's Restaurant News*, 33(19), 43.

Bon Appetite. (1999, January). The airline with a refrigerator you'd want to raid. 41.

Boone, J. (1997). Hotel-restaurant co-branding–a preliminary study. *Cornell Hotel and Restaurant Administration Quarterly*, 38(5), 34-43.

Borchgrevink, C. (Ed.). (1999). *Perspectives on the hospitality industry.* Iowa: Kendall Hunt Publishing.

Bowen, J. & Shoemaker, S. (1998). Loyalty: A strategic commitment. *Cornell Hotel and Restaurant Administration Quarterly*, 39(1), 12-25.

Buckley, J. (1999, February 19). Flying for peanuts carriers are tightening the belt on food fliers. *USA Today*, 1A.

Business Wire. (1998, January 27). Frequent flyers reveal recipe for success when it comes to airline food. Available: www.elibrary.com. File No. 1431726.

Cebrzynski, G. (1999). Restaurants may want to get wired to meet generation Y youths. *Nations Restaurant News*, 33(10), 14.

Coles, C. (1997, February 2). Flying the hungry skies. *Minneapolis Star Tribune*, 1G.

Cardello, A. (1994). Consumer expectations and their role in food acceptance. In H.J.H. MacFie and D. Thomson (eds.), *Measurement of food preferences.* Glasgow, England: Blackie Academic, 253-295.

Cardello, A., Maller, O., Masor, H., Dubose, C. & Edelman, B. (1985). Role of consumer expectancies in the acceptance of novel foods. *Journal of Food Science* 50, 1707-18.

Czepiel, J. & Rosenberg, L. (1977). The study of consumer satisfaction: addressing the "so what" question. In H. Keith Hunt, (ed.) *Conceptualization and Measurement of Consumer Satisfaction and Dissatisfaction.* Cambridge, MA: Marketing Science Institute, 92-119.

Earle, J. (1998, March 30). Eating on the fly: Terminal eateries see no threat. *Atlanta Journal and Constitution*, E07.

Fishbein, M. & Icek, A. (1975). *Belief, attitude, intention and behavior.* Reading, MA: Addison-Wesley.

Fitzsimons, B. (1997). BA pulls peanuts to avoid allergic reactions. *Onboard Services*, 29(5), 1.

Food Management. (1996a, March). Brands on the go III: A special report. 62, 66.

Food Management. (1996b, July). Brand rotation: spin control. 22.

Food Service Management. (1996, May). Brand sales. 30.

Garman, E. (1996) *Consumer economic issues in America.* Houston, TX: Dame Publications.

George, M. (1998, October 27). Right of reply. *Newspaper Publishing*, 5.

Grattan, M. (1998). U.S. airlines overall food spending soars 32 percent. *Onboard Services*, 30(9), 1.

Gundersen, M., Heide, M. & Olson, U. (1996). Hotel guest satisfaction among business travelers. *Cornell Hotel and Restaurant Administration Quarterly*, 37(2), 72.

Heskett, J., Jones, T., Loveman,G., Sasser, W. E., & Schlesinger, L.A. (1994, March 1). Putting the service-profit chain to work. *Harvard Business Review*, April 1999, 164-175.

Kessler, J. (1998, March 30). Eating on the fly: Food for thought: In-flight meals: No one likes the experience. *Atlanta Journal and Constitution*, E6.

Lewis, R. & Chambers, R. (1989). *Marketing leadership in hospitality.* New York, N.Y.: Van Nostrand Reinhold.

Lieber, R. (1998). Smart Managing. *Fortune*, 137(3), 161.

Lombardi, D. (1996). Trends and directions in the chain-restaurant industry. *Cornell Quarterly Hotel and Restaurant Administration*, 37(3), 14-18.

Love, N. (1998). Customer service do you know its impact. Available: http://www.lblconsulting.com/. Online [1998, November 1].

Lundberg, D., Stavenca, M. & Krishnamoorthy, M. (1995). *Tourism economics.* N.Y: John Wiley & Sons.

Martin, J. (1998). Smart managing. *Fortune*, 137(3), 168.

Maxon, T. (1997, June 10). Return trip: Airlines bringing back food, other amenities. *Dallas Morning News*, 1D.

McCartney, S. (2000a, January 20). Airlines find a bag of high-tech tricks to keep income afloat. *Wall Street Journal*, CV(15), A1.

McCartney, S. (2000b, January 20). Hot lunches or snack sacks? *Wall Street Journal*, CV (15), A6.

McDermid, K. (1997, July). Brand power. *Food Service Management*, 21-22.

Miller, J. (1977). Studying satisfaction, modifying models, eliciting expectations, posing problems and making meaningful measurements. In H. Keith Hunt (ed.) *Conceptualization and Measurement of Consumer Satisfaction and Dissatisfaction.* Cambridge, MA: Marketing Science Institute, 72-91.

Moran, C. (1998) Industry forecast today's airport food needs to meet every traveler's needs. *Onboard Services*, 30(1), 10.

Muller, C. (1997). Redefining value. *Cornell Hotel and Restaurant Administration Quarterly*, 38(3) 62-73.

Nomani, A. (1999, January 24). Carry-on food really taking off as airlines cut back. *The Arizona Republic*, D12.

Pettijohn, L., Pettijohn, C. & Luke, R. (1997). An evaluation of fast food restaurant satisfaction: Determinant, competitive comparisons and impact on future patronage. Journal of *Restaurant and Foodservice Marketing*, 2(3), 3-20.

Pilner, P. (1982). The effects of mere exposure on liking for edible substances. *Appetite*, 3, 283-290.

Restaurants and Institutions (1998). Top 400: The ranking. Available: http://www.rimag.com/814/400rank.htm [1999, January 25].

Rosenberg, M. (1956). Cognitive structure and attitudinal effect. *Journal of Abnormal and Social Psychology*, 53, 367-372.

Schwartz, M. (2000, April 13). Airline customers air the ire customers hate delays, bad food. *Dallas Morning News*, 1C.

Sellers, P. (1998). Service: How to handle customers' gripes. *Fortune*, 118(9), 88.

Shubart, E. (1998). Investors target new locations. *Wall Street Journal*, CII(80), B22.

Slippery Enterprises (1994, September 1). October 11, Events day-by-day. Available www.elibrary.com. File No. 123746.

Spector, A. (1999). Catching a flight? No, just having dinner at my favorite airport. *Nation's Restaurant News*, 33(18), 84.

Sturgis, M. (1998, June 6). Food and drink: New culinary flight paths. *The Daily Telegraph*. Available www.elibrary.com File No. 767832.

Sundaram, D, Jurowski, C. & Webster, C.(1997). Service failure recovery efforts in restaurant dining: The role of criticality of service consumption. *Hospitality Research Journal*, 20(3), 137.

Thorpe, V. (1998, October 26). In-flight meal was a dog's dinner. *Newspaper Publishing*, 3.

Tu, A.(1997). Catering to passengers. *Onboard Services*, *29*(9),14.

Wetherill, G. (1986). *Regression analysis with applications*. New York: Chapman and Hall.

Williams, K. (1995, November 30). Better eat before you fly. *Capital Times*, 4C.

Woodyard, C. (1999, March 16). Serving sushi in the sky. *USA Today*, 5E.

Yuksel, A. & Rimmington, M. (1998). Customer-satisfaction measurement. *Cornell Hotel and Restaurant Administration Quarterly*, 38(6), 60-70.

A Study of Factors Influencing Parental Patronage of Quick Service Restaurants

Lisa R. Kennon
Johnny Sue Reynolds

SUMMARY. This study identified factors influencing parental quick service restaurant patronage. Results indicated that parents are influenced to patronize quick service restaurants most often by speed of service, menus that offered their childrens' favorite food items, and parents' lack of time for meal preparation at home. Children preferred French fries, pizza, hamburgers, cheeseburgers, and chicken nuggets at quick service restaurants. French fries, the favorite childrens' fast food menu selection, was ranked by respondents as one of the least healthy food items, while mashed potatoes, one of childrens' least favorite fast food menu selections, was perceived to be one of the healthier food selections by parents. Restaurateurs were encouraged to develop newer, healthier methods of preparing these popular menu items while offering more nutritious menu selections for children. *[Article copies available for a fee from The Haworth Document Delivery Service: 1-800-HAWORTH. E-mail address: <getinfo@haworthpressinc.com> Website: <http://www.HaworthPress.com> © 2001 by The Haworth Press, Inc. All rights reserved.]*

KEYWORDS. Fast food, quick service restaurants, children, parents, nutrition, patronage

Lisa R. Kennon, PhD, is Assistant Professor, and Johnny Sue Reynolds, PhD, is Associate Professor and Interim Associate Dean, Hospitality Management, both at the University of North Texas.

[Haworth co-indexing entry note]: "A Study of Factors Influencing Parental Patronage of Quick Service Restaurants." Kennon, Lisa R., and Johnny Sue Reynolds. Co-published simultaneously in *Journal of Restaurant & Foodservice Marketing* (The Haworth Hospitality Press, an imprint of The Haworth Press, Inc.) Vol. 4, No. 3, 2001, pp. 113-122; and: *Quick Service Restaurants, Franchising and Multi-Unit Chain Management* (ed: H. G. Parsa and Francis A. Kwansa) The Haworth Hospitality Press, an imprint of The Haworth Press, Inc., 2001, pp. 113-122. Single or multiple copies of this article are available for a fee from The Haworth Document Delivery Service [1-800-HAWORTH, 9:00 a.m. - 5:00 p.m. (EST). E-mail address: getinfo@haworthpressinc.com].

INTRODUCTION

In the next five years, quick service establishments are poised to win the race and capture the largest share of the growing restaurant market. The trade publications *Technomic* and *Takeout Business* predict quick service restaurants to gain $7.6 billion or 38 percent of the growth in the takeout market during the next five years (Vila, 1999). According to a recent survey by Zagat Survey LLC, a New York-based publisher of restaurant guides, the average American dines out 3.6 times per week (Vila, 1999, p. 14). Children have become the new food gatekeepers for the family, and eating out has become an important family activity (Reynolds & Kennon, 1993). Children are just as comfortable eating at restaurant tables as they are in their own homes (Foodtrends 2000, 1999).

REVIEW OF LITERATURE

Eating at quick service restaurants is no longer a novelty while dining out, as opposed to a couple of decades ago (Jekanowski, 1999). It has become a mainstay in many Americans' diet and daily routine. We are entering a new age for the definition of quick service restaurants, which is no longer being limited to burgers, tacos and pizza (Vila, 1999). Foodtrends 2000 (1999), a survey of consumers, found that nearly 71 percent choose commercial foodservice because they are not in the mood to cook for themselves. Forty percent of the Foodtrends 2000 survey respondents said not being in the mood to cook, celebrating a special occasion and spending time with family and friends are always a part of the reason for eating out.

Parents with young children enjoy the conveniences of eating out, and they often take their young families to quick-serve restaurants. They especially like to take their young families to those restaurants that are equipped with playgrounds or play areas, and those that offer give-aways to their children (Reynolds & Kennon, 1993).

Lifestyles of Families Today

The increase in single parents and dual earner families has affected the amount of time families have to plan, purchase, and prepare meals at home. A consumer on a constrained time budget will likely favor small shops over large ones, spend less time comparing prices, use technology to reduce transaction time, and patronize businesses that make life easier (Miller, 1999, p. 63). Today's parents are busier than any other generation of parents. Dual income families usually have more discretionary income to spend on food, less time and energy to prepare food at home, and a strong desire and a need to spend time together as a family unit. Eight out of ten working mothers agree that eating together is a preferable way to enjoy quality family time (Vila, 1999, p. 14). With more women working and more two-career families, more people will eat out or opt for takeout (Allen, 1991, p. 42). Between 1982 and 1997, there was an increase of the proportion of away-from-home food expenditures on quick service restaurant. Expenditures increased from 29.3 percent to 34.2 percent during this period (Jekanowski, 1999).

Factors Influencing Quick Service Restaurant Patronage

What consumers value in terms of products and services is time and convenience. They refuse to be a wait-in-line culture anymore. They are used to getting what they want and getting it immediately (Miller, 1999). American families are experiencing a time-famine (Stanton, 1990, p. 29). Due to this time-famine, more families are eating out, with an average of only 1.2 meals prepared at home each week (Allen, 1991; Stanton, 1990). One thing that will continue to contribute to the growth and success of restaurants is the evolving eating patterns and needs of consumers. Conflicting schedules, busy lifestyles, varied food preferences, a robust economy and a low unemployment rate are some of the reasons for the increase in dining out (Vila, 1999, p. 14). According to Stanton (1990), the increased time pressures resulting from Americans' lifestyles and demographic and environmental transitions impact the way we eat today and how we will eat tomorrow.

Being family friendly has become the motto of many quick service restaurants (Howard, 1997; Cebrzynski, 1998). While operators such as Legal Sea Foods are targeting families by emphasizing a new wholesomeness of the children's menu selections (Howard, 1998), quick serve operators are still using toys and promotions to draw children's patronage (Casper, 1998; Cebrzynski, 1998). Subway restaurants have started emphasizing its Kid's Pak and the message that Subway is a place where adults can get food they feel good about eating and feel good about feeding to their kids (Casper, 1998, p. 83).

Allen (1991) contends that the consumers' continuing concern in value and taste overshadows nutritional interest. American parents are raising some of the most unhealthy children in decades. The big factor in todays eating habits centers around convenience (Vila, 1999, p. 14). Parents look for quick service and restaurants where their children are well cared for, but may choose speed and convenience over nutritional considerations for their children (Reynolds, Kennon, & Palakurthi, 1995). Reynolds and Kennon (1993) found that parents ranked nutritional selection for childrens meals seventh out of twelve restaurant amenities. Nutrition ranked behind quick service, special prices, and children's activity packets.

PURPOSE OF THE STUDY

The researchers sought to conduct a study that differed from previous investigations into the quick service industry. Previous studies determined the reasons that parents frequented quick service restaurants, but they did not report health-related perceptions of popular quick service restaurant menu items, nor did they identify the foods those children select most often at QSRs. The primary purpose of this study was to identify the factors influencing parental quick service restaurant patronage; and the secondary purpose is:

1. to develop and administer a measurement tool to families with children under 12 years of age,
2. to report respondents health-related perceptions of popular menu selections at quick service restaurants, and
3. to identify the foods that are selected most often at QSRs by children.

METHODOLOGY

The population for this study consisted of parents/families with children under the age of 12 who were attending educational and child care facilities in a major metropolitan area in the Southwest. Administrators from these facilities were contacted to obtain permission to distribute the surveys. A test protocol was developed to assure that each of the test administrators gave the same instructions to the participants. Personnel such as teachers and teacher assistants from the selected child care centers and schools were trained by the researchers to assist in administering the instrument. These trained test administrators were utilized by the researchers to assure historical validity in the research process. Convenience sampling was used to obtain responses from 250 individuals. However, sixty-two responses were discarded because they contained incomplete information. This resulted in a useable sample of 188.

The researchers, following review of the literature and interviews with parents, developed the survey instrument. A panel of experts from the fields of Hospitality Management and Consumer Sciences evaluated the instrument for content validity. The panel of experts consisted of professionals who had a variety of age and ethnic characteristics and who were from different geographic locations. The members of the panel incorporated professionals working in the hospitality industry, secondary school educators, university educators and parents. In determining validity, the panel members were asked to critique the instrument intent and scope. The responses by the panel provided input for adding and deleting questions. The survey was then pretested to evaluate for clarity of instruction, format of questions, as well as ambiguity of questions.

The survey instrument contained four parts. Part One of the survey contained questions designed to ascertain demographic and background characteristics of the respondents. Participants were then asked about quick service restaurant menu items that were most frequently selected by their children in Part Two of the instrument. Perceptions of healthiness of popular menu selections at quick service restaurants were determined in Part Three and the factors that influence patronage of quick service restaurants were established in Part Four. Descriptive statistics were used to summarize the data.

RESULTS

Respondents ranged in age from 18 to 63 with a mean age of 35 years. The majority of the respondents were married (82%) and eighty percent of the subjects were female. The majority of the respondents were well educated with 58% having at least a bachelor's degree. Fifty-eight percent of the respondents were classified as dual earners. Forty-three percent of the respondents had two children.

To determine factors that influence patronage of quick service restaurants, parents were asked to select among 17 factors which they felt influenced them to patronize quick service restaurants (see Table 1). The number one factor that influenced the respondents to patronize quick service restaurants was speed of service (69%). This was closely followed by offer children's favorite food items (66%). Lack of time for meal preparation at home ranked third among factors that influenced quick service restaurant patronage (61%). Factors which influenced quick service restaurant patronage the least were discounts for groups with children at off-peak hours (7%) and the amenities such as high chairs or booster chairs (11%), coloring books or crayons (15%), and menu games and/or puzzles (16%).

From a list of 20 popular quick service restaurant menu selections, respondents were asked to select their child's three most favorite foods by rank order (Table 2). Parents believed that the most popular quick service restaurant items for children were (1) french fries (58%), (2) pizza (52%), (3) hamburger/cheeseburger (51%), and (4) chicken nuggets (43%). The least popular quick service restaurant menu item was mashed potatoes, followed by potato chips, diet soft drinks, milk, salads, frozen yogurt, and cookies.

TABLE 1. Factors Which Influence Parental Patronage of Quick Service Restaurants (N = 182)

Rank Order	Factor	%	n
1	Speed of service	69	126
2	Offer children's favorite food item/s	66	120
3	Drive thru service	61	111
4	Lack of time for meal preparation at home	61	111
5	No time to cook	51	93
6	Child's menu	48	87
7	Reduced priced promotion	45	82
8	Play ground	42	76
9	Offers give-aways such as toys to children	38	69
10	Child size portions	33	60
11	Not having to mess up my kitchen	32	58
12	Lack of time for grocery shopping	21	38
13	Birthday parties	19	35
14	Menu games and/or puzzles	16	29
15	Coloring books or crayons	15	27
16	High chairs or booster chairs	11	20
17	Discounts for groups with children at off-peak hours	7	13

Parents were asked to give their opinion about the healthiness of the same 20 popular quick service restaurant menu items. Participants were asked to respond to a 3-point scale (Healthy, Moderately Healthy, and Not Healthy) to indicate the perceived nutritiousness of the menu items (Table 3). Menu items that were perceived to be healthy by respondents were milk (88%), salads (85%), frozen yogurt (59%), and mashed potatoes (50%). The least healthy food items as perceived by respondents were french fries (2%), corn dogs (2%), cookies (3%), and soft drinks with sugar (3%).

DISCUSSION

This study revealed that traditional children's amenities offered by quick service restaurants were not as important to parents as were the service components of speed of service, children's favorite food items, and drive thru service.

TABLE 2. Children's Favorite Quick Service Restaurant Menu Items by Rank Order (n = 182)*

Ranking	Quick service restaurant Menu Items	(n)	%
1	French Fries	105	58
2	Pizza	94	52
3	Hamburger/Cheeseburger	92	51
4	Chicken Nuggets	79	43
5	Ice Cream	31	17
6	Soft Drinks (with sugar)	25	14
7	Hot Dog	15	8
8	Fried Chicken	13	7
9	Fish and Chips	12	7
10	Grilled Cheese Sandwich	11	6
11	Tacos	10	5
11	Corn Dog	10	5
12	Frozen Yogurt	9	5
12	Salad/Salad Bar	9	5
12	Cookies	9	5
13	Milk	7	4
14	Potato Chips	3	2
14	Soft Drinks (diet)	3	2
16	Mashed Potatoes	2	1

*Respondents could make three ranked selections.

Less than half of the respondents indicated that a playground or a give-away such as toys was influential in quick service patronage.

Amenities such as coloring books and menu games were not as valued by parents, in part, because utilizing these activities indicate that they are waiting for food or service. This is in support the study finding of speed of service and drive thru service as two of the top factors for quick service patronage.

Wendy's vigorously applied and developed the drive-thru concept to the quick service industry in 1974 (Jekanowski, 1999). This has caused a revolution in the accessibility to quick service restaurant by individuals on the go. Parents utilize quick service restaurants because of the speed and convenience

TABLE 3. Parental Perception of Nutritional Selections of Traditional Quick Serve Menu Items (n = 182)

Rank	Menu Item	Healthy %	Moderately Healthy %	Not Healthy %
1	Milk	88	7	4
2	Salad/Salad Bar	85	13	1
3	Frozen Yogurt	59	32	9
4	Mashed Potatoes	50	41	7
5	Fish and Chips	25	48	26
6	Grilled Cheese Sandwich	20	62	17
7	Pizza	17	59	22
8	Tacos	14	57	28
9	Soft Drinks (diet)	12	20	67
10	Hamburger	7	56	36
11	Chicken Nuggets	7	63	29
12	Ice Cream	6	50	42
13	Fried Chicken	6	51	43
13	Cheeseburger	6	53	39
15	Hot Dog	5	34	59
15	Potato Chips	5	18	77
17	Soft Drinks (with sugar)	3	19	77
17	Cookies	3	33	62
19	Corn Dog	2	29	65
19	French Fries	2	23	73

• Percentages do not total 100%, due to non-responses.

that is provided for their busy lifestyles. Why didn't respondent's favor discounts to quick service restaurants at off peak times? Parents want the convenience and time factor that quick service restaurants provide. They want this convenience now, not later.

Quick service restaurants have capitalized on this lifestyle and offer items that are portable and designed for eating inside the car. However, the classic hand held food items offered on children's menus have typically not been very nutritious. Whatever items are offered, they must be portable and easy to eat inside a car.

It is interesting to note differences in nutritional perceptions of similar foods. The favorite children's quick service restaurant menu selection, french fries, was rated by respondents as one of the least healthy food items, while mashed potatoes were perceived to be one of the healthier food selections. But, mashed potatoes were the least favorite choice of children. Both items are potato products and depending on preparation method may be similar in nutrient profile.

Parents have less time to cook at home and that is a big problem these days. If parents are going out and taking their children, they want good, wholesome, healthy food (Howard, 1997, p. 6). In an article by Lin and Guthrie (1996) concern is expressed that the increasing popularity of dining out will lessen the control over what our children and we eat and how it is prepared, and in turn impacting the nutritional quality of our diets. A study by Grazin and Olsen (1997) found consumers most involved with nutrition stay away from quick service restaurants. The study goes on to suggest that managers who seek to cater to these customers should feature menus that offer recognizable alternatives to the standard fare that is relatively high in fat, calories, and sodium.

As found in this study, the most popular food items selected by children (french fries, pizza, hamburgers, cheeseburgers, and chicken nuggets) are some of the foods that parents perceive as not being healthy. In addition, diet soft drinks were perceived as being healthier than soft drinks containing sugar. Even with information that has been disseminated about possible dangers of aspartame, parents feel that soft drinks containing sugar are less healthy for their children than diet soft drinks. Overwhelmingly, parents rated milk as a healthy choice, but as with other choices, milk was low on the children's list of favorite food items.

Some quick service restaurant operators have a unique marketing opportunity to reach families with children utilizing a healthy approach. For example, some pizza restaurants have a salad bar that is included as a part of the whole meal package, including meals for children. This encourages children to select healthy food items such as salads or fresh vegetables in addition to traditional pizza selections.

In the future, healthy options for to-go kids meals could include carrot sticks instead of french fries, hand held yogurt sticks such as Yoplait's Go-Gurt, and flavored milks in aseptically packaged containers. These options meet the demand for hand held to-go food, but also provide a healthy alternative to traditional quick service restaurant menu items. The 1992 Census of Retail Trade reports that roughly 23 percent of all quick service restaurant establishments do not have seating for on-premise dining, catering instead exclusively to consumers who eat on the run (Jekanowski, 1999, p. 11). Therefore, it is important for all food items to be portable.

Future research should focus on the brand loyalty of children and their parents. How much influence do parents have over children in the selection of quick service restaurant items? When given a choice of a healthier menu or healthier children's meal options, will parents purchase the healthier items or allow children to again be the gatekeeper of food selection? Casual family restaurants are positioning themselves to reach the family dining market by spotlighting menus that appeal to parents who are concerned about their children's diets. Quick serve restaurants continue to have the advantage of convenience and speed, but will increase customer satisfaction by creating healthier menu choices.

REFERENCES

Allen, R. (1991). Demographics changing, shaping industry. *Nationals Restaurant News*, 25 (42), 42.

Casper, C. (1998). Happy meals. *Restaurant Business*, 97(20), 83.

Cebrzynski, G. (1998). New Dominos ad campaign targets parents through kids. *Nationals Restaurant News*, 32(19), 8.

Foodtrends 2000. (1999, December). Ground-breaking: Consumer study reveals what drives dining decisions. *Nations Restaurant News*, 33, 4-5.

Grazin, K. & Olsen, J.E. (1997). Market Segmentation for fast-food restaurants in an era of health consciousness. *Journal of Restaurant & Foodservice Marketing*, 2(2), 1-20.

Howard, T. (1997). New marketing promos insist it's a family affair. *Nation's Restaurant News*, 31(49), 6, 18.

Jekanowski, M.D. (1999). Causes and consequences of quick service restaurant sales and growth. *Food Review*, 22(1), 11-16.

Lin, B. & Guthrie, J. (1996). The quality of children's diets at and away from home. *Food Review*, 18, 45(6).

Miller, A. (1999). The millennial mind-set: Its here, its clear, get used to it! *American Demographics*, 21(1), 60-65.

Reynolds, J. & Kennon, L. (1993). The new food gatekeeper: The influence of children on the family's selection of a restaurant. *Texas Home Economist*, 59(1), 16-18.

Reynolds, J.S. Kennon, L.R., & Palakurthi, R.R. (1995). Parent's perceptions of quick service restaurant consumption by children. *Journal of Family and Consumer Sciences*, 87(4), 39-44.

Stanton, J. (1990). Future food for Americans. *Food and Nutrition News*, 62(5), 1-31.

Vila, G. (1999). Consumers shape future. *Food & Service News*, 1 (5), 14-15.

A REVIEW ARTICLE

White Castle: How Billy Ingram Made Hamburger "The America's Choice"

Dave Hogan

SUMMARY. Billy Ingram is the acknowledged father of American fast food. This article describes how, through White Castle system, Ingram created a whole new industry, an industry that employs millions of workers and accounts for billions in sales. *[Article copies available for a fee from The Haworth Document Delivery Service: 1-800-HAWORTH. E-mail address: <getinfo@haworthpressinc.com> Website: <http://www.HaworthPress.com> © 2001 by The Haworth Press, Inc. All rights reserved.]*

KEYWORDS. White Castle, fast food, entrepreneur, Billy Ingram

Extraordinary innovators and entrepreneurs are not necessarily true inventors or originators. More often, the significant entrepreneur takes existing ideas, perfects them, and then creatively applies them to a specific venture or a

Dave Hogan, PhD, is Professor, Department of History, Heidelberg College, 310 E. Market Street, Tiffin, OH 44883.

[Haworth co-indexing entry note]: "White Castle: How Billy Ingram Made Hamburger 'The America's Choice.' " Dave Hogan. Co-published simultaneously in *Journal of Restaurant & Foodservice Marketing* (The Haworth Hospitality Press, an imprint of The Haworth Press, Inc.) Vol. 4, No. 3, 2001, pp. 123-135; and: *Quick Service Restaurants, Franchising and Multi-Unit Chain Management* (ed: H. G. Parsa and Francis A. Kwansa) The Haworth Hospitality Press, an imprint of The Haworth Press, Inc., 2001, pp. 123-135. Single or multiple copies of this article are available for a fee from The Haworth Document Delivery Service [1-800-HAWORTH, 9:00 a.m. - 5:00 p.m. (EST). E-mail address: getinfo@haworthpressinc.com].

purpose. Business history is full of such successes, individuals who honed an existing idea and converted it into a fortune. Henry Ford did not invent either automobiles or the assembly line and young Mrs. Fields certainly did not bake the first chocolate chip cookie. Even the revered Edison, popularly credited with boundless originality, often synthesized earlier ideas and inventions into practical and marketable machines. Innovators in virtually every industry seized upon unformed or unexploited ideas and concepts and transformed them into gold. In the restaurant industry, one such innovator was Billy Ingram.

THE EARLY BEGINNINGS OF HAMBURGER

Edgar Waldo "Billy" Ingram is the acknowledged father of American fast food. He single-handedly did more to advance this industry than all the Krocs (McDonald's), Sanders (KFC), Thomases (Wendy's), and Carneys (Pizza Hut) who would later follow. Ingram successfully sold his hamburger to America, effectively convincing the consuming public to buy both the product and the concept of fast food. Prior to his creation of the White Castle System in 1921, hamburger meat was held in very low regard by most Americans, and was rarely eaten while eating in public places. Ground beef was often referred to as "the poor man's steak." In fact, during the early 20th Century many meat processors were selling high quality beef to retailers and restaurants and lower quality beef to food processors for making ground beef. This practice has its roots in Western Europe where old meat was often processed and sold as ground beef before it is spoiled. Since it is less than prime quality it is sold at discounted prices and often bought by poor people. Consequently, ground beef became a cultural symbol of lower classes and resented by the upper classes. Since the cultural roots of the American society extend into the traditions of Western Europe, many Americans during the early 20th century were less than enthusiastic about consuming ground beef products while dining out.

Ingram the Innovator

But within a decade, Ingram magically transformed the lowly hamburger into the most popular food item of America, spreading his chain of restaurants from Wichita, Kansas to New York City. His initial success in the 1920s spawned countless other entrepreneurs in the hamburger business, who quickly spread this humble sandwich to every city corner and small town in the nation. Though the founder of fast food, Billy Ingram did not invent either the hamburger or the fast food concept. His true genius lay in adapting and combining existing products, concepts, and marketing techniques into a successful restaurant chain, which served as the prototype for the industry. Similar to Ford, Ingram converted undeveloped ideas into profitable ones, building a thriving organization, a substantial personal fortune and a place in business history.

Creating a successful restaurant chain, founding a major American industry, introducing a new style of eating and introduction of a single most popular restaurant menu item is each historically significant in itself. Most entrepreneurs certainly never accomplish even one of these goals, much less all four. During his forty-five years in the hamburger business, Billy Ingram did succeed in all of these areas and much more. The historical focus of this essay, however, is not to extol Ingram's many remarkable achievements, but rather to analyze how he achieved these feats.

By blending imitation, true invention, and a creative synthesis of both, Mr. Ingram built a burger chain, which transformed American culture. He opened his first White Castle restaurant in Wichita, Kansas in 1920 and built a restaurant empire, the oldest restaurant chain in America. A shrewd businessman, Ingram initially invested all of his capital in a dubious product that he wholeheartedly believed in, but most others still scoffed at. He packaged his restaurants to represent positive virtues, both in name and appearance, refining and combining proven ways to instill customer confidence. Ingram proudly borrowed many of his most effective marketing strategies from more established industries, such as newspaper coupons used by grocery stores and corporate hostesses employed by food manufacturers. As a result of his resourcefulness, White Castle survived the peaks and valleys of recent history, continuing to thrive through the economic hardships of the Great Depression, the severe commodity shortages of World War II, and the intense fast food competition in postwar America.

INGRAM AND ANDERSON–"THE HAMBURGER TEAM"

Ingram's initial entry into the hamburger business was the most dramatic example of how he improved on an existing idea. In 1916, a fry cook and self-described "ne'er do well" Walt Anderson opened a makeshift food stand on a Wichita street corner. Selling borrowed ground beef on his first day in business, Anderson used a flat piece of iron as a griddle, and hawked his hamburger sandwiches up and down the sidewalk, proclaiming them to be "just a nickel a piece." Rather than offering a round meatball on a slice of bread, which was the popular norm of the era, Anderson broke with tradition and served a flat patty between two halves of a roll. Legend claims that he came upon this new serving twist in a rather unexpected manner, simply by one day growing impatient with the cooking time of a meatball, and exasperated, finally smashing the patty flat with his spatula. Regardless of his sandwich's precise origin, Anderson's new version of the hamburger was an immediate hit on the streets of Wichita. It was very affordable at five cents each, and easy to hold in the hand and eat. In fact, Anderson's hamburger sandwich proved so popular that he soon expanded his operation, adding three more food stands in other Wichita neighborhoods. Each of these new stands similarly thrived, allowing Walt Anderson to purchase a handsome house in 1920, located in an affluent section of the city. Despite his financial success, however, Anderson's street-corner operations were still only patronized by hungry working-class customers, and generally held in contempt by Wichita's middle class. As Prohibition was getting underway, much of the public believed a popular notion that "food joints" were often front operations for bootleggers selling illicit spirits. Thus, in 1921 when Anderson sought to rent space to open a new restaurant in a more affluent district of the city, the dentist who owned the building was very reluctant about granting him a lease. For assistance in securing the lease, Anderson turned to his fellow Rotarian, real estate and insurance agent Billy Ingram. The dentist relented when Ingram, an established and respected member of Wichita's business community, agreed to co-sign Anderson's lease.

In the process of these negotiations, Ingram learned the details of Walt Anderson's burger business, became impressed by his rapid financial success, and envisioned ideas for even greater profit potential. Though still very profitable, Anderson's existing operation already saturated Wichita's working class market, and its profit growth potential appeared to be stagnant. Ingram realized immediately that Anderson's good idea just needed new management style and innovation. The two men agreed to form a new company, with Anderson as the titular president. Ingram, the risk taker and entrepreneur, quickly sold his other interests and invested these proceeds into his new enterprise. From the beginning of their new company, Ingram primarily guided the growth and direction of the business. Anderson remained on board until 1933, but he only supervised day-to-day operations, with little real input into marketing strategies and territorial expansion.

The First Fast Food Restaurant Chain: White Castle

Billy Ingram's first decision was to reorganize and recreate the entire business. He saw the potential for business expansion beyond Wichita. In fact, from the very beginning of their new company, Ingram professed his very optimistic goal of creating "a national institution." With this goal in mind, Ingram set out to build his new business from the ground up, beginning with the actual buildings. It was obvious that many potential buyers objected to the questionable appearance and image of the street-corner stands and especially distrusting the cleanliness and quality of the food. In 1921 many people in America still remembered the lurid details of Upton Sinclair's *The Jungle,* which depicted the unsanitary conditions in the meat industry. Compounding these popular doubts was the common belief during that era that ground meat was the old, and sometimes spoiled remnants from the butcher's display case, ground up and mixed with chemicals for prolonged salability. Though the working-class public of Wichita was warming to the idea of hamburgers by 1921, the rest of the country ate them only sparingly, if at all. This distrust of the meat product was further heightened by the fact that most hamburger stands appeared to be unhygienic and temporary operations, often housed in either ramshackle buildings or portable stands on street corners.

To counter what he referred to as "the prejudice against ground meat," Ingram planned a campaign to legitimize both the burger and the business. In fact, Ingram was thoroughly convinced that Anderson's little hamburger sandwich could easily become a gold mine, if only marketed more effectively. First, he wanted a name for the company name that could communicate clearly to its customers in a positive way. Ingram chose the name "White Castle" for the company banner to convey the dual image of high level of cleanliness ("White") and permanence ("Castle") (not another street corner vendor). To give greater meaning to the new name of the company, White Castle, he constructed sturdy buildings in a castle motif with a crenelated tower and roof reflecting the intended strength and stability message. Consistent with his theme, Ingram even whitewashed his new mini-castles and installed gleaming stainless steel fixtures for a clean, hygienic appearance. Setting a precedent for his future management decisions, Ingram borrowed the whitewashed facade concepts from leading luncheonettes of the day and the castle motif from nothing less than the famed Chicago Water Tower. Though neither concept was new, combining towers and white paint together into a small hamburger restaurant was quite original. Interestingly, however, this somewhat hokey, and certainly unorthodox combination would go on to become the archetype for the next twenty years of fast food stands across America.

This new White Castle image, however, did not stop at the bricks and mortar. Ingram boldly tried to develop a whole new approach to restaurant eating. He offered little seating in his castles, forcing his patrons to carry away their food in insulated paper bags. Ingram even taught them how to purchase burgers in bundles. His marketing philosophy can be best described as "higher profits through higher volumes." He coined the marketing slogan "Buy 'em by the Sack!" To his great pleasure, the customers happily complied, giving birth to America's first fast food chain.

BILLY INGRAM AND MARKETING OF HAMBURGERS

Selling high quality meat sandwiches for only five cents in unique white castle motif restaurants was proved to be a great success. His first attempt at expansion was very short-lived in the small town of Eldorado, Kansas. Quickly realizing that large cities are better suited to his need for high volume business, he closed Eldorado and opened his first restaurant in Omaha, Nebraska, a relatively larger town, where Ingram had roots from prior employment. Soon more bustling White Castles appeared in Omaha, quickly followed the next year by additional Castles in Kansas City, and then only months later in St. Louis. As the company expanded, Ingram established strict corporate rules to guarantee uniformity and quality throughout the chain. This principle became the industry standard for other fast food chains to follow.

During these growth spurts, Ingram also resolved that his restaurants would always remain company-owned, feeling that franchising meant a loss of control, and thus unpredictable quality. Closely emulating the company-owned chain concept from earlier successful restaurant entrepreneurs, he devised a tiered management structure, a communication system, and a supply network with which he could guide the company from his Wichita office. This more systematized approach became increasingly crucial as the company expanded even further to the east, opening White Castles in urban centers such as Chicago, Minneapolis, Cincinnati, Louisville, Columbus (Ohio), and New York City by the end of the decade. After less than nine years in the burger business, Ingram's company consisted of over two hundred restaurants, extended over a territory of fifteen hundred miles. To more effectively supervise this rapidly growing company area, Anderson and Ingram bought airplanes to fly from city to city. An even more effective control device, however, was his management team, consisting of early Wichita employees, who Ingram assigned to oversee key markets. These managers enjoyed significant power over their respective areas, yet remained fiercely loyal for decades to the strict guidelines set by Ingram.

White Castle grew steadily throughout the 1920s, but Billy Ingram actually only opened restaurants in twelve cities. His goal was to fully saturate his chosen markets, yet still averse to franchising; he only opened Castles at a rate that he could afford to build with his own capital on hand.

With the popularity of White Castles in twelve cities and over 200 units, many substitutes and look-alikes started opening up across the country, making hamburger "America's Choice Food." Virtually every fast food chain that appeared in the 1920s closely resembled the White Castle model in architecture, marketing, service, and food offerings. Many of the "copy cat" chains chose suspiciously similar company names, such as White Tower, White Palace, White Clock, or Red Castle. These name variations of either "White" or "Castle" numbered to over a hundred, and usually adorned small-whitewashed buildings that were almost identical to Ingram's original design. For a while during the 1920s, in fact, the term "White Castle" became an almost generic term used for both the hamburgers and the small white restaurants. This wave of imitators served to flatter and amuse Ingram, who usually only challenged overt, illicit use of the White Castle name. He did object, however, to a direct challenge from the Milwaukee-based White Tower chain that opened restaurants in established White Castle markets, used the slogan, "Buy 'em by the Bag," and then sued White Castle for trademark infringement. After years of fierce litigation, White Castle was vindicated, with a federal judge finding that White Tower had copied almost every aspect of Ingram's chain. Despite these many examples of shameless imitation, or maybe perhaps because of it, the White Castle eating style and hamburgers soon caught on everywhere in America. Ingram's actual White Castle chain never truly became a "national institution," but the hamburger that he first marketed took the nation by storm. The concept of fast food chains took firm hold in American society. Probably, this is the major contribution of Ingram to America and the hospitality industry.

Marketing Innovations–Hamburger Coupons

As the Great Depression hit America in the early 1930s, the unemployment rate rose to high 20s. This depleted the buying power of the working class, the core customer base of White Castle. So to survive and expand his business, Ingram decided to attract more of the middle class to his White Castles. Convincing the middle-class public, however, to accept his hamburgers required more than just interesting architecture and clean restaurants. Middle class America still distrusted ground meat and convincing them proved to be a challenge.

Observing other retail operations, Ingram marveled at how grocery stores attracted new customers by offering discount coupons in their local newspapers. Introduced years earlier by grocer Bernard Kroger, these coupons served both as advertisements and enticements. Shoppers simply had to clip out the coupons, and then bring them to the grocery to receive lower food prices. Needless to say, grocers intended this couponing to attract customers away from competing stores, thus securing a greater market share. There was a cost to this approach, of course, both for the newspaper advertising and for the alluring price break given to customers on the foods items. Grocers found that their greatly improved sales volumes quickly outweighed these costs. Realizing the many parallels that exist between his White Castle System and grocery business, Ingram experimented with newspaper coupons offering a sack of ten hamburgers for a discounted twenty-five cents. This price represented a fifty percent price reduction–essentially two and a half cents per sandwich–that was actually less than the cost Ingram paid for the raw materials. Customers responded to the ads with great enthusiasm, often winding long waiting lines far around the block at many of the White Castles offering discounted burgers.

Following the first round of coupons, large crowds overwhelmed many castle operators, who quickly ran out of food to serve. Subsequent ad runs, however, found the Castles stocked with ample supplies and additional counter help. Ingram proclaimed the coupon campaign a success. Similar to the experience of the grocery chains, hungry customers bought their discounted sacks of burgers and purchased additional high-profit food items, such as coffee, Coca-Cola, and pie. The bottom line of the balance sheets showed a growing profit, proving that Ingram's couponing strategy was sound. Instead of being a regular weekly occurrence, however, he only used the newspaper coupons on a sporadic basis. Ingram's intention was more to introduce his product to the uninitiated, fostering their habitual patronage, rather than to consistently sell burgers at a significant loss. Profits stabilized, then continued to rise, containing steadily upward throughout the darkest years of the Depression. In fact, Ingram saw such success with couponing that he made it the centerpiece of his company's marketing plan for the next thirty years. Since coupons were published in newspapers, many of the coupon users were from the middle class. Through the usage of coupons Ingram was able to attract middle class America to at least try hamburgers while eating out.

Hamburger and America's Middle-Class

While he promoted middle-class patronage through discounted prices, Ingram also wanted to further reassure new buyers that his product was healthy and safe. Negative sentiment towards ground beef continued to thrive in the early 1930s and many diners were still reluctant to eat hamburgers. One publication of that era described hamburgers "*to be just about as safe as walking in a garden while the arsenic spray is being applied, and about as safe as getting your meat out of a garbage can standing in the hot sun.*" Though by then the most popular sandwich in America, the hamburger still held working-class connotations and was not yet standard middle-class fare. For assistance in convincing the often well-educated middle class, Ingram turned to scientific research and the university community, commissioning a study to prove both the safety and nutritional value of his hamburgers. Researchers at the University of Minnesota fed a student only White Castle hamburgers and water daily for sixty days to determine the health effects of such a diet. At the end of the study, they proclaimed that the student was in ideal health, and hence, that White Castle hamburgers were good and nutritious food. Finally gaining the scientific legitimacy for what he long espoused, Ingram naturally proclaimed these results to every possible listener. With research proof of quality in hand and a stable empire of over two hundred Castles, his original claims of "purity and strength" seemed quite valid.

Though couponing and proving quality were enormously successful, Ingram still looked for other ways to expand his customer base. During the 1920s, many food manufacturers, and even other industries, created the role of "corporate hostesses" to serve as marketing ambassadors for their products. Rather than just being sales representatives, however, these hostesses were expected to be the epitome of refined womanhood, displaying the positive virtues which the manufacturers wanted the buying public to associate with their products. Typically these women were assigned pseudonyms in an attempt to more fully mold their image, and to possibly heighten their acceptance. Such corporate hostesses routinely visited middle-class women's clubs across the nation to introduce their employer's products, demonstrating how these products benefited families or otherwise improved household efficiency. Most memorable of these corporate hostesses was General Mills' Betty Crocker, who gained such popular acclaim that the company eventually applied her name to a new line of ready-to-bake mixes.

Unbeknownst to most General Mills consumers, "Betty" was not one, but actually many different women, operating either successively or simultaneously under this same pseudonym. Despite the mild deception, the use of corporate hostesses proved to be a great asset in marketing food. Billy Ingram first took notice of this practice during the mid-1920s, but he did not create a White Castle hostess until 1933, when he hired Ella Louise Agniel to fill that role. Under the pseudonym of "Julia Joyce," Agniel's primary task was shuttling between company cities to visit women's clubs and garden clubs in an effort to introduce White Castle's hamburgers to middle-class women. Her approach was to demonstrate how these hamburgers when accompanied by a vegetable and a potato dish could make a nutritious and satisfying family meal. Agniel may not have convinced everyone in her audiences that White Castle burgers were healthy mainstream fare, but her trips usually correlated with increased sales during a period when many of White Castle's competitors closed down. Ingram believed that the very polished and gracious Agniel presented a positive public face for White Castle, providing legitimacy for his products among middle-class buyers. She remained with White Castle as both hostess and a valued advisor to Ingram until she died suddenly in 1952. By this time, however, Agniel's work was long completed, with the hamburger meal already firmly embedded in the growing post-war middle-class culture. Her successful efforts twenty years earlier, combined with Ingram's other marketing strategies, certainly convinced the middle class to "Buy 'em by the Sack," further changing consumption patterns and guaranteeing White Castle's continued profitability during the Depression. Miraculously, Ingram's White Castle System finished the Depression in a slightly better financial position than it enjoyed in 1929. By direct comparison, most of the franchised chains of White Castle imitators collapsed and faded away long before.

HAMBURGERS DURING WORLD WAR II

Industry leadership and good fortune in any business does not last forever, and Ingram's twenty-year reign over fast food abruptly ended with the onset of World War II. White Castle was immediately wrecked by a labor shortage as their countermen (employees) departed for the war duties, necessitating hiring of female and teenage employees for the first time. Soon food shortages, especially those of meat, coffee, and sugar, severely limited the types and quantities of items for sale in the Castles. In a matter of just a few months, Ingram went from creative and proactive to reactive in his management style. Crises seemed to arise daily, with little relief in sight. With often little food to sell and an unstable workforce, almost fifty percent of Ingram's restaurants closed before the end of the war in 1945. The realities of the wartime economy, flush with cash but limited in commodities, caused the damage that neither fierce competition nor the ravages of the Depression could inflict. Remaining Castles often kept limited hours, closing at night for the first time in twenty-two years. Ingram feared that quality would fall below acceptable levels and he optimistically predicted a positive rebound after the war. He was further frustrated by a storm of new government regulations that he perceived them as far too restrictive. His only solace was that White Castle remained marginally profitable, primarily buoyed by income from its subsidiary companies, while many of his longtime competitors folded.

When the much-anticipated victory over the Germans and Japanese finally came, ongoing labor problems, sporadic shortages, and a changing postwar marketplace further frustrated him. Rather than setting the pace of his industry and making the rules of the game, Ingram discovered that the entire game had suddenly changed. Though his remaining White Castles were still on traditionally busy city street-corners, much of the traffic now bypassed these streets in favor of new high-speed expressways and interstate highways. Even worse, these new freeways were carrying much of Ingram's longtime customer base to their new homes in the suburbs, where the newest fast food chains immediately staked out their territory. Stranded on both low-volume thoroughfares and in the declining cities themselves, White Castle became stagnant in the 1950s.

Unable to recapture his leadership role in the industry, Ingram could only watch as a new generation of franchised fast food chains claimed the allegiance and dollars of the growing suburbs. Slow to react, he finally began building a few new Castles further out for the center city, and altered the menu to include items such French fries to suit changing customer tastes. White Castle's progress was most hindered during this era by Ingram's staunch belief in the ideas and practices that succeeded in the early years of the company. He still preferred rebuilding his Castles in the city, rather than expanding to the suburbs, and he long ignored the collective advice of his area managers to advertise on television and radio. The rules of the 1920s did not apply in the world of burgeoning Burger Kings and McDonald's, which were created to serve an increasingly mobile population. Yet Billy Ingram's reluctance to change his cardinal principles may have ultimately helped his business to survive. Franchised fast food restaurants competed head-to-head with each other on virtually every suburban commercial strip, and finally only the strongest chains survived. Ingram's preference to stay out of that direct fray may have enabled White Castle to successfully outlive chains such as Burger Chef, Red Barn, Gino's, and Borden's Burgers. Conservative and fixed in his approach to business, he stayed comfortably close to his roots, both in philosophy and geography.

PASSAGE OF LEADERSHIP: END OF AN ERA

When Billy Ingram died in 1966, thousands of McDonald's dotted the American landscape, but White Castle still had fewer restaurants than it did before World War II. Considered very tiny by the fast food chain standards of the 1960s, Ingram's company was still quite profitable. His son Edgar had a strong base to work from and immediately set out to strengthen it even further. Six years later he passed the reins to his own son, Bill Ingram, who set out to modernize and expand the chain. Though still adhering to his grandfather's business philosophies of cautious expansion, high quality, and strict central control, Bill successfully tripled the size of the company, extending out into suburban areas and along major interstate highways. Selling both anachronistic charm and tasty little hamburgers, Ingram cemented a stable position for White Castle in the fast food market.

Billy Ingram ranks as one of the most adaptive and inventive entrepreneurs of the twentieth century. Using time-tested strategies and by assessing the marketplace he succeeded in promoting fast food on numerous levels. On the most basic and tangible level, Ingram founded a remarkable chain of restaurants and led it for forty-five years. To achieve this success, he adopted Walt Anderson's little burger sandwich, adapted the marketing techniques of other industries, and usually lived up to high moral standards. White Castle is the first restaurant chain to offer benefit packages to its employees that were considered to be very liberal by many standards. Though never the biggest, his White Castle chain is the sole survivor of three generations of burger wars, having outlived many thousands of competitors. In a much larger sense, Ingram succeeded in creating a whole new industry, which now directly and indirectly employs millions of workers and annually accounts for billions in sales. In the extreme, Ingram's innovations created a new style of eating, and quite possibly the primary American ethnic food. His hopes for creating a "national institution" bore fruit–not as a whitewashed, castle themed burger business, but rather as an integral facet of American culture.

White Castle Restaurant, Wichita, KS circa 1921.

Creating Consumer-Driven Demand Chains in Food Service

Roger D. Blackwell
Kristina S. Blackwell

SUMMARY. The foodservice industry, as well as many others, is experiencing growth and increased competition. In a cycle of growth, both in terms of number of restaurants and number of consumers eating out on a regular basis, foodservice marketers need to monitor the changes occurring in the marketplace and adapt their strategies accordingly. The evolutionary changes occurring in businesses today focus on two intertwined primary areas: supply chain management and consumer (human) behavior. In the foodservice industry, consumers are moving from their traditional passive position at the end of the linear supply chain, to one in which they drive the direction of the channel. Thus, these two key topics will affect the restaurant and foodservice areas in how they do business with partners and with consumers.

KEYWORDS. Demand chains, consumer behavior, restaurants, foodservice, marketing

---

Roger D. Blackwell, PhD, is Professor, Dept of Marketing, Fisher College of Business, the Ohio State University.

Kristina S. Blackwell, MBA, is Director, Daymark Corp. and Vice-President, Blackwell Associates, 3380 Tremont Road, Columbus, OH 43221.

[Haworth co-indexing entry note]: "Creating Consumer-Driven Demand Chains in Food Service." Blackwell, Roger D., and Kristina S. Blackwell. Co-published simultaneously in *Journal of Restaurant & Foodservice Marketing* (The Haworth Hospitality Press, an imprint of The Haworth Press, Inc.) Vol. 4, No. 4, 2001, pp. 137-154; and: *Quick Service Restaurants, Franchising and Multi-Unit Chain Management* (ed: H. G. Parsa and Francis A. Kwansa) The Haworth Hospitality Press, an imprint of The Haworth Press, Inc., 2001, pp. 137-154. Single or multiple copies of this article are available for a fee from The Haworth Document Delivery Service [1-800-HAWORTH, 9:00 a.m. - 5:00 p.m. (EST). E-mail address: getinfo@haworthpressinc.com].

Commerce has entered a new cycle of transformation. Some experts are characterizing the changes to business brought on by the Internet and new technology as *revolutionary*–insisting that business will be forever changed. However, we believe that as the Internet will continue to change the rules of the game, the nature of the game remains the same–providing products and services customers want to buy, in the form in which they want to buy them, and at a profit. Furthermore, we like to characterize commerce as being in an evolutionary stage. Just as electricity, telephones, and computers have aided and even fostered the evolution of commerce, so too has the Internet.

The foodservice industry, as well as many others, is experiencing growth and increased competition. In a cycle of growth, both in terms of restaurants and number of consumers eating out on a regular basis, foodservice marketers need to monitor the changes occurring in the marketplace and adapt their strategies accordingly. The evolutionary changes occurring in businesses today focus on two intertwined primary areas: supply chain management and consumer (human) behavior. They will affect the restaurant and foodservice areas in how they do business with partners and with consumers.

THE CHANGING NATURE OF COMPETITION: A GENERAL DISCUSSION

The hyper-competitive business arena of today is causing suppliers, manufacturers, wholesalers, retailers, restaurants, and facilitating organizations alike to rethink their strategic initiatives with both suppliers and customers. The best supply chains are formulating alliances up and down the channel to develop flexible, fast, and efficient partnerships that can respond to changes in the marketplace.

In the past, foodservice supply chain members furthest removed from the end consumer were primarily concerned with the changes directly affecting their organizations. Among these forces are *technology and information exchange, quality and customer service concerns,* and *efficiency and speed to market*. Survival in the future will require foodservice organizations to have vision *beyond* how these forces affect firms and their supply chain partners and *toward* how these forces are affecting end-consumer markets. For insight into how this change in vision might affect foodservice firms and restaurants, it is necessary to examine the realities businesses in other industries are facing.

Changes in Competition

In the past, retailers fought retailers for dominance in the marketplace, while wholesalers and manufacturers followed suit. Each entity was strong enough to take on its own competitors. Whether these companies survived and flourished or languished and folded was largely a result of how they performed in relation to their peers.

In the new millennium, the rules of battle will be rewritten. Organizations will still compete in the marketplace, not against individual competitors in their field, but fortified by supply chain alliances. In essence, *competitive dominance is achieved by an entire supply chain, with battles fought supply chain versus supply chain.*

This change in philosophy has affected marketing strategies for many leading firms in a variety of industries. They realize that effective marketing strategies can no longer be built upon the activities of a single organization. Strategy is shifting from the scope of a single organization to partnerships and alliances involving many organizations, all focused on customer satisfaction as well as consumer satisfaction.

Large firms with a lot of clout in the channel have capitalized on their position by demanding lower prices, dictating operating systems, and guiding strategy. As more large firms begin to devour small, independent firms, this type of behavior will likely continue. However, clout will no longer be determined just by size; it will be determined by information about, alliances with, and loyalty from consumers. As smaller organizations begin to use their inherent strengths–customer contact, personal service, and one-on-one interaction–they will become more valuable in the channel. However, size still counts, and the smaller, independent firms found in fragmented industries may band together for collective bargaining power. This type of cooperation among competitors is facilitated by technology, for the purpose of collective buying and collective marketing, as discussed later in this paper.

For example, Daymark, a family-owned transportation and logistics company based in Russelville, Arkansas, recently changed from a supply focus to a demand focus. The Daymark Demand Circle represents a concept it is developing to identify and source a wide range of services its customers might need, ranging from transportation to marketing. Developing relationships with these "preferred providers," Daymark can expand its range of services to its foodservice and floral clients by partnering with other supply chain partners to deliver integrated logistics solutions. One of its successful alliances occurred when it bought and implemented technology from i2, the up-and-coming logistics software giant. With this new technology, Daymark is able to provide total logistics solutions for its clients. Daymark, however, received more than just software when it became an i2 customer–it became an i2 partner and was introduced to other i2 customers with whom it might do business.

Traditional Supply Chain Management

Traditional supply chains begin at the point of manufacture and end with the sale to consumers, with retailers selling products that manufacturers conceive and wholesalers supply. Ironically, it is the manufacturers and food processing firms, the firms most removed from the consumer market, which traditionally dictated what products moved through supply chains. More often than not, these products were derived not from demonstrated need or consumer preference, but from the manufacturer's historic strengths, resources, and best instincts. This process has evolved, to some degree, to a more collaborative process between retailer and manufacturer, leading to retail-friendly product adaptations, such as packaging that is easier to stack on shelves, collaborative forecasting, and various inventory replenishment models. Yet even with closer collaboration between channel members, the traditional supply chain represents a linear, left-to-right progression in which the consumer stands passively at the receiving end of the chain (Figure 1).

Supply chains are efficient when it comes to delivering new products to consumers. Thousands of new products are conceived, designed, and produced by the manufacturer, delivered to the marketplace and made available to consumers. Most move through the supply chain in reasonably efficient and cost-effective ways, due in part to innovative streamlining of the distribution process, improved logistics strategies, and technological advancements. Technology will play an increasingly important role in supply chain management, making the functions of inventory control and purchasing more efficient. Large foodservice organizations may have to change their systems to integrate with their largest supplier and vise-versa, depending on which entity has the most power in the channel.

Limitations of Existing Supply Chains

Traditional supply chains usually operate on a transactional basis; that is, each entity involved is hired to perform a specific task, with its responsibilities

FIGURE 1. Traditional Supply Chain

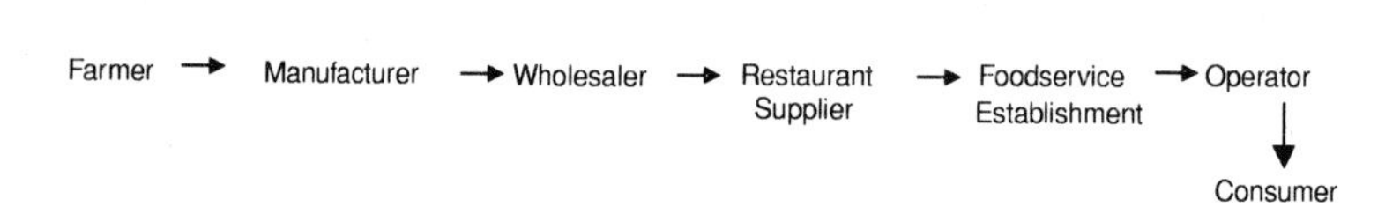

determined by the functions it has historically performed in other supply chains. Traditional supply chain members receive a variety of data regarding market needs, logistics efficiencies, profit margins, costs, and a host of other topics, but do not share data readily with other members of the channel. Supply chains are brought together piece by piece and may work to create a channel one year, only to have the entities involved change the next. Such changes in supply chain members limit the amount of time, money, and other resources channel members are willing to invest in making the channel as efficient as possible. This model does not satisfy well the ultimate need of the customer–a long-term solution to its supply chain needs. Relationships that are formed on trust and shared data last longer than those that are created to fulfill short term contracts.

The passive role of the consumer is also a limitation in the traditional supply chain. It is focused primarily on delivering systems and products that are most efficient from the channel's perspective rather than the consumers' perspective.

DEMAND CHAINS

Taking the supply chain several steps closer to the consumer is the *demand* chain, which gets its name from serving the needs of consumers. As Figure 2 shows, a demand chain represents a circular process which flows from the mind of the consumer through the various supply chain entities to the market. The key is that consumer behavior and consumer analysis fuel the direction of the demand chain. Rather than building and operating their supply chain from manufacturer to market, the best firms are creating alliances with channel partners best able to fulfill consumers' needs, wants, problems, and lifestyles.

The focus on the end-user is creating the supply to demand paradigm shift, causing foodservice and other partners to examine their roles in the supply chain. Among today's emerging demand chains, the players are the same as in the traditional supply chain, but the rules of the game have changed. The roles and responsibilities of each demand chain member are not based on historical strengths or traditional roles. Rather, responsibilities are assumed by the demand chain as a whole. The firm best able to complete a role does so, even if it means breaking the mold. In a demand chain, products can be developed at any point and by any player in the chain–based upon consumer research and information gathered by any entity and shared with all partners.

With whom does the responsibility of understanding consumer demand in the restaurant and foodservice arena lie? The restaurant, the institutional or hospitality food service organization, the wholesaler or other supplier, the advertising agency, or the manufacturer? Or does this responsibility shift according to the situation? The answer is that it depends on which entity is best able to monitor consumer research on eating patterns and preferences to guide collaborative marketing and design efforts.

The best scenario exists when the entire supply chain keeps a finger on the pulse of the consumer market. While each member might not conduct consumer research, each member should be given information on relevant consumer trends. All members involved in designing, creating, marketing, and selling a specific product or service should be privy to consumer information regarding the product, making it easier for any member of the chain to identify food usage and eating patterns, product and packaging improvements, marketing opportunities, or brand extensions. Participation from all members of the demand chain in the transfer of products from mind to market is essential for the success of the product and all channel members. Whether information is obtained from POS data bases, focus groups and quantitative surveys, or from newer methods of "shadowing" consumers and in-home research on food consumption, the data must be shared, analyzed and acted upon by all channel members acting as one entity, if supply chains are to be reinvented as demand chains.

Trends in Foodservice Marketing

Foodservice marketing will experience changes in the future, namely in how items are marketed to consumers and buyers and which entity in the sup-

FIGURE 2. Product and Information Flow in the Mind-to-Market Chain

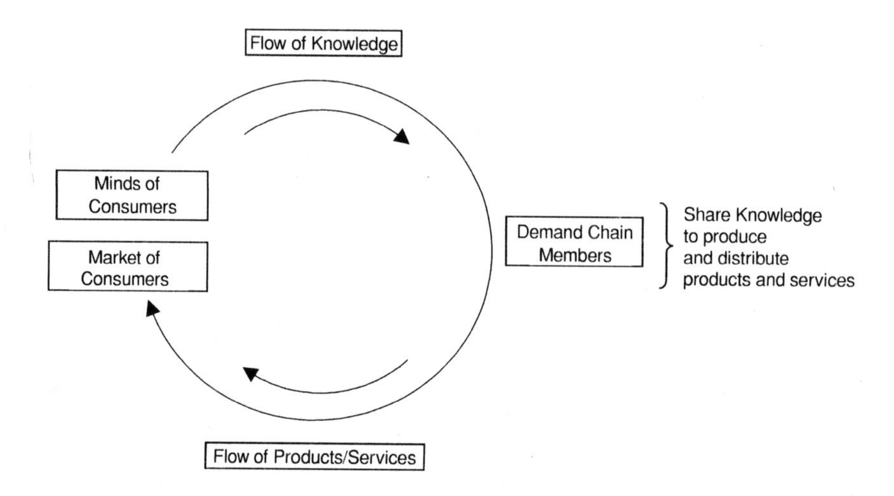

ply chain is responsible for the various marketing functions. Many activities call for cooperation between supply chain partners, as mentioned earlier, with a goal among all partners to operate for the "good" of the end-user.

Shifting and Sharing the Role of Marketing

The roles of marketing and branding in the foodservice arena have traditionally belonged to the party closest to the particular activity at hand. In the past, a restaurant was responsible for conducting its own marketing and advertising, and the distributor did its own marketing of the brands it represented. Similarly, the manufacturer was responsible for promoting the food brand to consumers in hopes that consumer demand would then draw the product from the retailer and restaurant through the chain.

But who is responsible for marketing Pierre's Café, a small independently owned French café? Traditionally, it has been Pierre's responsibility. However, as a small, independently owned restaurant, how can he invest the resources to compete with national chains, such as Olive Garden or Chili's–advertising during prime-time television shows? Could Pierre and some of his "fellow restaurateurs" band together and promote the nuance of dining at a one-of-a-kind restaurant, rather than a "me-too" national chain? If Pierre grows to have several restaurants in a city and prepares to enter a food festival and serve specialty pork, can he approach the National Pork Producers Council to sponsor an ad for him featuring the "Other White Meat" slogan or develop a reliable source of specialty pork cuts with accompanying recipes? Perhaps.

Small independent restaurants will struggle to adapt to the systems suppliers will demand them to have in order to integrate into the supply chain. While one independent store may not be able to invest in the hardware and software to comply, a cooperative may be a solution.

As the small independents look for ways to increase economies of scale, cut costs, and increase efficiencies, large national chains will search for ways to become more personal with local patrons. Local-based promotions and database marketing will continue to be important marketing tools for all foodservice firms. These types of promotions may feature brand name ingredients found in menu entrees, such as Absolut vodka, Hunt's tomato paste, or Morningstar veggie burgers. Co-branding activities between restaurants and manufacturer brand products have led to the appearance of restaurant foods appearing on grocery shelves. Restaurants and institutions may explore the reverse form of co-branding by featuring branded ingredients to signal freshness and quality.

Institutions will need to compete with restaurants as well. University dining halls need to expand and add variety to their menus to keep students from buying meals off-site. Unlike chain restaurants, which may change their menus once or twice per year, institutions can change their menus more frequently and may be more open to change and responsive to supplier promotional ideas. Suppliers should strive to help in developing new recipes or menus that maximize variety with given ingredients. In turn, suppliers may ask institutions to test new products or menu items, which can then be sold to other entities.

From Buying Customer Patronage to Earning Customer Patronage

Foodservice marketers will need to explore new ways to reach and retain consumers in the future. Though many marketing models still focus on giving discounts to "buy" consumers' patronage, other models will be developed to "earn" consumers' patronage. Some restaurants and food manufacturers will attract new customers with coupons and discount offers, but they will phase them away from this model to one that is based on other, more constant, consumer benefits–such as preferred seating, special menu items, or personalized service. More marketers are beginning to ask–why give customers free food when they are willing to pay full price? The best restaurants and foodservice marketers will develop a "menu" of customer benefits to promote rather than price.

The Human Factor

With all of the talk about technology, there are few industries in which *human interaction* will be more important. It will not go away any time soon; in fact, with the increased time spent on computers at work and communicating through e-mail and voice-mail, the human factor may become even more important than it is today. Many retailers, restaurants included, will make the mistake of not training (and rewarding) their staff sufficiently to help customers and make them feel important and appreciated. In a time when retail associates could be a store's greatest asset, they are often their greatest liabilities. Rude, unkempt, and perhaps excessively pierced or tattooed associates probably don't make most people feel good about their foodservice experience.

Limited service restaurants take note. This is not just a rule to be followed by full-service or fine dining establishments. If there are McDonald's, Burger King, Wendy's, and KFC all positioned at an intersection, poised to serve the lunch crowd from nearby office buildings, price and convenience are not the only determinants of which restaurant is chosen. In fact, in these establishments, price and convenience differences are negligible. Other factors, such as accuracy of order, cleanliness of restaurant, and friendliness of employees (all dependent upon the individuals working in the store) determine consumer choice.

THE DEMAND CHAIN'S FOCUS ON CONSUMER BEHAVIOR

If consumer information and understanding is the fuel of successful firms and demand chains, foodservice and restaurant marketers need tools to analyze changes among consumers and their behaviors. In order to identify some of these trends and strategies, marketers must examine the entire food purchase process. When and how consumers decide which foods to buy and which restaurants to patronize can guide marketing efforts.

How Consumers Make Purchase Decisions About Food

In order to thrive in the hypercompetitive climate of the future, foodservice companies of all sizes and types must focus on the first order of any business transaction–namely, understanding how consumers make food purchases, restaurant, and other food choice decisions. Quite simply, without understanding the *minds of consumers,* it is difficult to formulate strategies to keep existing customers and entice new ones. By understanding why consumers buy certain foods from certain restaurants instead of others, how they buy them, and how lifestyle, demographic, and environmental factors affect their choices, foodservice marketers can determine how to meet customers' needs and wants. Sometimes meeting these needs means introducing new products, reformulating existing ones, or altering communication strategies to reach new consumer segments. And sometimes it means removing items from menus and closing unsuccessful restaurant concepts. Sometimes it means developing a deeper understanding of how home meal replacement (HMR) interacts with traditional food consumption patterns.

In the disruptive, discontinuous markets of the foodservice industry, marketers often refer to a "road map" of how consumers make purchase decisions to guide their marketing strategies. The Consumer Decision Process (CDP) model, a simplified version of which is shown in Figure 3, represents a roadmap of consumers' minds that marketers and managers can use to help guide product development, communication, and sales strategies (Blackwell, Miniard and Engel 2001).

The CDP model, developed by James Engel, David Kollat, and Roger Blackwell in 1969, shows how people solve the every day problems in life that cause them to buy and use products of all kinds, including food. As the model shows, consumers typically go through seven major stages when making pur-

FIGURE 3. How Consumers Make Decisions for Goods and Services

Source: Roger Blackwell, Paul Miniard, and James Engel, *Consumer Behavior* (Harcourt; Fort Worth), 2001

chase decisions: need recognition, search for information, alternative evaluation, purchase, consumption, post-purchase evaluation, and divestment. The study of consumer behavior focuses primarily on these seven stages and how various factors influence each stage of consumers' decisions. By understanding the stages in the consumer decision making roadmap, foodservice marketers can discover why people are or are not buying certain foods and flavors and what to do to get them to buy more or from a specific retailer. Marketers can examine how consumers choose between the existing 465,000 restaurants that operate in the U.S. today. As you progress through the discussion of the model, please note the role that brand plays in almost every stage of the decision process.

This model can be used to map-out the buying process for suppliers and retailers as well as consumers. The stages are the same for each entity, but the influences and effects of each are different depending on the entity making the purchase and the situations in which decisions are being made. Regardless of whether a supplier or a consumer is making a purchase decision, each behaves as a human being, laying common ground between institutional and retail buying. Using this model can aid in identifying the effects of consumer trends and technology on decision making and purchase behavior.

Recognizing the Need

The starting point of any purchase decision is a customer problem, need, or wish. The recognition of a *need* occurs when an individual senses a difference between what he or she perceives to be the *ideal* versus the *actual* state of affairs. Consumers do not just walk into a restaurant and say, "I notice you have food to sell. I have some extra money I would like to spend so just pick something out and charge it to my credit card." Consumers buy a meal to satisfy a variety of needs–including hunger, desire to celebrate and have fun, and entertainment. Increasingly, time constraints have played a major role in the success of carryout food, currently expanding twice as fast as in-restaurant dining. The problem being solved in this instance is not just "need to feed the family"; it is lack of time to prepare the meal to feed the family. And as fewer children grow up in households where home-cooked meals are commonplace, the less they learn to cook, making the next generation even more dependent upon foodservice and such concepts as HMR and carry-out.

Marketers must *know consumers' needs*–if they know where consumers "itch," they have a better idea of where to "scratch" with new and improved products, more effective communication programs, and more user-friendly distribution channels. Knowing consumers' desires and feelings, often leads to an insight, which can be used to guide further product or concept development.

When Folgers began to interview consumers and talk to them about what they like about coffee and why they drink it, consumers indicated not only was taste important but so was aroma. Further in the discussion, they mentioned how coffee made them feel, which led to an insight. Coffee reminds people of being home with their parents and feeling safe and comforted by childhood memories. This insight led to a new ad campaign focusing on family moments, and it revitalized the brand. Starbucks acknowledges that it does not just want to sell coffee and snacks, it strives to be a place people want to be–following a similar consumer insight.

Retailers and manufacturers alike must monitor consumer trends because as consumers change, so do their problems and needs. The influences most likely to alter the way consumers look at food-related problems and the ways to solve them are *family size, health, age, income, health,* and *reference groups.* Spotting *changes* in these variables is often the key to new marketing opportunities. Thirty year old consumers with families need to buy more food items (usually in larger quantity packages) than do consumers in their seventies, who may be living single and in smaller homes with less storage space. Yet, both families are likely to exhibit similar frugal behavior, stemming from a variety of needs, some similar and some different. Today, consumers are looking for a combination of wellness and indulgence, bold flavors, and convenience in their food choices.

As consumers move through different *life-stages,* their needs and buying habits for food and dining can be expected to change. Marketers must ask questions, including "how do changes in income, employment status, health consciousness, and age change what people expect from food and restaurants, and how they solve their food needs?" The answers to these questions are more likely to be found by the demand chain rather than any of its individual entities. With hectic lifestyles, off-premise (or take-away) food is growing at two times that of in-store sales. Factors such as "easy to eat," temperature retention, presentation, and portability are important in guiding consumers' purchase decisions in this area.

Marketers often communicate a need, thereby raising consumers' awareness of unperceived needs or problems. Many years ago Listerine mouthwash used advertising to increase the awareness of "halitosis" and in the process dramatically increased mouthwash sales. Listerine did not create the problem of bad breath; it simply raised awareness of the problem. And Scope still does today with ads that make people more aware of "morning breath." Can marketers create needs? Not really, but they can show how a product meets unperceived needs or problems consumers may not have considered before. That is often the goal for groups that conduct industry-wide promotions, such as the National Pork Producers Association, the National Pasta Association, the American Dairy Association, and other food related trade organizations.

Search for Information

Once need recognition occurs, consumers begin searching for information and solutions to satisfy their unmet needs. Consumers may retrieve knowledge from memory, instinct, and past experiences or by collecting information from peers, family, and the marketplace. Sometimes consumers search passively by simply becoming more receptive to information around them, while at other times they engage in active search behavior, such as talking to friends about the latest restaurants, paying attention to ads, searching the Internet, or venturing to shopping malls and other retail outlets.

The length and depth of search is determined by variables such as personality, social class, income, size of the purchase, past experiences, prior brand perceptions, and customer satisfaction. If consumers are delighted with the *brand* of restaurant they frequent or product they consume, they may repeat the behavior with little if any search behavior, making it more difficult for competitors to catch their attention. This increases the need to monitor satisfaction continuously with comment cards, focus groups, and other research. That's why restaurants place a high priority on keeping customers satisfied. When consumers are unhappy with current products or brands, search expands to include many alternatives, and unfortunately, they tell others about their bad experience.

Increasingly, information search for a variety of products and services is occurring on the Internet. While some searches on the Internet take longer, others are shorter, depending on how the Web site is designed (Moorthy, Ratchford and Talukdar 1997). Some researchers indicate that if online retailing decreases the cost of price information search, consumers will become more price sensitive (Hogue and Lohse 1999). Other studies have shown that by altering Web design and making it easier to search for and compare quality information, consumers became less price sensitive and were more likely to purchase quality products (Alba, Weitz, Janisqewski, Lutz, Sawyer, and Wood 1997).

Some restaurant managers may have ignored the Internet proclaiming that "you can't sell food on the Internet," but consumers and suppliers alike will rely more on the Internet for information, whether items are purchased using the technology or not. Restaurants can provide information about catering, hours, daily specials, locations, and menu items, thereby, making it easier for consumers to "shop" for the perfect dining experience. It is possible today to make reservations and indicate seat preferences electronically for some progressive restaurants. How widespread this practice becomes depends on how many consumers want to communicate electronically for these purposes in the future.

While some consumers prefer the "old-fashion" approach to search, called "shopping," others would prefer to have the information come to them. Email marketing is beginning to capture the attention of some restaurants and consumers alike. Small, independent restaurants that sometimes lack the technology to reach current and potential customers may turn to specialized e-marketing firms, such as HometownOffers.com and CoolSavings.com, to become their e-marketing partners.

Evaluating Alternatives

In this stage consumers seek answers to questions such as "What are my options?" and "Which is best?" when they compare, contrast, and select from various restaurants and the menu items of each food service alternative. Customers go through the same process weighing alternative suppliers. People compare what they know about different products and brands with what they consider most important and begin to narrow the field of alternatives, and finally they resolve to buy one of them. Menu engineering has made great strides in recent years to influence how consumers evaluate items on a menu and how they put together a meal. Guiding consumers to make selections with higher price points and order more items a la carte increases the check total and profits per turn.

How do consumers evaluate alternatives? They rely on specific criteria to select what and where they buy food and where they dine and make the selections that will most likely result in their satisfaction. Different consumers employ different *evaluative criteria*–the standards and specifications used to compare different products and brands. How individuals evaluate their choices is influenced by both individual and environmental influences. As a result, evaluative criteria become a product-specific manifestation of an individual's needs, values, lifestyles, and so on. They use evaluative criteria to decide both where to buy and what to buy.

Marketers must conduct research to monitor how consumers' evaluative criteria are changing, and in which situations certain criteria are more important than others. For example, how important are criteria such as food preparation, freshness of ingredients, time of wait, and speed of service to customers? Understanding how important these criteria are to certain patrons and in certain situations, can help develop effective in-store and external marketing strategies.

Some attributes upon which alternatives are evaluated are *salient* in nature, while some are *determinant,* yet both affect marketing and advertising strategy. Consumers think of *salient attributes* first and consider them most important. In the case of purchasing a car, these attributes would include price, vehicle type (sedan versus sport utility) and quality ratings. But surprisingly, it's the *determinant attributes*–details such as color, display panels, and even the number and type of cup-holders–that usually determine which brand or store consumers choose, especially when they consider alternatives to be equivalent on salient attributes. Foodservice marketers need to examine which salient and determinant attributes are most influential in consumers' evaluations of alternative food and dining options. Beyond the salient attributes of "filling the stomach," how do consumers trade off taste versus fat content or versus hedonic attributes such as ambiance, environmental music, and food entertainment (table side service or serving the "sizzle" with a steak)?

Recent research has focused on the elements affecting the choice process for experience goods, such as entertainment. Similar to restaurants, movies serve as a good category to research since they are experience goods that are hard to evaluate prior to viewing (Lynch and Ariely 1999). While variables including word-of-mouth, and critic reviews have been identified as critical influencers on consumer choice of movies (Sawhney and Eliashberg 1996), other psychological variables, such as emotional expectations and latent product interest, also play an important role in choice. The findings that new movie choice is influenced by emotional expectations, not by cognitive assessment of product attributes, recognizes the role of emotions in certain areas of consumer behavior (Eliashberg and Shugan 1997). In fact, when evaluating alternatives and making trade-offs between product attributes, emotion-laden trade-offs complicate how trade-offs are made and what value is assigned to various attributes during the choice process (Neelamegham and Jain 1999). These are probably the attributes relevant in evaluating dining choices, as well.

Purchase

A consumer might move through the first three stages of the decision process according to plan and intend to purchase a particular product or brand. But consumers sometimes buy something quite different than intended or opt not to buy at all because of what happens during the purchase or choice stage. A consumer may prefer one restaurant but choose another because of a promotional event, location, traffic-flow problems, or special menu items. Inside the store, consumers may talk with someone who changes the decision, see a table tent display, use a coupon or price discount, or lack the money or right credit card to make the purchase. The best restaurants manage the overall attributes and image to achieve preferred patronage among the market target and to manage, in micro detail, all aspects of the dining experience.

Consumption

After the purchase is made, consumption can occur–the point at which consumers use the product. For most products, consumption can either occur immediately or it can be delayed, even food bought in restaurants. For example, what proportion of food served at a meal is consumed at the restaurant and how much is left on the plate? If entrees are too large, the desire for dessert and appetizers is diminished, yet if portions are too small, then consumers feel they are not getting value. How much consumers want to eat, expect to receive, and want to take home with them, depends on situations and overall consumer trends. Trends that promote healthy eating and low calorie, light dining may warrant smaller portions, while consumers with large families may look for larger portions to consume there and take home. The quality of packaging for these "leftovers" may influence perception of the quality and the loyalty that leads to return visits to that restaurant.

Post-Purchase Evaluation

The final stage of consumer decision-making is post-purchase evaluation, in which consumers experience either a sense of satisfaction or dissatisfaction. *Satisfaction* occurs when consumers' expectations are matched by perceived performance; when experiences and performance fall short of expectations, *dissatisfaction* occurs. The outcomes are significant because consumers store their evaluations in memory and refer to them in future decisions. If the consumer is highly satisfied, subsequent purchase decisions become much shorter. Competitors, for the most part, have a hard time accessing the minds and decision processes of satisfied customers because they tend to buy the same brand at the same store. But consumers that are dissatisfied with products they buy or stores from which they buy are ripe for picking with the marketing strategies of competitors who promise something better.

Emotions also play a role in how someone evaluates a product or transaction. An emotion can be defined as a reaction to a cognitive appraisal of events or thoughts; is accompanied by physiological processes; is often expressed physically (e.g., in gestures, posture, or facial expressions); and may result in specific actions to cope with or affirm the emotion (Luce, Payne and Bettman 1999). Just as consumers compare price and evaluate the fairness of exchange in the alternative evaluation stage, so do they revisit these issues during post-purchase evaluation. Some research indicates that how consumers view the fairness of the exchange *over time* affects current and future usage behavior. Price and usage also affect their overall evaluations of the fairness of the exchange. In turn, these evaluations affect overall satisfaction and future usage (Bagozzi, Gpoinath and Nyer 1999).

Divestment

After consuming food, consumers must dispose of leftovers and packaging. Marketers must also consider the packaging they offer consumers who take home leftovers. Can the containers be microwaved? Are they recyclable, easily disposable, or reusable in the home?

Repeat Purchase Behavior (Otherwise Known as Brand Loyalty)

One goal of the CDP model is to help increase the overall satisfaction level associated with the purchase of a product. The more marketers understand how consumers make decisions, the more likely they are to formulate strategies to affect all levels and increase customer satisfaction. And with satisfaction comes the increased likelihood of repeat purchases and brand loyalty (Bolton and Lemon 1999).

Brand loyalty affects the amount of time and effort consumers put into the decision making process. If patrons are happy with a particular restaurant or brand of food, they are likely to spend less time on the search and evaluation stages and skip to the purchase stage. As such, one of the greatest contributions of brand loyalty to consumers is a saving of time. McDonald's and Wendy's, which have both succeeded in building their brands in the minds of consumers and delivering food and dining experiences that meet consumers' expectations, not only offer their customers consistency; they deliver time savings. When consumers see these brands, they know what to expect and do not have to "shop around" for other options if they are satisfied with what they think the brand will deliver.

Multiple branding takes that phenomenon one step further. When consumers are faced with having to make food choices several days of the week, they may not want to go to the same restaurant each time. Instead, clusters of restaurants (such as Tricon Food's Pizza Hut, Taco Bell, and KFC) which offer multiple food choices to consumers mimic the Wal-Mart one-stop-shopping strategy that limits the search process for consumers.

CONSUMER BEHAVIOR AS THE DEMAND CHAIN DRIVER

Creating winning demand chains in the foodservice industry depends on how each member rates on the following attributes:

- Ability to recognize, monitor, and work together to meet the changing demands of customers and consumers
- Willingness to communicate and share information with channel members
- Ability and willingness to invest in growth, in order to grow with other members
- Ability to foster trust among members, which is often dependent on sharing similar values
- Willingness to communicate honestly and admit what they can and cannot do
- Ability and willingness to be proactive in addressing changes in the marketplace

Consumer behavior, analyzed best with the CDP model, provides the foundation for creating demand chains in food service. Understanding the minds of consumers, gathering and disseminating information to all demand chain partners, communicating about trends and business limitations, and working toward the same goal of delivering consumer-desired products summarizes the process taken-on by foodservice demand chains. The goal of all of these activities is to increase consumer loyalty by delivering products that are desired by consumers and driven by their needs, wants, and lifestyles.

REFERENCES

Alba, J., J. Lynch, B. Weitz, C. Janisqewski, R. Lutz, A. Sawyer, and S. Wood (1997), "Interactive Home Shopping: Consumer, Retailer, and Manufacturer Incentives to Participate in Electronic Marketplaces," *Journal of Marketing*, 61 (July), 38-53.

Bagozzi, R.P., M. Gopinath, and P. Nyer (1999), "The Role of Emotions in Marketing," *Journal of the Academy of Marketing Science*, 27 (Spring), 184-206.

Blackwell, R., P. Miniard, and J. Engel (2001), *Consumer Behavior*, 9th Edition, Fort Worth, TX: Harcourt Publishing

Bolton, R.N., and K.N. Lemon (1999, "A Dynamic Model of Customers' Usage of Services: Usage As An Antecedent and Consequence of Satisfaction," *Journal of Marketing Research*, 36 (May), 171-186.

Eliashberg, J. and S. Shugan (1997) "Film Critics: Influencers or Predictors?" *Journal of Marketing*, 61 (April), 68-78.

Hogue, A.Y. and G. Lohse (1999), "An Information Search Cost Perspective for Designing Interfaces for Electronic Commerce," *Journal of Marketing Research*, 36 (August), 387-394.

Luce, M.F., J. W. Payne, and J. R. Bettman (1999), "Emotional Trade-Off Difficulty and Choice," *Journal of Marketing Research* (May), 143-159.

Lynch, J. and Dan Ariely (1999), "Interactive Home Shopping: Effects of Search Cost for Price and Quality Information on Consumer Price Sensitivity, Satisfaction with Merchandise Selected and Retention," working paper, Marketing Department, Duke University.

Moorthy, S., B.T. Ratchford, and D.Talukdar (1997), "Consumer Information Search Revisited: Theory and Empirical Analysis," *Journal of Consumer Research*, 23 (Fall), 263-277.

Neelamegham, R. and D. Jain (1999), "Consumer Choice Process for Experience Goods: An Econometric Model and Analysis," *Journal of Marketing Research*, 36 (August), 373-386.

Sawhney, M. and J. Eliashberg (1996), "A Parsimonious Model for Forecasting Gross Box Office Revenues of Motion Pictures," *Marketing Science*, 15, (Spring), 113-131.

The Restaurant Industry in China: Assessment of the Current Status and Future Opportunities

Jinlin Zhao
Lan Li

SUMMARY. This paper examines the impact of political, economic, sociocultural, technological and ecological environments on China's restaurant business and discusses the current status and competition of the restaurant industry. The study indicates that liberalization of economic and legal policies of China provided the impetus for growth and development of franchised western restaurant chains as well as local restaurants. Multinational companies have introduced various new foods and service methods with a positive and permanent impact on Chinese eating habits. China's restaurant industry is experiencing the rapid growth stage of its industry life cycle. The future of the restaurant industry in China is very bright and is partly tied to the continuing liberal policies of the government. *[Article copies available for a fee from The Haworth Document Delivery Service: 1-800-HAWORTH. E-mail address: <getinfo@haworthpressinc.com> Website: <http://www.HaworthPress.com> © 2001 by The Haworth Press, Inc. All rights reserved.]*

KEYWORDS. Restaurants, China, quick service restaurants, restaurant chains, foodservice

Jinlin Zhao, PhD, is Associate Professor, School of Hospitality Management, Florida International University, North Miami, USA.

Lan Li, PhD, is Associate Professor, Food Systems Administration, School of Family, Consumer and Nutrition Sciences, Northern Illinois University.

[Haworth co-indexing entry note]: "The Restaurant Industry in China: Assessment of the Current Status and Future Opportunities." Zhao, Jinlin, and Lan Li. Co-published simultaneously in *Journal of Restaurant & Foodservice Marketing* (The Haworth Hospitality Press, an imprint of The Haworth Press, Inc.) Vol. 4, No. 4, 2001, pp. 155-172; and: *Quick Service Restaurants, Franchising and Multi-Unit Chain Management* (ed: H. G. Parsa and Francis A. Kwansa) The Haworth Hospitality Press, an imprint of The Haworth Press, Inc., 2001, pp. 155-172. Single or multiple copies of this article are available for a fee from The Haworth Document Delivery Service [1-800-HAWORTH, 9:00 a.m. - 5:00 p.m. (EST). E-mail address: getinfo@haworthpressinc.com].

INTRODUCTION

As early as the thirteenth century, in his *Il Milione*, Marco Polo introduced to the world the importance of food in Chinese culture. Tannahill (1988), a food historian, wrote:

> *Marco Polo was intrigued by the vast abundance of food. None of countries at that period were as advanced and varied as China was. As well as ordinary eating houses, fast food restaurants, tea houses, noodle shops and wine shops, all with their own Chef's specials–chilled fruits or honey fritters, steamed pork and beans, won tons, barbecued meat, fish soups and so on. Every morning between 1 A.M. and dawn the proprietors hurried off to specialist food markets for pork or silkworms or shrimp from which they made pies to serve with their drinks, or oysters, mussels, or bean curd that nourished the poorer classes.*

While China is known for its rich culture and cuisine, food consumption at home has changed dramatically from meeting basic physiological needs to an occasion for enjoyment. Since the significant economic and political reforms of 1978, China's restaurant industry has developed at a rapid pace. During the last 20 years, the number of restaurants has grown 25 times from 117,000 in 1978 to 2,588,000 in 1996 (Ling, 97; Yan, 1998). Likewise, employment in the restaurant industry has increased sevenfold from 1,044,000 in 1978 to 7,753,000 in 1996 (Yan, 1998). In 1997, the Hong Kong Trade Development Council estimated that China's fast-food sector alone was valued at more than $3.6 billion and expected to grow at an annual rate of more than 46 percent over the next 3 years (Cee & Theiler, 1999).

The current paper discusses the macro forces driving the changes and the impact of the reforms in the development of China's restaurant industry. The data for the report were collected during in 1998 and 1999 through personal interviews with Chinese officials from the National Tourism Administration, the State Administration of Internal Trade, directors from the hospitality and tourism associations, editors of hospitality and tourism publications, industrial professionals and hospitality educators. Related information from newspapers and magazines as well as Internet sources in both English and Chinese were also analyzed as part of the research.

DRIVING FORCES OF CHINA'S RESTAURANT INDUSTRY

Many factors have influenced the rapid development of the restaurant industry in China. Factors such as deregulation, liberalized new laws and rules, economic development, sociocultural changes, technological and ecological considerations have all had an impact. The following sections will discuss these issues and their importance to the Chinese restaurant business environment.

Political Impact

The Chinese restaurant industry has benefited from political reform. Many government policies, laws, and regulations established in the last 20 years have made a positive impact on the industry.

The Open-Door Policy. One of the first reform policies initiated in 1978, the open-door policy has encouraged international investors and companies to enter Chinese markets in the form of joint ventures, franchising, management contracts, and sole ownership, allowing China to absorb over $236.8 billion in direct foreign investment in the last 20 years (Peng, 1998). The multinational restaurant companies found their way to China in the later 1980s. KFC pioneered the fast-food market in 1987; Pizza Hut, as a part of PepsiCo, followed KFC and entered the Chinese market in the early '90s. At present, there are some 350 KFC and Pizza Hut restaurants in 90 cities with 23,000 employees (Zhang, 1999). McDonalds entered China in 1990 and now has more than 241 outlets (Guo, 1999). U.S. based companies such as Pizza Hut, A&W, Subway, Domino's, TGI Friday's, Kenny Rogers Roasters, Schlotzsky's Deli, Dairy Queen, and Hard Rock Café are all competing in China's restaurant industry, along with companies from Japan, Korea and Singapore (Wang, 1997; Hillis, 1998).

The concepts of quality, service, cleanliness, and value brought by the Western fast-food chains have had a dramatic impact on local restaurant operators. After observing the success of international restaurant chains in China, many local entrepreneurs developed a growing interest in restaurant quality control and efficient store management. In 1989 the State Administration of Internal Trade allocated $1.2 million for a campaign to encourage Chinese entrepreneurs to take advantage of its traditional cuisine and industrialize the fast-food sector with efficient production and service systems.

To encourage foreign foodservice companies to enter China's foodservice market, the Chinese government provides tax incentives including tax exemptions for the importation of food operations equipment. Also, under the open-door policy, the importation of fruit has increased nine times from 83,611 tons in 1993 to 764,748 tons in 1997, with a spending of $235 million on fruit importation in 1997 alone (*Economic Daily*, 1998).

The Coexistent Economic Policy. The flourishing development of the restaurant industry has mainly relied on the economic policy of coexistence, which allows the coexistence of state, collective, and private ownership of restaurants. Before reforms, a majority of the restaurants were owned or controlled by state and/or local governments. During the reforms, the restaurant industry was the first sector to allow private ownership and partnership. By 1997, more than 90 percent of the restaurants were in the private sector (Yang, 1997; Yan, 1998). Consequently, entrepreneurs from the private sector represent the most dynamic force in today's Chinese economy. Entrepreneurs from the private sector are expected to serve as the engine pushing China's continuous economic growth and to create new employment opportunities.

Legal Environment. To promote the development of the restaurant industry, the Chinese government has passed numerous laws and regulations. The 1995 "Law of Food Sanitation and Safety," for example, regulates the practice of food production, processing, and transportation; employee hygiene; and food additives in order to protect the health of consumers and enhance the quality of food production and consumption. The "Regulation on Hog Slaughter," passed by the State Council in 1997, mandates that only a licensed slaughter house can process pork products, and that it must follow strict sanitation and safety practices (*People's Daily*, 1997). The "Implementation Plan for Improving Human Nutrition Intake" (Sinanet.com, 1998), issued by the State Council in 1998, provides guidelines to promote balanced diets and encourage nutritious food. To regulate the Chinese-style fast-food sector, the State Administration of Internal Trade issued a document in 1997 analyzing the restaurant business environment and establishing the goals and objectives of Chinese fast-food development (State Administration of Internal Trade, 1997). In the first part of 2000, the Ministry of Public Health issued "Food and Beverage Safety and Sanitation Measures" to safeguard public food safety.

Economic Impact

During the past two decades, the Chinese economy has been one of the fastest-growing economies in the world with Gross Domestic Product (GDP) per capita increasing about 5.4 times (*People's Daily*, 1998). Despite the recent economic crisis in Asia, China still enjoys a 7.8 percent GDP growth rate in 1998 (O'Neill, 1999), outstripping most other large economies. Its currency has maintained stability without devaluation for many years. The disposable income per urban resident has increased 15 times, from 343 yuan in 1978 to 5,188 in 1997 (*People's Daily*, 1998). The increase of disposable income has made a great contribution to the rapid development of the restaurant industry. According to the State Statistics Bureau, 46.4 percent of the disposable income in urban areas is used for food expenditures (Wu, 1998). However, there is a large gap in disposable income between the rich and poor and the gap seems to be growing ever wider.

One of the major economic issues in China is its large unemployed workforce. Because of the restructuring of state-owned enterprises, it is estimated that 13-15 million people became jobless by the end of 1998 (SCMP, 1998). Job creation for the unemployed is the number one task for the government and local officials. The government has budgeted two to three billion yuan to help laid-off workers keep their basic living standards; it has also made some effort in re-employing the unemployed, including encouraging the laid-off workers to explore re-employment opportunities in the restaurant industry. Lured by its great growth potential, a new generation of Chinese entrepreneurs has emerged in the restaurant sector. With an improved image of themselves and vastly different from that of the traditional mom and pop family restaurant owners, new entrepreneurs have been carefully studying the operation system of international restaurant chains in China and started developing indigenous restaurant chains. Restaurant entrepreneurs surely help to absorb some of the unemployment pressure but more efforts are needed to stimulate the national economy in order to address the prevailing unemployment problem.

Socio-Cultural Impact

China has a population of 1.2 billion people. Three-fourths of the population live in the countryside and one-fourth or about 360 million are urban residents. Although the rural population has more or less maintained its lifestyle and eating habits, the urban Chinese residents have dramatically changed theirs. Because of the faster pace of life in the cities, the urban residents are gradually moving away from the tradition of cooking and eating at home to a trend of eating out with family and friends. People used to think that eating out wa s a waste of money. Now they go out to eat in order to save time, have fun and/or socialize. With the gradual increase in disposable income, the consumers have changed their focus from indiscrete product purchase to the demand of high-quality products, friendly services, comfortable eating environments, and even live entertainment.

Generally, food prices are affordable for the average citizen. Office and city workers may purchase a box lunch delivered for 6-12 yuan ($0.75 to $1.50) per meal. The box lunch usually consists of stir-fry meat and vegetables, rice, and soup. As a symptom of urbanization most city dwelling Chinese enjoy trying new things, and they are increasingly interested in tasting foreign foods. The rapidly changing lifestyles and recently acquired new eating habits have resulted in a booming fast-food industry that offers foreign foods and Western style service. Among the foreign foods, Western burgers and fried chicken have gained the most popularity in cities, particularly among youth and working urbanites. According to some estimates, frequent restaurant-goers may visit Western-style fast-food restaurants approximately three times a month, while children prefer to go to McDonald's up to 10 times a month. One survey indicated that kids aged 12 and under in 5 major cities (Beijing, Shanghai, Guangzhou, Chengdu, and Xian), totaling 5.21 million, spend about 3.5 billion yuan a month, with 54.3 percent of their expenditure spent on food. As is the case in America, kids often influence a family's dining-out choice: about 2/3 of parents often ask their children's preferences before most purchases. The survey also found that 26.4 percent of families prefer to eat foreign foods while dining out (France-presse, 1998). Another survey indicated that 72.7 percent of the patrons in foreign fast-food restaurants are 15 years old or younger, while 70 percent of the patrons in Chinese fast-food restaurants are adults aged 35 and over (Yang, 1998). Today, for most Chinese, quick service restaurants represent more than just a different taste, they stand for a new lifestyle and social status.

Technological Impact

Unlike many other industries, the Chinese foodservice industry has not changed much in its traditional food preparation techniques. Although gas has become the major energy source for restaurants in urban centers, coal is still the energy source for many mom and pop restaurants. Traditional cookware is still the main cooking equipment. Except for restaurants in joint-venture hotels, computer application in most restaurants is minimal and limited to cash registers, despite the fact that the number of Internet users in China is estimated to increase. To modernize and industrialize the foodservice industry requires extensive innovation as Chinese cuisine is so diversified with distinctive regional tastes and local preferences. As a result, standardization in mass production is more challenging in China than in the Western countries with predominantly nondiversified cultures. Standardization of equipment, food preparation techniques and service levels are critical to achieve any reasonable level of success in the restaurant industry.

Ecological Impact

Pollution of air, water and soil has become one of the major by-products of the country's economic modernization. The central, regional, and local governments have begun to realize the negative impact of pollution and made it a major priority issue. The State Environmental Protection Agency was upgraded to a ministry-level organization in 1998 (Hon, 1998), and it monitors and evaluates environmental protection measures in the country's 570 cities (Ni and Dai, 1998). Since the State Council approved the Natural Environmental Protection Plan in early 1999, environmental efforts have been made to clean the rivers and lakes, establish standards to limit automobile exhaust, switch from coal to natural gas for cooking, and convert internal combustion engines to those powered with electricity or natural gas.

The Ministry of Agriculture has encouraged farmers to adopt environmental-friendly practices to produce "Green Foods," less-or not-pollution prone. Since 1990, about 700 kinds of such foods have been produced, with an annual production rate of 3.6 million tons (Wu & Pu, 1998). To reduce solid waste, the country plans to stop using polystyrene food boxes and plastic food containers by October 30, 2000 (Bi, 1999). Instead, "green food boxes" made from biodegradable materials, such as rice straw, are being encouraged (France-press, 1998). The National Tourism Administration has recently adopted the principle of "thinking environmental protection before tourism development" (Ni & Dai, 1998). For the restaurant industry, the environmental protection issues on its agenda include reducing waste, controlling cooking fumes, and recycling

Impact of Travel and Tourism Industry

Travel and tourism industry in China has grown at a rapid pace in the past 20 years. The tourism industry is contributing more than 4 percent of GDP annually. With significant growth in disposable income combined with a high number of national holidays and non-working days (114 days in total), domestic travel has become the first choice for many Chinese looking for ways to spend their leisure time. Profit from domestic tourism is increasing more than 15 percent annually (*China Daily*, 1999). Traditional festival celebration has moved from home gatherings to tourism sites. The restaurant industry has benefited 50 billion yuan from both overseas and domestic tourists in 1998 (*China Daily*, 1999).

Clearly, the above driving forces in the Chinese society have had a very positive impact on the rapid development of the restaurant industry. As China prepares to enter the World Trade Organization, its restaurant industry will surely be influenced by global economic development.

STATUS OF THE RESTAURANT INDUSTRY IN CHINA

The restaurant industry in China consists of both domestic and overseas markets. The domestic market can be further categorized into local patrons and domestic travelers. In 1996, food sales from domestic travelers was 202.5 billion-yuan, 37 times the sales of 5.5 billion yuan in 1978 (see Figure 1). The number of foodservice outlets has increased 22 times from 117,000 to

2,558,000 between 1978 and 1996 (see Figure 2). Figure 3 shows that the restaurant industry employed 7.75 million people in 1996, seven times that in 1978 (Yan, 1998). In 1998, local patrons contributed a lion's share of the domestic market, about three-fourths of the total sales of the restaurant industry. The overseas tourists spent $1.586 billion on food, or 13.1 percent of $12 billion of the total overseas tourism receipts in 1997 (China Tourism Statistics, 1998). The overseas travelers have mostly spent their food dollars in Chinese hotels and classified restaurants.

CLASSIFICATIONS OF THE CHINESE RESTAURANT INDUSTRY

Because China is a large country with various ethnic and cultural groups, the food preparation techniques, food ingredients and food preferences vary greatly from one region to another. Traditionally, Chinese restaurants are categorized by the different types of cuisine they serve. Four types of cuisine are classified by geographic region: Shandong, Huaiyang, Sichuan, and Guangdong. *Shandong cuisine,* originating in the northern region, has a strong and a comparatively salty taste. *Huaiyang cuisine,* originating in the southeast region, is known for its many soup dishes, and food from this region is light in color and plain in taste. *Sichuan cuisine,* mostly popular in the southwest, is spicy and hot. *Guangdong cuisine,* found in the southern regions, tastes sweeter than the other cuisines and utilizes a wide variety of ingredients. Fast development of the restaurant industry and the significant liberal changes in the general environment that took place recently have permitted the offering of inter-regional foods and foreign foods.

Many sub-regional foods are also found in various areas in China. As a result of fast development of the restaurant industry, many restaurants now feature combinations of various cuisines, new styles of service, and foreign foods.

Classification Based on Prices

Generally speaking, there is not much difference in the price of breakfast versus lunch. However, there is great price variation in the dinner market. Restaurants in China target various different groups of patrons depending on the levels of disposable income. For the dinner market, low-end restaurants are priced at 20 to 40 yuan ($2.5-$5) per person. Most restaurants are moderately priced with dinner costing about 50 yuan ($6) per person. Expensive locations are often priced at 150 to 200 yuan ($18-$25) or more per person. The quality of food and services, along with the ambience, vary greatly based on menu

FIGURE 1. Restaurant Sales China

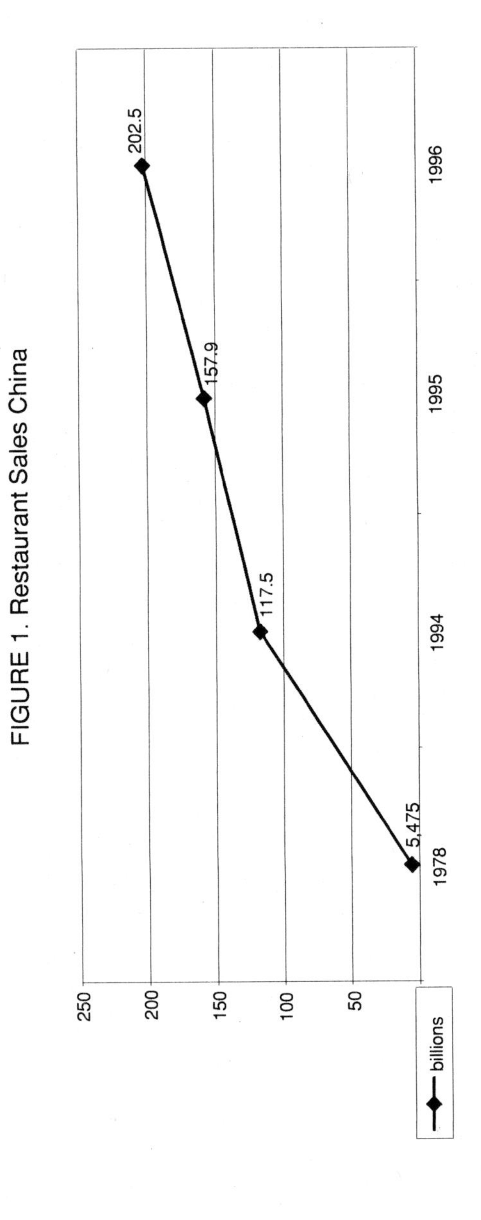

FIGURE 2. Total Restaurant Numbers in China

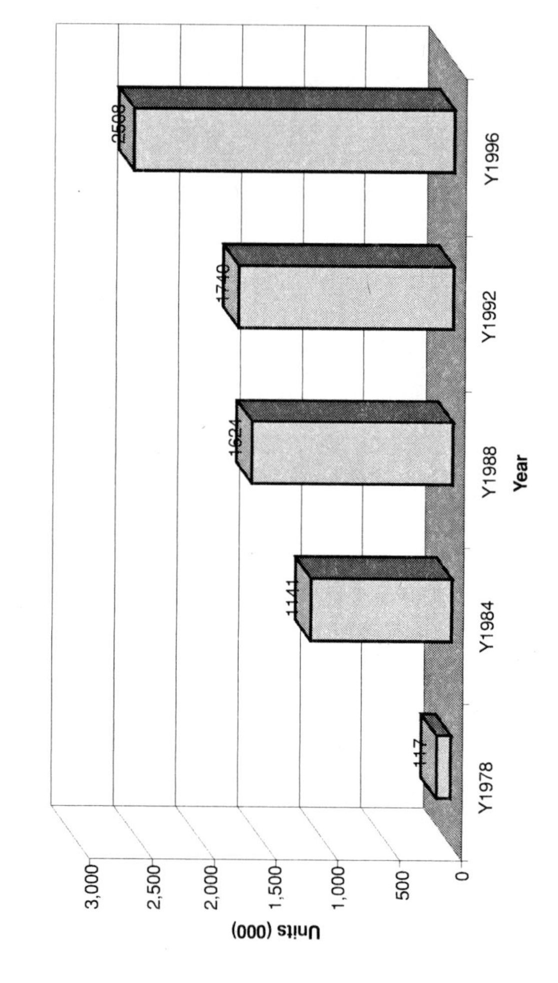

FIGURE 3. Restaurant Employment in China

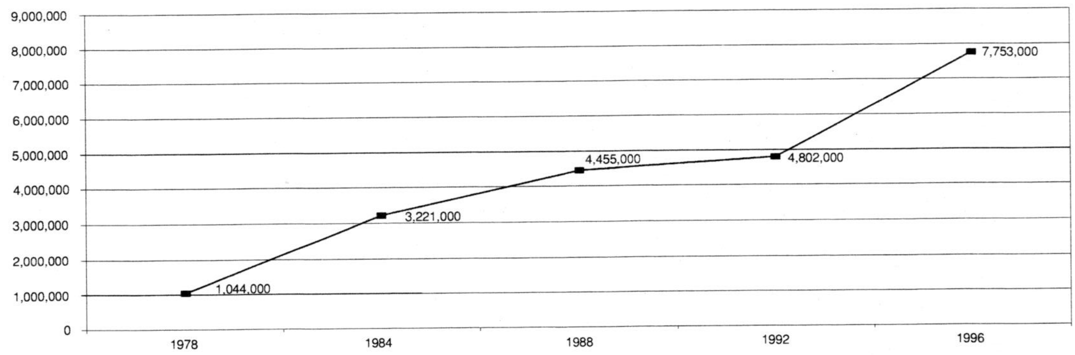

prices. As an added feature, restaurants in the more expensive category may provide live or karaoke entertainment.

Governmental Classifications of Restaurants

Unlike many countries where restaurants are classified by trade organizations, classification of Chinese restaurants is one of the responsibilities of the State Administration of Internal Trade (SAIT), which has established guidelines for restaurant classifications and evaluations. Table 1 shows a breakdown of different classification requirements. If a restaurant fits into a certain category, the SAIT will issue a classified certification, which subjects the restaurant to frequent evaluations by the organization. The SAIT has identified four classes of restaurants: Prime, First, Second, and Third Classes, each with clearly spelled-out requirements based on reputation, quality of service, facility, atmosphere, knowledge level of the chefs and cooks, and sanitation and safety practices.

Type of Establishment

Traditional, full-service restaurants are the backbone of the industry. Many of them feature any one of the four major Chinese cuisines. Some restaurants specialize in only one kind of food, such as Beijing Roast Duckling or Tianjin Steamed Dumplings. Others are family restaurants serving combinations of

TABLE 1. Governmental Classification of Restaurants

Class	Service area and guests	Atmosphere	Kitchen and Menu	Employee Qualifications
Prime	500 Sq. meters serve 100+ international guests	Excellent in Décor, light, carpeted, air conditioned, stereo system	Well equipped, one of major cuisines with 12 famous dishes	Prime class chef certification, 40% second class chef, 50% service staff are bilingual
First Class	250 Sq. meters, serve international guests	Good décor and facility	More than 10 well-known dishes from traditional cuisine	One prime class chef, 30% second class chefs, 30% service staff with third class service professionals
Second Class	Mainly serve domestic guests	Good décor and facility	More than 10 well-known dishes from traditional cuisine	One first class chef, 15% second class chefs, 20% third class service staff
Third Class	Mainly serve domestic guests	Good décor and facility	More than 6 well-known dishes from traditional cuisine	One first class chef, 15% third class chefs, all serving staff must be trained

various cuisines. Although the sizes and classifications are different, they all offer a la carte menu style and cook food to order. In recent years, new designs for food facilities have been introduced into China, such as food courts in malls, buffets, supermarket meals and quick service restaurants, food deliveries and catering services. Most food courts are in shopping malls and serve a variety of foods, mostly Chinese local dishes or snacks. In newly built malls Western foods have been added. Foods are either pre-prepared or cooked to order. Prices are usually affordable at food courts.

With their emphasis on self-service, buffets, supermarket meals, and quick service restaurants have all become popular, each with its own characteristics and advantages. Restaurants that offer buffets, similar to the USA, charge a set price for a buffet meal. Choice of foods includes cooked or uncooked items, and those choosing an uncooked item must prepare it themselves. Supermarket meals, similar to the commercial cafeteria in the USA, offer about 200 or more food items, cooked or uncooked, arranged on shelves or counters, and food items are priced individually. Customers pay for whatever dishes they select. For uncooked dishes, they can choose the chef on duty to cook for them or they may cook for themselves. The Shaoezai group established its first supermarket meal restaurant in 1993 and today it has more than 40 such restaurants in over 10 cities.

Quick service restaurants and home delivery are gaining popularity, along with the faster pace of life, among China's urban residents. While Western-style quick service restaurants are still considered somewhat of a novelty by Chinese children, Chinese adults find the food less tasty or varied than Chinese foods. This leaves room for development of Chinese-style quick service restaurants offering authentic foods or foods with native flavors. Inspired by the success of international quick service restaurant chains, more and more local entrepreneurs have become interested in developing their own brands. Emerging Chinese quick service restaurant brands include Malan Noodle, Huarong Chicken, and Red Sorghum. Food delivery companies serve boxes of food for lunches or dinners for the convenience of "blue collar" as well as "white collar" workers, to school lunch programs, and to parks and recreation facilities. Both Chinese-style quick service restaurants and home delivery restaurants charge very reasonable prices. For example, Lihua, a food-delivery company in Changzhou and Beijing, offers its box lunch or dinner–rice, two to four stir-fry meats and vegetables, and soup–for only 6-12 yuan ($.75-$1.50).

WESTERN IMPACT ON THE CHINESE RESTAURANT INDUSTRY

For centuries, the mom and pop noodle shops on local street corners met the basic fast service needs in China. The industry began to change in the late 1980s, when KFC and McDonald's introduced Western quick service restaurants. With a strong brand image, these chains are well perceived by Chinese consumers for their quality control and efficient store management. As a result, Western quick service restaurants earned more than 20 percent of the total sales in China's quick service market (Cee & Theiler, 1999). The success of Western-style quick service restaurants is serving as a catalyst for developing and improving the Chinese fast-food sector, where cumulative annual sales have reached 40 billion yuan, one-fifth of the annual revenue of China's restaurant industry. According to one report, there were 800 Chinese quick service restaurant companies in 1996, of which 400 were chain operations with 400,000 outlets (Wu & Pu, 1998). Despite the increase in number of units and sales, the performance of Chinese quick service restaurant operations do not compare well with their Western-style counterparts, which have, though with fewer outlets, captured millions of patrons eager for a new taste experience. Recently both KFC and McDonald's units of China have won the one-day-sales record in the world for their respective companies. McDonald's, with fewer than 20,000 employees, has recently earned 2.5 billion yuan annually.

Besides financial success, Western foodservice companies have influenced the Chinese food operations and management in four areas. First, foreign food companies have brought new tastes and service, quality, ambience, sanitation, and safety standards that enhanced the Chinese food service operations, and they have also influenced the cooking and eating habits in China. Second, by demonstrating the power of brand names, technology, and equipment, these companies have stimulated the Chinese entrepreneurs to improve their own operations and management, and to implement new technology in processing and production. Third, these companies have brought franchise operation systems and methods of chain management and training that can serve as a model for Chinese restaurant industry. Last, international restaurant chains with operational superiority and efficiency have challenged the traditional Chinese restaurants that relied on the advantages of low-tech, labor-intensive, and traditional foodservice methods.

Competition in the Restaurant Industry

The rapid development of the Chinese restaurant industry has intensified competition among industry sectors. The most common competitive approach utilized by most food operations is low pricing to attract high volume. Some large hotels, taking advantage of their more elegant eating environment, have joined the fast-food competition by lowering their prices to attract local residents. As a result, quality ambiance and service have become strong competitive tools. New specialty foods, service styles, concepts, and themes also come into play. In order to attract patrons, restaurants offer bonuses, awards, sales discounts, and dining club memberships. Many join in sales wars to capitalize on holidays and special events. They offer special party packages on site, deliver special food packages to patrons' homes, or even send their chefs to cater holiday cooking and special events. To compete with their Chinese counterparts, with their advantages of low price and traditional taste, foreign quick service restaurant operations have tried to adapt to local tastes and culture while capitalizing on brand name image, good quality, fast service, clean eating environment, new technology, and management know-how.

FUTURE CHALLENGES

Just as the emergence of a new Chinese restaurant industry converged with the political and economic reforms of the last 20 years, its continuing development is dependent on political and social stability of the country. A fast-growing economy and an increasing domestic consumer demand are seen as favorable forces to shape the industry. Since the industry largely exists in the private sector, protection of private ownership and financial support from the government are imperative.

As competition in the restaurant industry accelerates, merger and acquisition activities are anticipated. These activities will help establish brand-name operations, virtually nonexistent in China a decade ago, and chain management concepts and methods. As chain operations continue to grow they are likely to enjoy economies of scale. Emerging restaurant chains may have greater financial resources to industrialize food production, establish preparation centers, and computerize management operations. Chain management of Chinese restaurants may follow the model of Western restaurant chains while accommodating regional food preferences. The restaurant industry could establish its own trade association to create a forum where members can exchange ideas and management techniques. The association may establish its own research organizations to monitor and analyze the changing environment and provide its members with useful marketing information. It should introduce management know-how and new technology, help the industry establish its professional image, and encourage its members to adhere to high standards of food sanitation and safety practices.

For multinational foodservice companies, the growing Chinese restaurant industry provides enormous opportunities. It is an ideal time to introduce and develop Chinese brand names and chain operations through joint-ventures, and to introduce new management technique and production technology to their partners. Investment in training for the Chinese restaurant industry will have a long-term positive effect.

In 1997, there were only 221,500 students enrolled in hospitality and tourism programs in vocational high school and two-year and four-year colleges (*China Tourism Statistics*, 1998). The ratio between current enrollment and industry employment is 2.6 to 100, a great impetus to develop hotel and restaurant management training programs.

CONCLUSIONS

The Chinese restaurant industry has developed rapidly during the last 20 years in part due to the liberalized political and economic reforms. Societal changes are also a major driving force for that development. The restaurant industry of China faces many new challenges in the future: applying new technologies, learning new management strategies, establishing chain operations, and training new generations of foodservice professionals.

With great growth potential of the restaurant sector, a new generation of Chinese entrepreneurs will emerge from the private sector so long as growth and reform in China's economy continues. Young and aggressive, the new generation of entrepreneurs may perceive themselves differently from the traditional mom and pop noodle shop owners. By learning from the international restaurant chains emerging in China, they may constantly test new production and service methods, forging the frontiers in developing chain operations in China's restaurant industry. Some of them may well seek global market opportunities. They represent the future of the Chinese food industry.

REFERENCES

Anon. (1998). *The yearbook of China tourism statistics*. National Tourism Administration, Beijing, China.

Anon. (1998). Import fruit quantity and cost table. *Economic Daily*, July 5, 5.

Anon. (1997). Hog slaughter management regulations. *People's Daily*, December 25, 1.

Anon. (1998). GDP figures between 1978-97. *People's Daily*, November, 18. http://www.peopledaily.com.cn/item/20years/news/newflies/d1020.html

Bi, F. (1999). *Polystyrene food boxes are not allowed to be used by 10/30/2000*. November 11. http://dalilynews.sina.com.cn/china/1999-11-4/28842.html

Cee. J. & Theiler, S. (1999). US French fries heat up China's fast food industry. *AgExporter*, 11(2), 19-20.

China Daily (1999). http://www.chinadaily.com.cn./cndydb/1999/10/d2-6tour.a05.html

France-presse (1998). *"Little emperors" lap it up*. June 23. http://www.scmp.com/news/template/template . . . ina&template=Default.htx&maxfieldsize=1183

France-presse, (1998). *Polluting food boxes face chop*. December 15. http://www.scmp.com/news/template/China . . . a&template=Default.htx&maxfiedlsize=1162

Guo, Y. et al. (1999). *McDonald's plans to develop franchisees in China*. http://202.84.17.11/info/9910111450082.htm

Hillis, S. (1998). *Feature–U.S. fast food thriving in China*. http://www.infobeat.com/stories/cgi/story.cgi?id=2554800844-155

Hon, C.Y. (1998). *Green body upgraded to ministry*. April 2. http://www.scmp.com/news/template/templat . . . na&template=Default.htx&maxfieldsize=1566

Lateline News (1998). *China to cut a quarter of government workforce*. http://lateline.nuzi.net/cgi/lateline/news?p=4997

Ling, D. (1997). The new year thought. *Chinese Cuisine*, No. 1, 4-5.

Ministry of Public Health (2000). *Food and beverage safety and sanitation measure*. http://dailynews.sina.com.cn/china/2000-2-17/62584.html

Ministry of Domestic Trade (1997). The principle of Chinese fast-food development. *Chinese Cuisine*, No. 12, 42.

Ni, S. & Dai, Y. (1998). *Five principles of promoting environment protection and tourism development*. October, 17. http://www.peopledaily.com.cn/199810/17/col_981017001921_jryw.html

O'Neill, M. (1999). *Dark days ahead for economy, says Beijing*. January 7. http://www.scmp.com/news/template/Front . . . t&template=default.htx&maxfieldsize=2987

Peng, F. C. (1998). *GDP target requires US$30 billion investment*. http://www.scmp.com/news/tempate/templa . . . a&template=Default.htx.&Maxfiel-dSize= 2629

Scmp. (1998). *Unemployed predicted to hit 18 million*. http://www.scmp.com/news/template/China . . . a&template=Default.htx&maxfieldsize=1058

Sinanet.com (1998). *Mainland issued the first document of improvement of human nutrition*. http://dailynews.sinanet.com/china/cip/1998/06/16/8.html

Sinanet.com (1998). *China's children market is huge*. June 1 http://dailynews.sinanet.com/business/zhongguo/1998/06/01/1.html

Tannahill R. (1988). *Food in History*. New York: Crown Publisher, 757.

Yan, Y. (1998). The status and development of China food and beverage industry, *Chinese Cuisine*, No. 1,

Yang, M. (1997). The status and development trends of China restaurant industry. *Chinese Cuisine*, No. 3, 4-6.

Yang, M. (1998). The thought of Chinese fast-food development. *Chinese Cuisine*, No. 2, 4-6.

Wang, W. (1997). American fast-food in China. *Chinese Cuisine*, No. 10, 7-9.
Wu, J. & Pu, L. (1998). There is a great potential in food-table economy. *People's Daily*, July 20, 2.
Wu, J. & Pu, L. (1998). Food table leads all the businesses. *Economic Daily*. July 22, 2
Zhang, J. (1999). *Harvard University studies the success of KFC and Pizza Hut*. http://dailynews.sina.com.cn/world/1999-12-7/39258.html

Impact of Mentoring and Empowerment on Employee Performance and Customer Satisfaction in Chain Restaurants

Jerrold Leong

SUMMARY. The purpose of the current study was to determine the viability of mentoring and empowerment practices to enhance employee performance and customer satisfaction in chain restaurants. The manager's demographic profile was related to mentoring. According to the present study, there were significant effects relative to employee success, goal setting, personal initiative, self-confidence, task achievement, character building, trust and respect, networking, customer loyalty, bonding between supervisor and employees, and rapport between employees and customers. The respondents indicated strong agreement with the fact that mentoring contributed to employee success, goal setting, maturity, self-confidence, and initiative. Furthermore, the respondents indicated a preference for empowerment, as it contributed to task achievement, character development, networking, and providing additional value to enhance customer satisfaction. The respondents indicated that mentoring and empowerment contributed higher value and quality to products. In turn, employees exceeded customer expectations, built rapport, and stabilized customer loyalty and return patronage. *[Article copies available for a fee from The Haworth Document Delivery Service: 1-800-HAWORTH. E-mail address: <getinfo@haworthpressinc.*

Jerrold Leong, PhD, FMP, is affiliated with the Oklahoma State University, Stillwater, OK 74078.

[Haworth co-indexing entry note]: "Impact of Mentoring and Empowerment on Employee Performance and Customer Satisfaction in Chain Restaurants." Leong, Jerrold. Co-published simultaneously in *Journal of Restaurant & Foodservice Marketing* (The Haworth Hospitality Press, an imprint of The Haworth Press, Inc.) Vol. 4, No. 4, 2001, pp. 173-193; and: *Quick Service Restaurants, Franchising and Multi-Unit Chain Management* (ed: H. G. Parsa and Francis A. Kwansa) The Haworth Hospitality Press, an imprint of The Haworth Press, Inc., 2001, pp. 173-193. Single or multiple copies of this article are available for a fee from The Haworth Document Delivery Service [1-800-HAWORTH, 9:00 a.m. - 5:00 p.m. (EST). E-mail address: getinfo@haworthpressinc.com].

KEYWORDS. Mentoring, empowerment, chain restaurants, employer performance, customer satisfaction

MENTORING

Mentoring is an important process for professionals entering the foodservice industry. A formal mentoring process is defined as a transfer of organization philosophy, beliefs, values, and standards of performance from an experienced mentor to a new protégé within an organizational structure. This period of indoctrination and orientation will allow the new employee to become accustomed to the organizational culture, hierarchy, policies, procedures, protocols, and work environment. The mentoring relationship encourages sharing between individuals and alliance building among peers. In most cases, mentors are usually assigned to new professionals based on common interests and levels of expertise of the individual protégé. This relationship may be temporary, or may be for an extended period of time.

New professionals are keen to find alliances with those peers that have mutual interests and expertise levels. From a managerial perspective, effective mentoring relationships could enhance employee's self-efficacy and professional development in the hospitality industry. The business and hospitality industry literature has provided greater evidence that employee-mentoring programs have beneficial effects on leadership and organizational performance (Cunningham, 1999; Henley, 1999; Mitchell, 1998; Walsh & Borkowski, 1999; Walsh, Borkowski, & Reuben, 1999; Hayes, 1998). Pfeffer and Veiga's (1999) study indicated that there is a positive relationship between financial success and management's commitment of treating their employees as assets. Moddelmog's (1999) study indicated that 46 percent of the women business owners reported having a mentor or a role model that offered advice and encouragement. Edmundson (1999) concluded that the success of law enforcement recruit's career was based on an effective mentoring program. Results of the police recruits' mentoring program fostered self-esteem, empowerment, access to resources and sense of satisfaction. Olesen (1999) contended that companies are realizing the value of training as a recruitment tool. According to him, an employee's contribution to corporate effectiveness is a function of training in computer technology, independence in job structure, multiple skill training and mentoring. Cashman (1999) concluded that coaching and mentoring facilitated optimal growth, affected bottom-line, improved retention of essential people and enhanced managerial change. Mentoring in a college environment is effective when student's persistence to learn the theoretical concepts is enhanced through a formal student and faculty interchange process (Tinto, 1997). Employees that experienced a structured mentoring program reported more control in their jobs, and confidence in accomplishing the tasks related to their respective positions.

EMPOWERMENT

Empowerment (delegation) is defined as a process of granting decision-making power to worthy employees. Employees that have exhibited managerial potential during the mentoring process are empowered to facilitate effective team performance. Empowerment gives nonmanagerial individuals an opportunity to make their own decisions and develop their abilities and job performance skills (Kreitner & Kinicki, 1992). Empowerment gives authority and responsibility to members to enhance the design of their work environments so that they may better meet customers' needs. Thus, when team members are empowered the work teams become more effective in achieving organizational goals (Hellriegel, Jackson, & Slocum, 1999). In empowerment, autonomy comes along with authority and responsibility. Each individual is accountable for the outcome and its impact on the organizational structure. Empowerment offers a sense of encouragement to the employee, who seeks to provide optimal performance and exceed customer's expectations.

The Ritz-Carlton Hotel Company, 1992 Malcolm Baldridge National Quality Award recipient, empowers its employees to spend up to $2,000 to satisfy a guest and decide upon the acceptability of products and services. According to Woods and King (1996), organizational structure that allows empowerment of employees results in lower production costs, reduction in operational inefficiencies, enhancement of customer satisfaction, and higher profitability.

Herrenkohl, Judson, and Heffner (1999) developed an operational definition of employee empowerment, created a measure of empowerment, and examined the measure's validity. They indicated that employee empowerment encourages formation of self-managed or self-regulated teams. The team concept is cohesive and facilitates natural work environments in which to optimize organizational performance. Gable (1999) has mentioned three principles in creating a positive work environment and enhancing employee retention: (a) open culture, (b) empowerment, and (c) teamwork. In addition, Willis' (1999) study revealed that companies that encourage their employees to be creative and innovative experience improved quality of life and organizational effectiveness.

This creates an environment that encourages employee involvement, commitment, and continuous quality improvement.

Empowerment process also encourages employee involvement and continuous quality improvement. Foegen's (1999) study demonstrates a need to push authority downward and giving ordinary employees a greater say. The results of his study showed that by generating highly effective and motivated employees in the workplace customer satisfaction could be enhanced. Covey (1999) suggested that a high trust corporate culture becomes the basis for both empowerment and quality enhancement. Organizations must empower their employees to realize their fullest potentials.

PERFORMANCE

Performance is defined as the outcome that is achieved as a result of using the right combination of past experiences, ability, skills, and efforts to accomplish the specific goal and objectives (Kreitner & Kinicki, 1996). Importantly, performance outcomes give employees a benchmark for future job performance challenges. The employee involvement process emphasized making the employee a stakeholder and strengthening employee motivation. Successful performance of the individual's job is directly related to the managerial willingness to mentor and empower the employee during their career progression. Employee development is enhanced when management nurtures the employee and provides a positive working environment. Empowered employees will then develop a sense of self-efficacy as they use their previous experience, learn how to perform tasks vicariously, receive encouragement from their peers, and develop a positive attitude to achieving a particular task (Kreitner & Kinicki, 1996).

Team performance is critical to individual development and organizational vitality in the business arena. Members of a team must have the skills, expertise, and ability to function effectively within a heterogeneous group of individuals. Herman and Reichelt (1998) noted that team building by first-line nurse managers resulted in patient satisfaction, improved relationship with other employees, and created a strong team spirit. Moreover, team building increased employee satisfaction and productivity. The development of team cohesiveness is premised on building trust and respect among the team members. Members who collaborate and share the authority and responsibilities of the job are involved in a joint decision-making process. The operations management decisions made by the multifunctional teams have resulted in process and quality improvement. This consensus decision process results in timely and rapid implementation of the decisions (Luebbe & Snavely, 1997). There is more accountability and ethical practices among members when there is a spirit of sharing, commitment, and cooperation. Individuals who are strategically integrated and allowed to work in a cross-functional work environment are inclined to embrace the corporate culture and provide optimal service to their clients.

CUSTOMER SERVICE

Customer service is the competitive advantage that sustains a foodservice organization's viability within the industry. Customer service embraces the notion of providing optimal product quality, value, and impeccable service in a timely manner. The customers' perception of optimal product and service quality is based on previous experiences, degree of satisfaction with products and/or services rendered.

Jogaratnam, Tse, and Olsen (1999) observed the linkage between business strategies of a restaurant and its performance. The highly effective and successful restaurants placed a premium on being proactive, were more innovative, and more committed to delivering quality service than their less successful counterparts. Furthermore, Spector (1999) indicated that casual dining restaurants must differentiate their brands to reach the sophisticated customer and improve their dining experiences. Moreover, Murphy (1999) noted that McDonald's success in the United Kingdom was premised on customer-focused marketing and emphasis on quick service. Papiernik (1999) indicated that restaurant patrons are looking for value, quality food, service, price, and ambiance. Girard-diCarlo (1999) suggested that highly successful healthcare organizations will be those with committed leaders who motivate each individual through their involvement in customer service programs.

PURPOSE OF THE STUDY

The purpose of the study was to determine the viability of providing employee mentoring and empowerment opportunities to enhance employee performance and customer satisfaction in casual dining and quick service restaurant chains. The study will address the following questions (Figure 1 & 2):

a. Does manager's perception of mentoring impact employee's performance?
b. Does manager's perception of empowerment impact employee's performance?
c. Does mentoring affect customer satisfaction?

d. Does empowerment affect customer satisfaction?
e. Is there an interaction effect of mentoring and empowerment on customer satisfaction?

Management's willingness to develop an employee mentoring and empowerment program to improve overall operational effectiveness may enhance customer satisfaction (Figure 1). Mentoring programs are used to improve employee performance and effectiveness in meeting organizational goals. Empowerment of employees is meant to improve employee confidence in the decision-making process. Performance can be expressed as individual and group performance or as financial outcomes.

An effective performance criterion creates a benchmark relative to mentoring and empowerment to facilitate employee growth and development to think creatively and develop strategies to meet or exceed customer expectations.

METHODOLOGY

Survey Instrument

The questionnaire was developed to determine to what extent managers participated in mentoring relationships with their employees, empowered their employees, and their impact on the level of customer satisfaction. The Likert-type scale was used to measure the respondent's evaluation of the attributes related to mentoring, empowerment, and customer satisfaction on a five-point scale (1 = strongly disagree; 5 = strongly disagree). In addition, they were asked to indicate their overall managerial perceptions of mentoring,

FIGURE 1. Mentoring, Empowerment, and Performance Paradigm

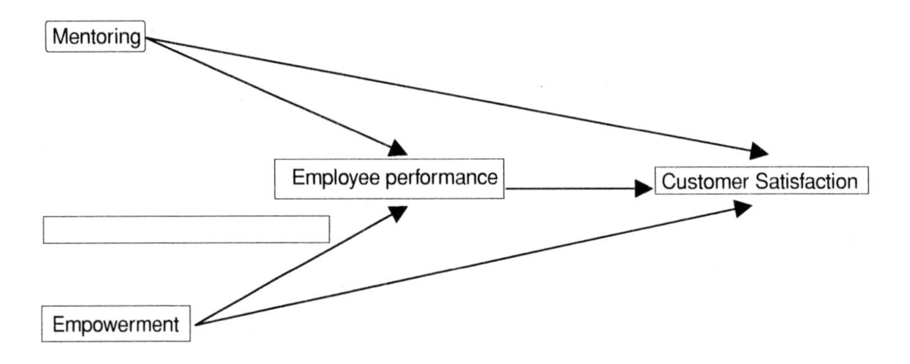

FIGURE 2. Interaction Effects on Customer Satisfaction, Mentoring, and Empowerment

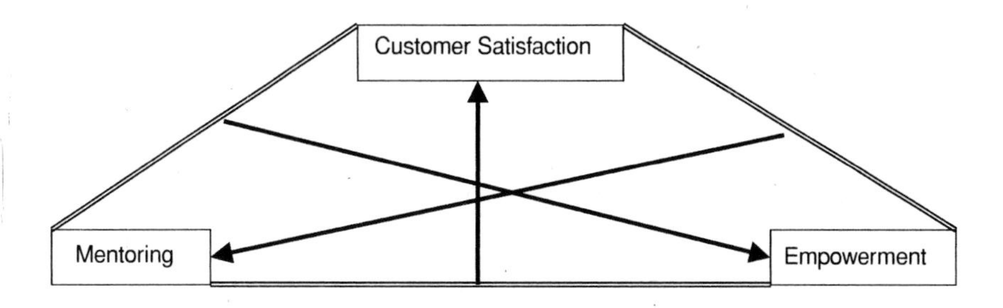

empowerment, and customer satisfaction. The respondents were then asked to provide personal demographic and company information. The demographic information would provide the study with an overview of the sample.

The first section of the questionnaire inquired about the following: (1) how mentoring affected employee performance, (2) how empowerment affected employee decision making process, and (3) how empowerment and mentoring influenced customer satisfaction. The respondents were asked to express to what degree they agreed or disagreed with the statements using a five-point Likert scale. The Likert scale would measure the respondent's perceptions relative to mentoring, empowerment, and customer satisfaction.

The second section inquired about the manager's demographic background, empowerment opportunities and compensation for mentored and empowered employees. The last section included information on type of establishment, type of ownership, total food and beverage sales, food cost percentage, labor cost percentage, check average, square footage, customer count, seating capacity, and turnover rate.

Sample Design

According to the Foodservice Data Base Company, the casual dining and quick-service restaurant industry consisted of over 3,100 business enterprises. This population was selected for the following reasons: (1) these food service managers commonly practice mentoring and empowerment in their restaurant units; (2) the organizations were dispersed throughout the United States and working with a diverse employee population; and (3) these managers have the ability to develop employees to meet the needs of their customers. A simple random sample was used to select 650 restaurant owners.

Data Collection

The survey was conducted from 10 February 2000 through 3 March 2000. A cover letter, two-page self-administered questionnaire, and a self-addressed business reply envelope was sent to 625 managers of chain foodservice corporations. The managers of the chain restaurants were asked to take ten minutes to complete the instrument. The actual number of usable questionnaires was 46 with an effective response rate of 7.36 percent. A possible explanation for the low response rate may be: (a) the managers' inability to devote time to review incoming correspondence, (b) their inability to spend time apart from their business activities to complete the questionnaire and return it within a three week interval, and (c) the methods used to contact the respondents, single mailing.

The data was analyzed using the Statistical Package for the Social Sciences (SPSS) software to determine one-way frequency, means, standard deviations, one-way analysis of variance (AVOVA), independent sample t-test, and Cronbach's Alpha (reliability coefficients).

RESULTS

Managers' Demographic Profile

The managers were asked to provide personal information related to their job status and their participation in the mentoring process. This information allows the researchers to determine how these independent variables affect the managerial perceptions of mentoring and empowerment attributes (see Table 1). Over eighty percent of the respondents were male and a large majority (65.2%) of the respondents was married. Most of the respondents (58.7%) were between 40 and 49 years of age. The majority (43.5%) of the respondents indicated they were college graduates, while an additional 13% stating they held postgraduate degrees.

The respondents indicated that they had technical expertise in the top ten areas: (1) operational experience (93.3%), (2) quality assurance (78.3%), (3) labor scheduling (73.9%), (4) food production (73.9%), (5) customer relations (71.7%), (6) purchasing (69.6%), (7) sanitation (63.0%), (8) marketing (63.0%), (9) human resources (60.9%), and (10) safety (58.7%).

A large majority (65.2%) of the foodservice mangers participated in a formal mentoring program during different intervals of the employees' development process. Most of the respondents (65.2%) were mentoring between one and five employees, while others were mentoring between six and ten employ-

TABLE 1. Manager's and Company Demographic Profile (n = 46)

Gender:	**Frequency**	**Percent**	**Technical Expertise**	**Frequency**	**Percent**
Male	37	80.4	Operations	42	91.3
Female	09	19.6	Quality Assurance	6	78.3
			Labor Scheduling	34	73.9
Marital Status:			Food Production	34	73.9
Married	30	65.2	Purchasing	32	69.6
Single	16	34.8	Sanitation	29	63.0
			Marketing	29	63.0
Age:			Human Resources	28	60.9
20 to 29 Years	01	02.2	Safety	27	58.7
30 to 39 Years	11	23.9	Equipment Buying	6	56.7
40 to 49 Years	27	58.7	Computer Expertise	22	47.8
50 to 59 Years	04	08.7	Accounting	21	45.7
60 Years	03	06.5	Bar Management	14	30.4
Education:			**Education**		
High School	07	15.2	Some College	13	28.3
College B.S.	20	43.5	Postgraduate	06	13.0
Manager's Mentoring Opportunities:			**Number of Employees Mentored:**		
Are you currently serving as a Mentor?			0 to 5 employees	30	65.2
Yes	30	65.2	6 to10 employees	12	26.1
			11 to15 employees	01	02.2
			20+ employees	03	04.3
Age of employees mentored.:			**Compensation for Mentored Employees:**		
19 years	17	37.0	More Compensation	29	63.0
20 to 29 years	09	19.5	Same Compensation	17	37.0
30 to 39 years	14	30.5			
40 to 49 years	06	13.0	**Compensation for Empowered Employees:**		
			More Compensation	30	65.2
			Same Compensation	16	34.8
Type(s) of Establishment:			**Type(s) of Ownership:**		
Casual Dining	24	54.3	Chain Owned	18	41.3
Quick Service	21	45.7	Franchised	15	32.6
			Independent	12	26.1
Total Annual Food Sales:			**Total Annual Food Sales:**		
Over $50,000,000	13	28.3	$25,000,000 to $49,999,999	04	08.7
$10,000,000 to $24,999,999	12	26.1	$5,000,000 to $9,999,999	05	10.9
$1,000,000 to $4,999,999	09	19.6	$500,000 to $999,999	02	04.3

Gender	Frequency	Percent	Technical Expertise	Frequency	Percent
Financial Information:	**Mean**	**Std. Dev**	**Financial Information:**	**Mean**	**Std. Dev.**
Food Cost Percentage	30.45	6.598	Labor Cost Percentage	24.50	5.154
Gross Profit Percentage	44.73	6.900	Average Customer Count	378.7	240.9
Guest Check Average	$11.33	13.63	Total Seating Capacity	153.3	263.4
Square Footage per Unit	2857	2128	Dining Room Seat turnover	3.4	.9409
Annual Sales per Location	$1,135,254	$1,009,746			

ees (26.1%). Thirty-seven percent of the respondents were mentoring employees that were less than 19 years old, while 30.5% were mentoring employees between the ages of 30 to 39 years. A large majority (63.0%) of the managers indicated that a higher compensation was afforded to mentor employees. A large majority (65.2%) of the managers indicated that they paid their empowered employees a higher wage rate than the non-empowered employees. Overall, the managers' demographic profile represented experienced managers, with a strong expertise in operations management, with a willingness to mentor and empower the younger foodservice professional. Managers consistently rewarded employees that participated in mentoring programs and those who accepted additional responsibilities with a higher compensation (see Table 1).

Demographic Profile of Participating Restaurants

The managers were asked to provide company characteristics and financial information. The majority (54.3%) of the managers worked for casual/family/dinner house/white tablecloth establishments, while forty-six percent of the managers indicated their affiliation with quick-service restaurants.

Over 51.0% of the respondents indicated that the type of company ownership was chain owned, followed by franchise operations (32.6%) and independent operations (26.1%). Over 28.0% percent of the establishments had sales more than $50 million, while 26.0% percent of the companies reported annual sales between $10 million to $25 million. The average food cost percentage was 30.45% and the average labor cost percentage was 24.50%. The average gross profit percentage was 44.73%. The average sale per customer was $11.33, average customer count was 378, and the unit sales were $1,135,254. The average seating capacity was 153 persons and the average seat turnover was 3.4 times per meal period. Overall, the company characteristics and financial information reflected industry norms of high-margin operations (see Table 1).

Mentoring and Employee Performance

For the purpose of comparison, Table 2 provided overall means of the manager's responses to mentoring, empowerment, and customer satisfaction attributes. The first group of questions were related to the managers' perceptions of the influence of mentoring to enhance employee's training, motivation, performance, and effectiveness in meeting organizational goals (see Table 2). The managers were asked to evaluate to what extent each mentoring attribute affected their employees.

Overall, the reported management's mean perception (4.15) indicated that mentoring improved employee performance and effectiveness in meeting organizational goals. Notably, managers perceived that mentoring enhanced employee self-efficacy or successfulness (4.19), assisted in goal setting (4.19), encouraged employee self-confidence (4.15), developed employee maturity (4.08) and builds employee self-esteem (3.91). Managers indicated that mentoring afforded personalized training and individual initiative in learning profession skills. The study calculated the reliability coefficient (Cronbach's Alpha) as .7818, which is considered acceptable.

Empowerment and Employee Performance

The second group of questions was related to the managers' perceptions of the impact of empowerment on the employee's decision-making process (see Table 2). The managers were asked to evaluate to what extent each empowerment attribute affected their employees performance. Overall, the reported management's mean perception (4.04) indicated that empowerment has improved employee confidence in decision-making process. Importantly, managers perceived that empowerment (delegation of authority) builds trust and respect among employees (4.06), allow employee latitude to enhance customer value (4.04), increased profitability (4.00), builds character and self-motivation (3.95), and builds employee self-confidence (3.89). The study calculated the reliability coefficient (Cronbach's Alpha) as .7770, which is considered acceptable.

Employee Performance and Customer Satisfaction

The third group of questions was related to the managers' perceptions of customer satisfaction resulting from mentoring and empowerment (see Table 2). Overall, the reported mean value for customer satisfaction (4.17) indicated that customer service has improved as the result of mentoring and empowering employees. Specifically, managers in general perceived that the level of cus-

TABLE 2. Measurement of Mentoring, Empowerment, and Customer Satisfaction

MENTORING	**Overall Mean**	
Mentoring has improved employee performance and effectiveness in meeting organizational goals.	4.15	
Mentoring Attributes		
Mentoring helps the employee **believe** in success.		4.19
Managers can **help** employees in professional **goal setting**.		4.19
Mentors help the employee feel **confident** about job.		4.15
Mentoring can **encourage** maturity in employees.		4.08
Mentoring **builds employee** self-esteem.	3.91	
Mentoring **drives** employee empowerment, and the combination contributes to bottom line profits.		3.84
Mentoring is **training** the employee about the work.		3.82
Mentoring teaches **initiative to** provide customer value.		3.80
Managers sometimes **refuse** to mentor employees.	3.76	
Reliability Coefficient = 0.7818 (Cronbach's Alpha)	N of Cases = 46	N of Items = 10
EMPOWERMENT	**Overall Mean**	
Empowerment has improved employee confidence in the decision-making process.	4.04	
Empowerment Attributes		
Empowerment builds **trust** and **respect**.	4.06	
Empowerment employee latitude to for **customer value**.		4.04
Empowerment is **affected** by mentoring and the combination contributes to bottom line profits.	4.00	
Delegating authority builds **character/self-motivation.**		3.95
Empowerment builds **self-confidence** in employees.	3.89	
Empowerment develops creativity in **task achievement.**		3.79
Empowerment leads to high **group involvement.**		3.76
Empowerment provides direction to **accomplish tasks.**		3.67
Empowerment develops **networking** among employees.		3.50
Reliability Coefficient = 0.7770 (Cronbach's Alpha) N of Cases = 46	N of Items = 10	

CUSTOMER SATISFACTION	Overall Mean	
Customer service has improved as a result of mentoring and empowering employees.	4.17	
Customer Satisfaction Attributes		
Customer loyalty and **return patronage are important to us.**		4.65
Mentoring and employee empowerment **can improve** the perception of customer value and satisfaction.		4.23
There is **rapport building** between employees and guests.		4.17
We are determined to exceed customers' **expectations.**		4.15
Customers perceive a **higher value and quality** of product.		3.80
There is a close **bond** between supervisors and employees.		3.71
The employees say **"Thank you and please come again."**		3.65
Employees **perform** in a highly professional manner.		3.60
Employees do whatever it takes to **please** customers.		3.56
Reliability Coefficient = 0.8077 (Cronbach's Alpha)	N of Cases = 46	N of Items = 10

tomer satisfaction (4.23) was improved, rapport building between employees and customers was improved (4.17), desire to exceed customer expectations was improved (4.15), and customers perceived a higher value and quality of product (3.80) (Table 2). The study calculated the reliability coefficient (Cronbach's Alpha) as .8077, which was most favorable.

Interaction Effects on Mentoring

The analysis determined whether the manager's overall perceptions of empowerment and customer satisfaction had a significant effect on mentoring attributes. The ordinal data from the survey responses was analyzed by using one-way Analysis of Variance (ANOVA). In all cases, the null hypotheses assumed that the population means were equal and the level of significance (alpha) was $p < 0.05$.

The study found that the manager's empowerment perceptions had a significant effect on two mentoring attributes, i.e., employee success (sig. = .019) and self-confidence (sig. = .018). The significance level was less than alpha $p < 0.05$. The null hypothesis of equality of means is rejected, that is, managers rated highly the effectiveness of mentoring as a managerial process that contributed to employee success and improved self-confidence (Table 3). Put another way, managers saw effective mentoring as a mutually beneficial process affecting employee success, goal setting, training, maturity, self-confidence, personal initiative, and mentoring as it drives employee empowerment (Table 3).

Interaction Effects on Empowerment

The study used one-way Analysis of Variance to determine whether the manager's overall mentoring and customer satisfaction perceptions had a significant effect on empowerment. The study found that the manager's mentoring perceptions had a significant effect on task achievement (sig. = .005), builds character (sig. = .024), employee direction (sig. = .042), trust and respect (sig. = .003) (Table 3). The null hypothesis of equality of means is rejected, meaning that managers agreed that the effectiveness of empowerment facilitates task achievement, character building, direction, trust and respect, networking, and empowerment, and they all are affected by mentoring. The null hypothesis of equality of means is rejected, which means that managers rated highly the effectiveness of empowerment to enhance task achievement, character building, employee direction, trust and respect, employees' abilities to providing value, and empowerment driven by mentoring.

Interaction Effects on Customer Satisfaction

The study again used one-way Analysis of Variance to determine whether the manager's overall mentoring and empowerment perceptions had a significant effect on customer satisfaction. The study found that the manager's mentoring perceptions had a significant effect on customer loyalty and return patronage (sig. = .009), rapport between employees and guests (sig. = .005), and bonding between supervisors and employees (sig. = .035) (Table 3). The null hypothesis of equality of means is rejected, meaning that managers perceived that customer satisfaction is improved by customer loyalty and return patronage, rapport between employees and guests, and a closer bond between supervisors and employees.

Furthermore, the manager's perceptions of empowerment have a significant effect on customer loyalty (sig = .001), return patronage (sig = .034), and customer satisfaction (sig = .002) (Table 3). The null hypothesis of equality of means is rejected, which meant that managers rated the effectiveness of empowerment as a measure of customer loyalty, return patronage, rapport between employees and guest, and improved perception of value.

TABLE 3. Relationship Between Mentoring, Empowerment, and Customer Satisfaction–Results of One-Way Analysis of Variance (ANOVA)

Interaction effects on Mentoring		
Empowerment	**F**	**Sig.**
Employees believe they can be successful	3.718	.019**
Mentoring helps employee feel confident	3.760	.018**
Customer Satisfaction	**F**	**Sig.**
Employees believe they can be successful.	6.694	.001**
Managers helping employee with goal setting	3.262	.031**
Training the employee in the work process	5.338	.003**
Mentoring encourages maturity in employees	6.885	.001**
Mentoring helps employee feel confident	7.705	.001**
Mentoring teaches initiative	3.482	.024**
Mentoring drives employee empowerment	3.131	.036**
Interaction Effects on Empowerment		
Mentoring	**F**	**Sig.**
Empowerment develops task achievement	4.874	.005**
Delegation builds character	3.488	.024**
Empowerment provides direction	2.979	.042**
Empowerment builds trust and respect	5.318	.003**
Empowerment builds networking	4.041	.013**
Empowerment is affected by mentoring	3.415	.026**
Customer Satisfaction	**F**	**Sig.**
Empowerment develops task achievement	3.746	.018**
Delegation builds character	2.847	.049**
Empowerment provides direction	3.514	.023**
Empowerment builds trust and respect	2.881	.047**
Empowerment allows employees to provide value	3.021	.040**
Empowerment is affected by mentoring	5.614	.002**
Interaction Effects on Customer Satisfaction		
Mentoring	**F**	**Sig.**
Customer loyalty and return patronage	4.402	.009**
Improve rapport between employee and guests	4.965	.005**
Close bond between supervisors and employee	3.132	.035**
Empowerment	**F**	**Sig.**
Customer loyalty and return patronage	9.374	.001**
Improve rapport between employee and guests	3.174	.034**
Employee's perceptions have improved	5.749	.002**

****p < 0.05 level of significance**

DISCUSSION

Mentoring

The results revealed that there were significant effects identified between mentoring attributes and the managers' overall perceptions of empowerment and customer satisfaction. The managers' perceptions emphasized important mentoring attributes: (a) employee success, (b) assisting in goal setting, (c) develops personal initiative, (d) training and skill development, (e) developing employee maturity, and (f) building self-confidence. These findings are consistent with self-efficacy theory (Cunningham, 1999; Hayes, 1998; Walsh, Borkowski, & Reuben 1999), which can be applied to employee success, employee involvement, and executive development.

The overall results confirm that the manager's willingness to mentor employees is positively related to the employee's skills, abilities, efforts, and success in accomplishing organizational goals. This has implications for managers who are contemplating implementing a mentoring process for their new employees. Employees must express a willingness to be involved in a mentoring process. The manager must be flexible and adaptable to meet the individual demands of their protégés. Furthermore, the mentoring process will alleviate any potential conflicts and misunderstandings about job performance expectations. However, the employee must be nurtured and be comfortable in participating in a variety of mentoring experiences depending on the skill and training that is desired.

Empowerment

The results revealed that there were significant effects identified between empowerment attributes and the managers' overall mentoring and customer satisfaction perceptions. The managers' perceptions emphasized important empowerment attributes: (a) creativity in task achievement, (b) character and self-motivation, (c) building trust and respect, (d) providing employee direction, (e) networking among employees, and (e) employees providing customer value. These findings are consistent with quality management environment (Foegen, 1999; Herrenkohl, Judson, Heffner, 19employee performance under total 99; Herman & Reichelt, 1998; Luebbe & Snavely, 1997; Reed, 1999), which linked participatory management to effective financial, lifelong learning, and human resource decision process to enhance corporate success.

The overall results confirm that the manager's willingness to empower employees is positively related to power-sharing that delegates decision making capabilities, and responsibility to non-supervisory employees. This has implications for managers who are unafraid to give up their power to seasoned employees. Individual employees and self-managed teams that are empowered continue to be more effective in dealing with operational and customer challenges. Managers must foster effective communications and feedback during the early stages of the empowerment process. Furthermore, the empowerment process will provide autonomy, meaningfulness, and social responsiveness for the employee when confronted with customer challenges.

Customer Satisfaction

The results revealed that there were significant effects between customer satisfaction attributes and the managers' overall mentoring and empowerment perceptions. The managers' perceptions emphasized important customer satisfaction attributes: (a) customer loyalty, (b) return patronage, (c) bonding between supervisor and employees, (d) employees' perceptions have improved, and (e) rapport between employees and customers. These finding are consistent with continuous improvement (CI) (Jogaratnam, Tse, & Olsen, 1999; Morrison, Mobley, & Farley, 1996; Tomas 1996), which can be applied to quality products, waste elimination, improve customer service, customer count, and monitoring performance for quality improvement programs.

The overall results show that the managers perceive that mentoring and empowering employees leads to higher customer satisfaction. Customers who exhibit loyalty to chain restaurants and continue to return time and again are personally recognized and much appreciated. This has implications for employees who deal with a steady clientele. The level of customer service is directly related to the level of job satisfaction experienced by the employee. It is important that managers provide a work environment that fosters employee well-being and team cohesiveness. Maintaining a proactive posture and anticipating the customers' needs and wants will lead to increased sales and ultimately increased market share.

CONCLUSION

The study provided baseline information related to the managers' overall perceptions and its influence on specific mentoring, empowerment, and customer satisfaction attributes. It is important to discern how the future impact of mentoring and empowerment will enhance employee and financial performance.

The manager's demographic profile indicated that the majority of the respondents were male, with expertise in operations, quality assurance, labor scheduling, food production and customer relations. Over forty-three percent had attained a baccalaureate degree and over fifty-eight percent were between 40 and 49 years of age. Over sixty-five percent mentored between one and five individuals who were less than nineteen years of age.

The respondents strongly agreed that mentoring contributed to employee success, goal setting, maturity, self-confidence, and initiative. Furthermore, the respondents indicated a preference for empowerment, as it contributed to task achievement, character development, networking, and providing additional value to enhance customer satisfaction. Finally, the respondents indicated that mentoring and empowerment contributed higher value and quality to products. In turn, employees exceeded customer expectations, built rapport, and stabilized customer loyalty and return patronage.

The study found that the manager's perception of empowerment had a significant effect on employee success and self-confidence. Similarly, the manager's perception of customer satisfaction facilitated employee initiative, goal setting, maturity, and success. The manager's mentoring process influenced their willingness to empower employees to build their self-efficacy, establishing trust and respect, networking, character building, and personal direction. The manager's perceptions about customer satisfaction had a significant effect on promoting rapport between supervisors and employees to build customer confidence. Finally, the effects of sales, prime cost percentages, and gross profit percentages influenced mentoring, empowerment, and customer service attributes.

The chain restaurant industry is continuously growing through merger and acquisition strategies, consolidation of present store location, new unit acquisitions, partnerships or traffic-boosting techniques (Liddle, June 26, 2000). Overall, the casual dining and quick-service restaurants are facing strong competitive forces and they seek to increase market share through optimal employee performance.

More importantly, managers who have advocated the use of mentoring should be encouraged to continue this coaching technique to gain a better understanding of the needs of their employees. Furthermore, managers that seek to empower their employees will create a flatter organizational structure and develop their employees' decision-making capabilities. Therefore, managers in the chain restaurant industry should not be reluctant to mentor and empower their employees, but be willing to provide a work environment that strives for continuous quality improvement and optimal performance in the employees' career development process.

Limitations

The survey results provided evidence of the manager's perception of the influences of mentoring and empowerment on performance. The results do not reflect how employees felt mentoring and empowerment enhanced their career development. In addition, the results do not reflect to what extent customers were satisfied with the service provided. As mentioned earlier, the low response rate is a major limitation of this study, which precludes any generalization of results to the entire industry.

This survey measures the perceptions of managers of chain restaurants about mentoring and empowering selected employees and how these activities affect employee performance. The survey results do not reflect the impact of mentoring and empowerment on performance from the employee's perspective.

REFERENCES

Cashman, K. (1999, September). Coaching from the inside out. *Executive Excellence*, 16(9), 14.

Covey, S. R. (1999, September). High-trust cultures. *Executive Excellence*, 16(9), 3-4.

Cunningham, S. (1999, July/August). The nature of workplace mentoring relationships among faculty members in Christian higher education. *The Journal of Higher Education*, 70(4), 441-463.

Dulen, J. (1998, December 1). Service masters. *Restaurant & Institutions*, 108(28), 67-76.

Edmundson, J. E. (1999, October). Mentoring program help new employees. *FBI Law Enforcement Bulletin*, 68(10), 16-18.

Foegen, J. H. (1999, April/June). Why not empowerment. *Business and Economic Review*, 45(3), 31-33.

Gable, B. (1999, July). Building and retaining staff–it's a whole new world. *Business Communications Review*, 29(7), 60-62.

Girard-diCarlo, C. B. (1999, November/December). The importance of leadership, *Healthcare Executive*, 14(6), 48.

Hayes, E. F. (1998, October). Mentoring and nurse practitioner student self-efficacy. *Western Journal of Nursing Research*, 20(5), 521-535.

Hellriegel, D., Jackson, S. E., & Slocum, J. W., Jr. (1999). *Management* (8th ed.). Cincinnati: South-Western College Publishing.

Henley, R. J. (1999, July). Mentoring helps financial professionals lead with integrity. *Healthcare Financial Management*, 53(7), 14-15.

Herman, J. E. & Reichelt, P. A. (1998, October). Are first-line nurse managers prepared for team building. *Nursing Management*, 29(10), 68-72.

Herrenkohl, R. C., Judson, G. T., & Heffner, J. A. (1999, September). Defining and measuring employee empowerment. *The Journal of Applied Behavior Science*, 35(3), 373-389.

Jogaratnam, G., Tse, E. C., & Olsen, M. D. (1999, August). Matching strategy with performance. *Cornell Hotel and Restaurant Administration Quarterly*, 40(4), 91-95.

Kreitner, R., & Kinicki, A. (1992). *Organizational Behavior* (2nd ed.). Homewood, IL: Richard D. Irwin, Inc.

Liddle, A. (2000, June 26). Top 100. *Nation's Restaurant News*, 34(26), 75-94,152.

Liddle, A. (1998, September 14). The millennium: A foodservice odyssey operations. *Nation's Restaurant News*, 32(37), 151-154.

Luebbe, R. L. & Snavely, B. K. (1997). Making effective team decisions with consensus building tools. *Industrial Management*, 39(5), 1-7.

Melnyk, S. A., & Denzler, D. R. (1996). *Operations Management: A Value Driven Approach.* Chicago: Richard D. Irwin, a Times Mirror Higher Education Group, Inc.

Messmer, M. (1999, October). Retaining your valued employees. *Strategic Finance*, 81(4), 16-18.

Mitchell, T. (1998, January/February). Mentorship as leadership. *Change*, 30(1), 48-49.

Moddelmong, H. (1999, November/December). Mentoring women in franchising. *Franchising World*, 31(6), 54.

Morrison, E., Mobley, D., & Farley, B. (1996, Fall). Research and continuous improvement: The merging of two entities? *Hospital and Health Services Administration*, 41(3), 359-364.

Murphy, C. (1999, February 18). How McDonald's conquered the UK. *Marketing*, 30-31.

Olesen, M. (1999, October). What makes employees stay? *Training & Development*, 53(10), 48-52.

Papernik, R. L. (1999, September 13). Customer Satisfaction: Value. *Nation's Restaurant News*, 33(37), 112-114.

Pfeffer, J. & Veiga, J. F. (1999, May). Putting people first for organizational success. *The Academy of Management Executives*, 13(2), 37-48.

Reed, M. K. (1999, April). Empowering lifelong learners. *Nursing Management*, 30(4), 48-50.

Spector, A. (1999, April 19). Brands sharpen focus on guest to garner dining distinction. *Nation's Restaurant News*, 33(6), 49-54.

Statistical Package for the Social Sciences (SPSS). (Computer Software). (1999). *Statistical Package for the Social Sciences*, Version 9.0. Chicago: SPSS, Inc.

Tinto, V. (1997, November/December). Classroom as communities. *The Journal of Higher Education*, 68(6), 599-623.

Tomas, S. (1996, November). Nothing is so good it can't be made better. *Hospital Material Management Quarterly*, 18(2), 48-57.

Walkup, C. (1999, August 9). Absence of training makes customer loyalty a matter of chance. *Nation's Restaurant News*, 33(32), 114.

Walsh, A. M., & Borkowski, S. C. (1999, Summer). Cross-gender mentoring and career development in the health care industry. *Health Care Management Review*, 24(3), 7-17.

Walsh, A. M., Borkowski, S. C., & Reuben, E. B. (1999, July/August). Mentoring in health administration: The critical link in executive development/practitioner application. *Journal of Healthcare Management*, 44(4), 269-280.

Wayne, S. J., Liden, R. C., Kraimer, M. L., & Graf, I. K. (1999, September). The role of human capital, motivation and supervisor sponsorship in predicting career success. *Journal of Organizational Behavior*, 20(5), 577-595.

Willis, A. K. (1999, May). Breaking through barriers to successful empowerment. *Hospital Material Management Quarterly*, 20(4), 69-80.

Woods, R. H., & King, J. Z. (1996). *Quality Leadership and Management in the Hospitality Industry*. Michigan: Educational Institute of the American Hotel and Motel Association.

Strategic Information Systems in the Quick Service Restaurant Industry: A Case Study Approach

Marsha Huber
Paul Pilmanis

SUMMARY. Information technology (IT) is increasingly needed to remain competitive. This article presents a framework for designing management information systems (MIS) from a strategic perspective for the quick service restaurant industry (QSR). We provide guidance for planning and managing systems strategically using two existing frameworks: critical success factors (CSFs) for implementing systems that are strategic, and "stages theory" for managing these systems in response to changes in technology and in the marketplace. While the frameworks are not new, they can be useful in suggesting ways to design strategic information systems. Two case studies, developed from interviews with MIS directors of two quick service chains, are presented to describe the firms in the context of these two frameworks. Implications useful for strategic planning and system development have been developed from the case studies. *[Article copies available for a fee from The Haworth Document Delivery Service: 1-800-HAWORTH. E-mail address: <getinfo@haworthpressinc.com> Website:*

Marsha Huber, MBA, CPA, is a Doctoral Candidate in Hospitality Management at Ohio State University.

Paul Pilmanis, MBA, is Information Technology Consultant, 91 High St., Butler, NJ 07405-1139.

[Haworth co-indexing entry note]: "Strategic Information Systems in the Quick Service Restaurant Industry: A Case Study Approach." Huber, Marsha, and Paul Pilmanis. Co-published simultaneously in *Journal of Restaurant & Foodservice Marketing* (The Haworth Hospitality Press, an imprint of The Haworth Press, Inc.) Vol. 4, No. 4, 2001, pp. 195-210; and: *Quick Service Restaurants, Franchising and Multi-Unit Chain Management* (ed: H. G. Parsa and Francis A. Kwansa) The Haworth Hospitality Press, an imprint of The Haworth Press, Inc., 2001, pp. 195-210. Single or multiple copies of this article are available for a fee from The Haworth Document Delivery Service [1-800-HAWORTH, 9:00 a.m. - 5:00 p.m. (EST). E-mail address: getinfo@ haworthpressinc.com].

KEYWORDS. Information systems, foodservice, restaurants, QSR, strategy, case study

INTRODUCTION

Management information systems (MIS) directors have been discussing the use of information systems to gain a strategic advantage since the mid-1980s (Gartner Group, 1999; Lederer & Salmela, 1997; Brancheau & Wetherbe, 1987). Constructing a system that has strategic advantages requires a firm to define the strategic needs of the company and to match information systems and applications to those needs. A strategic system aligns information systems with company goals, operations, products, and services, which creates an environment that helps firms gain a competitive advantage in the marketplace (Laudon & Laudon, 2000).

Whether information technology (IT) can be used to achieve a competitive advantage is debatable, but the impact IT has had on all industries, including the foodservice industry, cannot be denied. IT is often needed just to remain competitive, let alone for gaining a competitive advantage. There is a growing need to better manage technology and to respond to changes in technology and competitor actions.

In foodservice literature, there is a lack of direction on information systems planning from a strategic perspective (Kirk, 1998). Numerous articles in magazines, such as *Nation's Restaurant News,* focus on specific software packages or applications, not on strategic issues. Ansel and Dyer (1999) developed an IS planning framework that listed applications appropriate for a three-part model that included production systems, enterprise-resource-planning systems, and external systems, but provided little guidance on how to plan systems strategically.

We provide guidance for planning and managing IT systems strategically using two existing frameworks: critical success factors (CSFs), for implementing systems that are strategic, and "stages theory," for managing these systems in response to changes in technology and in the marketplace. While the frameworks are not new, they are useful for planning and maintaining strategic information systems. Two cases studies, developed from interviews with MIS directors of two quick service chains (QSR), are presented that describe QSR firms in the context of these two frameworks.

"Critical Success Factor" Approach to Developing Strategic Systems

The critical success factors framework (CSFs), originally proposed by Daniel in 1961, and refined by Rockart in 1979, was proposed as an objective and individualized way to determine the information needs of executives. Critical success factors are areas of activity in which companies must excel in order to be successful. CSFs highlight what is strategically important to a firm and allow the identification of key underlying activities with tactical significance. Proper execution of tasks within these areas helps organizations meet operational and strategic goals. Rockart determined that there were four prime sources of critical success factors: industry structure, competitive strategy, environmental factors and temporal factors. CSFs are unique due to the geographic, temporal and strategic environments (Rockart, 1979).

Critical success factors are broad categorizations of what a company must do to be successful. These factors are composed of critical activities, and once these activities are identified, an MIS director can determine if they are suitable for future systems development (Chu, 1995). How companies integrate their systems with these activities leads to competitive parity, advantage, or disadvantage. The fundamental reason for the failure of strategic systems occurs when systems do not support the truly critical activities (Chu, 1995).

In the foodservice industry, the most common critical success factors are: (1) cost reduction, (2) sales growth, (3) increased market presence, and (4) improved food and service quality. Providing a generalized list of critical activities for the QSR industry, however, is more difficult. Critical activities tend to be company specific and vary with a company's infrastructure, operating procedures, resources, and management practices. The purpose of this paper is not to generate a list of critical activities for the QSR industry, but to illustrate how IT can be used to perform critical activities. For example, for the cost reduction CSF, specific activities that can help reduce costs are demand forecasting and a shift in the power from suppliers to purchasers. By developing an IS-based forecasting model, waste could be reduced, or by developing an extranet that allows vendors to bid for contracts on-line, the power of suppliers might be reduced.

"STAGES THEORY" ROLE IN THE MIS PLANNING FUNCTION

"The Stages Theory is a framework for understanding the assimilation of information technology in business organizations" and "has been useful in understanding the effect of adopting new information technologies in organizations" (Nolan, Cronson, & Seger, 1993). Stages theory can be applied to individual organizations or to entire industries. "Stages theory" served as an important tool for planning the development of IT within a firm in the 1970s and 1980s. Although it was developed two decades ago, this framework is useful today.

Stages theory proposes four stages of systems development: initiation, expansion, formalization, and maturity (Gibson & Nolan, 1974). The initiation stage includes low-level operational applications such as data processing and accounting applications. The second stage, expansion, occurs when a proliferation of applications transpires to "get the job done" in various functional areas of the firm such as sales, purchasing, capital budgeting, or forecasting. The third stage, formalization, occurs when a moratorium is placed on the number of new applications installed. In this stage, a high level of control is exercised to promote greater efficiency and to encourage the permeation IT throughout the organization. Lastly, in the maturity stage, firms attempt to balance stability with innovation by installing applications that aid management in high-level decision-making. Examples of such decision support applications include simulations, financial planning models, and on-line queries.

Gibson and Nolan (1974) found that each stage of systems development also has potential problems. In the initiation phase, the IT function is often misplaced within the firm's organizational structure, and firms encounter employee resistance to the use of new computer applications. In the expansion phase, firms encounter problems such as of lack of control, an exploding budget, and a lack of alignment with an overall goal. In the formalization stage, to counter rapid expansion, firms often exert high levels of control, which can sometimes be excessive. When this occurs, budgets are cut, key personnel leave, and growth inevitably is stymied. In the maturity stage, firms are challenged to strike a balance between stability and innovation. Firms incorporate new technologies without losing the stability of existing systems. During this time, communication with top management is key to making a successful transition.

Herein lies the prescriptive nature of stages theory. Once a firm is identified with a particular stage, management can counter problems that typically occur. Nolan et al. (1993) also observed that while technology develops continuously, technological change in institutions is often discontinuous. During these periods of discontinuity, incumbent technology might suppress the adoption of new technology, and there often needs to be a transfer of knowledge from an experienced user group to an inexperienced user group. While technology constantly changes, these periods of transition remain a constant. Companies that better manage periods of discontinuity and the assimilation of information technology into the business activities may be more competitive than firms that do not (Nolan et al., 1993).

MIS AND COMPETITIVE ADVANTAGE

The use of information systems to achieve a competitive advantage has been a common and worthy goal for many organizations. However, instances where information systems provide an overpowering and sustainable competitive advantage are few. This is because technology is relatively easy for rival firms to copy (Clemons & Row, 1991) and competitive parity is usually reestablished within a short period of time (Kearns, 1995; Powell & Dent-Micallef, 1997).

Today, information technology is required to remain competitive in most industries. But there are organizations that manage IT better than others. Some organizations, all things being equal, can out-compete rivals. By integrating IT into its corporate goals and process on a tactical level, such advantages, though not gaping, might be obtained. The key to a sustainable advantage comes from identifying critical success factors and their underlying activities and integrating IT systems with these activities. Systems become strategic systems after they cross this threshold.

As early as 1985, Porter and Millar wrote that information technology has acquired strategic significance, and it affects competition by changing industry structures, by giving companies new ways to outperform rivals, and by creating new businesses (Porter and Millar, 1985). They observed that the information content in products and processes has increased in recent years. This is obviously truer today than when Porter and Millar made their observation in 1985. In the QSR, the amount of information that can be embedded in a product is limited. As such, sources of competitive advantage by differentiating products themselves is difficult, which leaves only cost control according to Porter's original framework.

Porter (1985) viewed companies as a collection of value-adding activities. The proper attention to critical activities and the careful management of linkages between activities could be a source of competitive advantage. While competitors can imitate activities of organizations, it is difficult to copy the linkages between activities (Porter, 1985). The ultimate goal in developing strategic information systems is to yield a sustainable competitive advantage that is not possessed or easily imitated by other firms (Mata, Fuerst, & Barney, 1995). Herein lies support for planning and developing strategic systems that support a firm's critical success factors.

CHARACTERISTICS OF THE FOODSERVICE INDUSTRY

Characteristic of the highly competitive QSR industry are low margins and a perishable product. Due to the low margins, control of food and labor costs is important. A strong market presence and the strategic placement of new stores are also important for long-term success. There are primarily five customer sales channels: delivery, dine-in, carryout, pickup window, and catering. In the QSR industry, IT is commonly used for order processing, accounting, purchasing, marketing, consumer behavior, and the location of new restaurants.

The foodservice industry as a whole is beyond the initiation stage with point-of-sale systems and accounting packages in place. The industry has entered a stage of proliferation, where a growing number of technologies are available. Conferences showcase technology and feature a full-range of applications, but few strategic systems.

Based on a series of actual interviews, two case studies illustrate how systems can planned and integrated within firms. The applicability of "critical success factors" and "stages theory" are examined in light of the case studies. While the following two case studies contradict some of the ideas of the "stages theory," they illustrate, how systems can be formulated around critical success factors, and how discontinuity and continuity can affect competitive advantage.

CASE STUDY OF ALPHA COMPANY

The first case study has been developed from interviews with the MIS director of a quick service chain with approximately 150 stores and $120 million in sales. For purposes of this paper, we will call this firm, Alpha. Alpha had been in the "initiation" stage for most of its existence. Its IT systems have been focused on its point of sale (POS) and accounting system.

Alpha began using IT began using systems more strategically in 1993, with the implementation of a proprietary POS system that was linked to an extensive database of financial information. During the "initiation" stage of POS, applications focused on accounting. The system was used for gathering data, tracking sales, determining food costs, placing both customer and inventory orders, scheduling labor, and tracking service delivery times. These early years served as an adjustment period. Each store had one server, a back-up server, and multiple workstations. In emergencies, the stores could convert to a paper system, in which orders could be taken by hand. The POS system had weaknesses: it lacked a touch screen and could not process credit cards.

In 1997, Alpha entered into the "expansion" phase by increasing in the number of applications used by many of the functional areas in the firm, including marketing, accounting, human resources, operations and management. The MIS director's objective during this time was to create reporting tools for both store managers and district managers. With top management support, he interviewed store managers and executives to find out their needs. The MIS director wanted managers to focus on operations and serving customers, rather than doing paperwork. In response to these discussions, the MIS director developed worksheets and database applications providing relevant control information and reducing paperwork. This information was placed on an intranet for easy access. District managers were given laptops and provided Internet access to Alpha's intranet.

During this time, IS maintenance was centralized, and individual stores paid the firm headquarters a maintenance fee monthly as well as for charges for additional work. Maintenance and programming were outsourced and computer equipment was leased. The use of third parties and leasing equipment allowed the IT department to concentrate on the tasks that supported the needs of the users.

The firm's systems development focused on its critical success factors. It was led by the MIS director, and supported by upper level management. They identified industry-wide CSFs as well as those unique to their company. The critical success factors Alpha identified were: (1) to increase market presence through brand awareness and growth, (2) to control of food and labor costs, (3) to control of food quality and service quality, (4) to select proper sites and control expansion. Alpha's systems supported the CSFs in the following ways.

CSF #1: An Increased Market Presence Through Brand Awareness and Growth

Information technologies are used extensively in marketing to support sales growth and to increase market presence. The primary use of MIS is to monitor brand awareness and growth. Sales information is used to evaluate the sales of new products and television promotions. Coupons, the most common type of promotion used by Alpha, are coded to identify the advertising medium from which they originated.

Alpha also uses IT to perform market analysis. "Market basket" analysis is used to determine the product preferred by each market segment. This practice has proven to be very valuable in the QSR industry, and is commonly used by both multinational franchises and small chains. This information helps Alpha decide what type of "value" meals and promotions to offer different customer segments.

Alpha also utilizes the web to increase market presence. Customers can gather menu information as well as place orders over the Internet. Alpha utilizes a third party, *www.Food.com,* to maintain this website. As of 1999, less than 1% of all weekly orders are placed over the website, but Alpha expects that to increase over time.

In response to customer demands for catering, a website was developed for catering orders over $100. This specialized web site targets a different market segment from the *Food.com* website. The catering market segment orders meals for parties, business luncheons, and other large gatherings. Catering orders go to a central location, and production orders are distributed to one or many stores for processing. Catering accounts for roughly 20% of all orders or $30 million in annual sales. Alpha plans to offer online menu planning at a later date, which will provide more interactivity and greater flexibility when placing orders.

CSF #2: Control of Food Costs and Labor Costs

Food and labor cost control is of vital importance to any foodservice establishment. In its "initiation" years, the POS system submitted daily purchase orders to headquarters. Headquarters then utilized electronic data interchange (EDI) to transmit orders directly to its primary suppliers. The automated ordering system was more efficient than manually preparing and submitting orders by phone.

Daily cost-of-goods percentages were available through the POS systems, but company-wide reports that could help district managers in managing area stores were not automated. An automated "real time" reporting systems was deemed too expensive in the mid-1990s. To provide "real-time" reporting, the MIS director estimated costs to be approximately $100,000 with an additional cost of $2,000 per store for a total of $300,000. Advances in technology in the late 1990s, however, enabled the Alpha to develop a new reporting system.

The new reporting system uses an intranet and two core technologies, Linux and OLAP. What made this so affordable is that Linux is a UNIX-based operating system available free of charge. Alpha's expense for these new technologies is less than $25,000 with costs of $700 to link Linux to SQL Server, $50 for site licenses per store, and $15,000 to link Linux to existing applications.

OLAP (on-line analytical processing) sped up the reporting process. Data is stored in OLAP cubes, which are faster than relational databases due to pre-defined dimensions. Prior to using OLAP it took two days to get product mix information using spreadsheet programs. OLAP now takes minutes. This reporting model was developed from the input of executive management.

District Managers could log into the Intranet from their homes and while on the road with their laptops. Linux was installed at all Alpha locations. Linux offered benefits in addition to the existing POS system and EDI ordering. These benefits included Netscape browser, Star Office–a collection of office applications similar to Microsoft Office–and advanced data processing capabilities.

The key to effective food cost control is reporting timeliness and relevance. Managers can access many types of reports, including weekly food cost percentage and variance reports for all the stores. Store and district managers can view not only their accounts, but also those of other stores.

Food cost problems often result from waste and the overproduction of food. Food costs and variances are determined weekly by comparing ideal food costs to actual food costs. These reports are generated on Excel spreadsheets. Each day Alpha's store managers estimated the amount of food needed, but fearing stock outs, they frequently overestimated.

Alpha developed a new application called the Food Estimator (FE) to reduce waste. Based on historical data, the FE forecasts the quantity of food needed for each store on daily basis. The FE indicates how much food to prepare and when to start preparation. This guarantees a fresh product and minimizes overproduction. According to the operations vice president, the annual savings attributed to the FE was a 50% reduction in waste. Annual savings are projected to be $1 million or 3% reduction in total food costs.

Alpha implemented *Peoplesoft,* a human resource software package that provides payroll, on-line benefits, direct deposit, and minor violation reports, to improve labor cost control in 1999. Minor violation reports track the hours worked by minors to avoid violating state laws. Prior to the installation of *Peoplesoft,* this was a time consuming process. The payroll processing was outsourced, and Alpha had to track minor hours separate from the payroll function. *Peoplesoft* integrates both the monitoring and payroll functions, and now all payroll processing is performed in-house, saving both time and money.

Additional labor savings resulted from the automation of cash management. Alpha is primarily a cash business and because of this security is important. The firm uses a bank clearinghouse to sweep money into an interest bearing account where all deposits are compared with sales data from the POS system. The POS system also tracks coupons and sales that have been voided. In some cases, vendor draws are taken from the account electronically. Reports generated by the system allow upper level management to monitor these accounts.

Savings in labor costs also resulted from posting of the Form W-4 to the intranet. The W-4 is the form that all new employees must complete for federal income tax withholding purposes. With the form easily accessible on-line, calls to the human resources to request forms have been minimized.

In the future, Alpha is investigating bandwidth, telecommunications, and vendor connectivity issues. Alpha plans to develop an extranet that lists product specifications allowing vendors to bid on supplier contracts. Alpha also plans to provide near real-time reporting, producing reports on a daily basis that now take a week to generate.

CSF #3: Control of Both Food Quality and Service Quality

Alpha prides itself on its food and service quality. The new intranet is also heavily utilized in these areas. Shoppers imitate customers and visit the restaurants at random to evaluate service quality. Secret shoppers rate the restaurants according to standardized measures such as food quality, store appearance, delivery times, and service quality. Shoppers dine out several times a week, and reports generated from these visits are placed on the intranet weekly. Alpha uses these service ratings and rankings to measure and improve service quality. The reports are posted to the intranet to give immediate feedback to store managers and the district managers. All the stores are listed so that store managers can see how their stores rank in relation to the other stores.

Service delivery reports and client defection reports are also reported to the intranet. Service delivery times are reported for each store. Client defection reports provide information about order frequency, the number of coupons redeemed, and service recovery attempts. This allows managers to benchmark their performance with other stores and company-wide averages.

CSF #4: Proper Site Selection and Controlled Expansion

An important part of future growth is establishing new restaurants. Some chains have encountered problems with uncontrolled growth, such as Boston Chicken, or poor site selection, such as the bankrupt G.D. Ritzys. Alpha uses demographics and IT to locate sites for new restaurants. The demographic and psychographics data is purchased from a third party. The information is used to profile Alpha's customers by factors such as education, age, income level, and geographic location. Actual customer data generated by the POS system is then integrated with the commercial database to develop a "best store" prototype. Potential sites are then scored against this "best store" prototype to predict the success of future locations.

Alpha also subscribes to National Decision Systems (NDS), another database with demographic database. This demographic information is used to make marketing decisions such as determining store dimensions and parking lot size. This is another tool Alpha utilizes regarding unit expansion and the construction of new stores.

CASE STUDY OF BETA COMPANY

The second case study is also a quick service chain, which we will call Beta. Beta has been in business for several decades with over 1,000 stores and sales of over $1 billion. Beta has invested millions of dollars in information technology. Beta has a state-of-the-art data center and uses similar types of IT applications as Alpha including cash management, food estimators, market basket analysis, human resource management, and electronic data interchange with stores, banks, and payroll.

Regarding the role of the stages theory and Beta, Beta did not conform to Nolan's "stages theory." Beta, with its many years of systems development, had some characteristics of the "formalization" stage with many applications are already developed. In place too was an IT steering committee and strict budget guidelines. On the other hand, with the implementation of new technologies, specifically an Enterprise Resource Planning System (ERP), Beta seems to have moved from the "formalization" stage back to the "expansion" stage with a focus on "getting the job done." The huge task of implementing the ERP seems to be characterized by a lack of alignment to an overall strategic goal.

As a result of these observations, the original "stages theory" with firms evolving from one stage to the next may be more dynamic than originally depicted. Our example implies that firms may move forward and backwards along a continuum, or might be in two stages concurrently, as impacted by innovations in technology (see Figure 1).

Beta's systems are several decades old and are mainly comprised of legacy systems that are not Y2K compliant. The Y2K crisis of the late 1990s forced Beta to update its systems. The MIS director felt the best solution was to replace the many legacy systems with an Enterprise Resource Planning (ERP) System. To make a smooth transition, it was necessary to strike a balance between stability and innovation. In addition to this, the MIS director wanted to exploit an existing database. To accomplish these goals, a step-by-step plan

FIGURE 1. Stages Theory Continuum

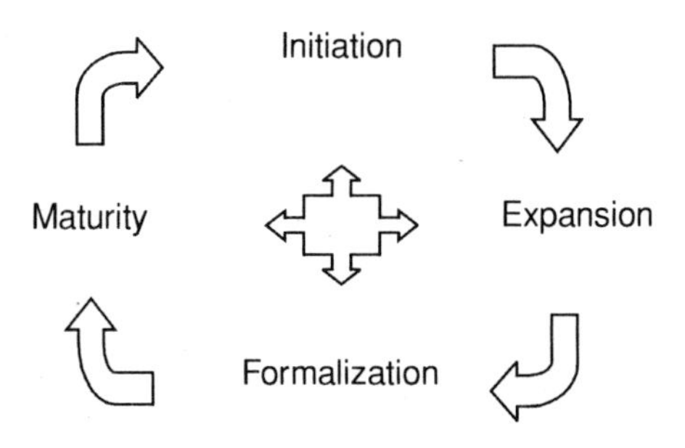

was developed, to develop infrastructure, upgrade networks, develop databases, and to install *Peoplesoft*.

In Phase One, technology infrastructure was updated. To better accommodate the demands of ERP and future information needs of the company, new mainframes, relays, robots, and subsystems were installed.

In Phase Two, *Peoplesoft* was installed. *Peoplesoft* is an enterprise resource planning system that integrates all the functional areas of a firm into one system. It operates from a central database that is shared by all the functional lines. The legacy systems for payroll, account payable, and the general ledger were the first systems replaced with *Peoplesoft*. The replacement of the legacy systems for billing, accounts receivable, and inventory control will follow in 2000/2001.

In Phase Three, web technology is to be updated, and its uses expanded to include intranets, extranets, and the Internet. The MIS director wants to get information faster to the end users through enhanced networked systems.

In Phase Four, data management will be improved to include data warehousing and mining. Also, the role of technology in the firm is being examined, and whether it should continue to be a support service or a future source of strategic advantage.

Beta's philosophy for application development differed from that of Alpha. Unlike Alpha that developed applications based on CSFs, Beta has many systems already in place, and was most concerned with updating its technology. Furthermore, Beta's upper management did not view the role of IT as strategic. The focus of Beta is more traditional: IT serves as a low-cost provider of data, whose primary focus is monitoring competitors and providing data for benchmarking. At Beta, IT held a supporting role and was not perceived to add value to the end product. The critical success factors identified for Beta are: (1) consistency, (2) quality food and service, (3) brand recognition, and (4) operations control.

CSF #1, #2, and #3: Consistency, Quality Food and Service, and Brand Recognition

Unlike Alpha, Beta did not use its systems to support these areas. A marketing research firm provides the reports on consistency of food preparation, service quality, and brand recognition of the company name and its products. Consumers are surveyed weekly and results are compared to its two nearest competitors on forty-three key performance attributes that include quality, value, taste and advertising recall.

Service consistency is measured by asking customers about drive-thru times, the accuracy of orders, what is considered the best value, and facility cleanliness. Questions about quality include the service and food quality, taste, and preparation. Questions about service ask whether or not the employees smiled, were courteous, and whether they seemed organized. Food quality questions inquire about which QSR had the best tasting food, freshest food, best quality meat, best French fries, and so on. Brand awareness questions centered on awareness of advertising campaigns and comparing sales levels with promotional efforts.

IT was not used to generate reports or to make report results available to all store managers. Due to the Y2K crisis, such reporting, if provided, has be delayed until Phase Three or Four of Beta's implementation process.

CSF #4: Operations Control

As mentioned earlier, an IT steering committee oversees Beta's MIS development. Top management views IT as a support service rather than a strategic partner. Budget and personnel limitations prevent Beta from developing truly effective MIS systems focused on its CSFs. The MIS department is not allowed to charge-back costs to the various departments nor are stores willing to pay for new systems. The MIS department's ability to purchase or develop new applications has been limited.

As a result of the lack of strategic understanding of systems, the focus of Beta's IT reporting system has been on transaction processing. Information needs for Beta center around daily sales reports, food cost reports, and weekly marketing reports. Although, these reports provide operational control, they are not as easily accessible to all of Alpha's critical personnel. The chief executives have reports accessible to them via desktop computers, but district and store managers do not since they do not have access to Beta's Intranet. Beta gave district managers laptops in 2000, enabling easier access to operational information. Prior to 2000, district managers must travel to a regional office to gather that information.

Whether Beta achieves competitive advantage, parity, or disadvantage will depend on how well it can integrate ERP with existing systems and applications, as well as its ability to assimilate new technologies.

IMPLICATIONS FOR PLANNING AND STRATEGIC DEVELOPMENT

The case studies of Alpha and Beta illustrate some interesting points related to developing strategic systems. First, using the CSF approach is one way to develop strategic systems. Alpha demonstrated this through its methodical development and alignment of its systems with firm critical success factors. Alpha personnel, when interviewed, felt "good" about their systems. In fact, Alpha's integration of IT with CSFs and critical activities may create boundaries to duplication that lead to competitive advantages.

Second, as originally stated by Nolan, old systems seem to be a hindrance to new systems. It is easier to develop new systems than to update old ones by rewriting thousands of lines of code. Alpha, the younger firm, assimilated new technologies quite easily into its firm. Beta, hindered by legacy systems and massive IT structure, found assimilation more difficult. Linking old systems to new ones often requires both extensive customization and changes to infrastructure. As a result of stymied growth, Beta users are reportedly "dissatisfied" with Beta's systems. For Beta, the implementation of ERP is a move that achieves competitive parity rather than advantage.

Third, the case studies illustrate how a company in the "formalization" stage can fall behind a company in the "expansion" stage. Alpha, an expansion company, utilized the latest technologies to gain competitiveness such as developing an intranet. The intranet provided fast and easy access to food cost information. Beta, on the other hand, is only beginning to explore this idea. In addition, Alpha uses an EDI link to order supplies from vendor that minimizes ordering time. Beta, on the other hand, requires managers to e-mail or call in orders to headquarters, which takes more time than EDI.

Fourth, it appears as though "stages theory" might be better depicted as a "continuum" with firms moving backwards or forwards among stages. Firms in fact could be in two stages at the same time, as it appears is the case with Beta. This complicates typical solutions to problems commonly faced by companies in the original "stages theory."

Fifth, top management's view of the role of IT in their firms has far-reaching effects on system development. Alpha's management supported the view that IT could serve as a strategic ally. Their support eased the assimilation of new technologies into the firm and supported the development of technology based on CSFs. Beta, on the other hand, faced difficulties in assimilation that might have been partly caused by management's view of IT as a low-cost provider of data.

CONCLUSIONS

Information technology is useful for automating transaction processing, providing information to decision makers, embedding technology into products and services, and bettering internal and external communications. Certain frameworks may be useful not only for those seeking an advantage, but also for those trying to remain competitive. This paper illustrates how two frameworks can guide MIS departments in planning and developing strategic systems. CSFs can be used to design systems that are strategic, and "stages theory" gives insight into the assimilation of new technologies.

The case studies of Alpha and Beta provide us with some striking contrasts. Alpha viewed IT as a strategic ally, and Beta viewed IT as a low-cost provider of data. Alpha focused on the informational needs of its users, and Beta concentrated on implementing ERP and updating its systems. Alpha better assimilated new technologies. Beta remained challenged with the task of maintaining competitive parity.

Our case studies illustrate some strategy and IT issues that are common to the food service industry. For example, in the pizza segment of the food service industry, Domino's, a company with 4,500 locations, has similar problems to Beta. Domino's has incompatible point-of-sale systems in its stores. Store managers must also call in orders to headquarters. Given the large volume of stores, upgrading and changing POS systems would be extremely expensive.

Another pizza franchise, Papa John's, is more similar to Alpha. Papa John's views systems as strategic and has a unified POS and ordering system. According to Tom Kish, systems vice president at Papa John's, "technology must align itself with a business' goals and strengths or else little synergy exists."

We suggested the use of existing two frameworks for implementing systems that are strategic and for managing such systems in response to changes in technology and in the marketplace. CSFs and critical activities highlight the tasks and areas of a company that can create greater return on IT investment and perhaps a competitive advantage. Stages theory emphasizes the need for a balance between new and old technologies and helps orient companies in what seems to a continuum of technological change.

This paper raises several unanswered questions. What other frameworks could help companies manage technological discontinuity? What critical activities are most common in the foodservice industry, and what are variations in critical activities attributed to? Do ERP packages stifle innovation by making the adoption of new applications and new technologies difficult? Do ERP packages trade stability for innovation? Future research could investigate the relationship of ERPs and competitive advantage. We also recommend a comprehensive examination of antecedents to successful system planning. This paper presented various antecedents including top management's view of MIS, age of the system, and stage of systems development. Empirical evidence to test these antecedents as well as the implications cited earlier would add to the knowledge base of the development of strategic information systems.

REFERENCES

Ansel, Daryl and Dyer, Chris (1999). A Framework for Restaurant Information Technology, *Cornell Hotel and Restaurant Administration Quarterly*, June 1999, 74-83.

Brancheau, J. C. and Wetherbe, J. C., (1987). Key issues in information systems management, *MIS Quarterly*, March 1987, 23-45.

Chu, P. C. (1995). Conceiving Strategic Systems. *Journal of Systems Management*, July/August, 36-41.

Clemons, E. and Row, M. (1991). Sustaining IT advantage: The role of structural differences. *MIS Quarterly*, Sept. 1991, 275-292.

Gartner Group. (1999). *The CIO Top Ten 1999-2003*. Retrieved May 26, 1999 from the World Wide Web: http://gartner6.gartnerweb.com/.

Gibson, C. F., and Nolan, R. L., (1974). Managing the four stages of EDP growth. *Harvard Business College Press*, product no 74104. p 232-244.

Kearns, G. S. (1997). Alignment of Information Systems Strategy with Business Strategy: Impact on the use of IS for Competitive Advantage. *UMI*, 1-260.

Kirk, D., & Pine, R. (1998). *Research in Hospitality Systems and Technology. International Journal of Hospitality Management*, 17, 203-217.

Laudon, K. C., & Laudon, J. P. (2000). *Management Information Systems* (6th ed.). New Jersey: Prentice Hall.

Lederer, A. L. and Salmela, H. (1996). Toward a theory of strategic information systems planning. *Journal of Strategic Information Systems*, 5, 237-253.

Mata, F. G., Fuerst W. L., and Barney, J. (1995). IT and Sustained Competitive Advantage: A Resource Based Approach. *MIS Quarterly*, December 1995, 487-506.

Nolan, R. L., Croson, D. C., and Seger, K. N. (1993). The Stages Theory: A Framework for IT Adoption and Organizational Learning. Harvard Business School, 9-193-141, March 19, 1-12.

Porter, M. E. and Millar, V. E. (1985). How Information Gives You Competitive Advantage. *Harvard Business Review*, July-August, 149-160.

Porter, M. E. (1985). *Competitive Advantage: Creating and Sustaining Superior Performance*. New York: Free Press.

Powell, T. C. and Dent-Micallef, A. (1997). Information technology as competitive advantage: the role of human, business, and technology resources. *Strategic Management Journal*, 18:5, 375-405.

Rockart, J. (1979). Chief Executives Define Their Own Data Needs. *Harvard Business Review*, March-April, 81-93.

A Conceptual Model for Quantifying the Effects of a New Entrant in a Competitive Restaurant Market

Radesh Palakurthi

SUMMARY. A challenge often faced by restaurant operators is to deal with the effects of a new restaurant opening up in their local market. The preexisting restaurants' management teams are invariably unsure about how the opening of the new restaurant will affect their established market share and revenue streams. The current paper has two major objectives. First, this paper proposes a conceptual model for quantifying the changes in the market shares of all the restaurants in a market, when it is known that a new restaurant will be opening up in the near future. Second, this paper shows, through the use of simulated data, how the model developed can be applied for making strategic decisions. A popular stochastic model known as the Markov Model is used to analyze the dynamics of market share when consumers switch brand.

KEYWORDS. Restaurant, foodservice, competition, Markov Models, new entrant, marketing strategy, market share effect

---

Radesh Palakurthi, PhD, is Associate Professor, Hospitality Management, San Jose State University, One Washington Square, San Jose, CA 95192-0058 (E-mail: Radesh.Rao@worldnet.att.net).

[Haworth co-indexing entry note]: "A Conceptual Model for Quantifying the Effects of a New Entrant in a Competitive Restaurant Market." Palakurthi, Radesh. Co-published simultaneously in *Journal of Restaurant & Foodservice Marketing* (The Haworth Hospitality Press, an imprint of The Haworth Press, Inc.) Vol. 4, No. 4, 2001, pp. 211-224; and: *Quick Service Restaurants, Franchising and Multi-Unit Chain Management* (ed: H. G. Parsa and Francis A. Kwansa) The Haworth Hospitality Press, an imprint of The Haworth Press, Inc., 2001, pp. 211-224. Single or multiple copies of this article are available for a fee from The Haworth Document Delivery Service [1-800-HAWORTH, 9:00 a.m. - 5:00 p.m. (EST). E-mail address: getinfo@haworthpressinc.com].

INTRODUCTION

A challenge often faced by restaurant operators across the world is to deal with the effects of a new restaurant opening up in their local market. The preexisting restaurants' management teams are invariably unsure about how the opening of the new restaurant will affect their established market share and revenue streams (Hamstra, 1998). With increasing competition in the hospitality industry, and with consumers perceiving little difference between the various brands and with decreasing customer loyalty (Warner, 1999; Yesawich, 1997; Koeppel, 1992), chance may play a big part in a restaurant purchase decision. The importance of the role of chance may be due to factors that are beyond an establishment's control such as when a travel agency randomly selects a restaurant for a client, or when a guest is forced to select another restaurant as his/her favorite restaurant has a long waiting time. In other instances, a customer's choice may truly be random–like a desire for a change that day, or, simply a whim or price discount–that affects the restaurant purchase decision. In such an environment, it becomes difficult for the preexisting restaurants to predict the effect of the new restaurant in their market.

This paper proposes a conceptual model for quantifying the changes in the market shares of all the restaurants in a market, when it is known that a new restaurant will be opening up in the near future. The methodology suggested could be applied by the preexisting restaurants months before even the new restaurant is built as long as the main characteristics are known, i.e., type of cuisine, target Per Person Average (PPA), and the amenities offered. With some effort such information can be obtained by various means including contacting the local Chamber of Commerce, meeting the building contractors or suppliers of the new entrant, communicating with the owners of the new restaurant.

Various researchers have offered many strategies for dealing with competition in tight markets. For example a recent article (Siguaw, Mattila, and Austin, 1999) suggests that restaurant managers should develop a brand- personality to successfully differentiate their restaurants from their competitors to maintain the trust and loyalty of their patrons. Brand personality refers to the set of human traits that consumers attribute to a brand. According to these authors, successful brands are differentiated from their lesser-known counterparts due to their robustness, desirability, distinctiveness and consistency. Thus, an important objective for brand management is the formation of well-defined brand personality.

THE MODEL

Different stochastic (or random) choice models can be used to predict the brand choices of consumers. These models are classified depending upon whether or not they consider brand choice, purchase timing, purchase quantity, and marketing variables such as advertising expenditures and pricing. In addition, the models also differ on how they handle heterogeneity–the differences between consumers. Some handle these differences by segmenting the population into groups that appear to behave similarly. Others assume that there are parameters within a given model (brand loyalty or price sensitivity), whose values vary across the population (Lilien, 1993).

One popular stochastic model used to analyze the dynamics of market share when consumers switch brands is known as the Markov behavior (Lilien, Kotler, and Moorthy, 1992). Much of the information used in the Markov Model can be extracted from what is called the brand-switching matrix (probability matrix). The elements in this matrix are obtained from purchase information on two purchase occasions that are separate in time, where the time between the two purchases may differ considerably for different consumers. For example, a local dine-out customer may eat at a specific restaurant today but may want to eat at another restaurant the next time a similar dine-out occasion occurs.

There are many types of Markov Models depending on the assumptions made about consumer purchase behavior (Lilien, Kotler, and Moorthy, 1992). The "zero order" models assume that a consumer's current brand choice (purchase) is unaffected by his/her previous brand choice (purchase). In other words, each consumer has a constant probability of buying each of the brands on each purchase occasion. A sort of mental roulette wheel determines the choice the consumer will make on the next purchase occasion (Lilien, 1993). For example, a consumer may have a brand purchase probability vector (Brand A = 0.6, Brand B = 0.4) from which he/she does not deviate.

On the other hand, Markov Models (of the first order) assume that only the last brand choice affects the current purchase. These models assume that all consumers that purchased a given brand last to have the same probabilities of either remaining loyal or switching the next time (Lilien, 1993). If all such information needed to characterize the consumers' brand choice is contained in the probability matrix, denoted by a P, then mathematically,

$$P = \sum p_{(i \mid j)}$$

Where,

$p_{(i/j)}$ = Conditional probability that a customer will choose to dine at restaurant "i" on the second dine-out occasion given that he/she eats at restaurant "j" on the first occasion.

And,

$$0 \leq p_{(i/j)} \geq 1$$

(Probability of dining at any restaurant range from 0 to 1)

$$\sum_{i} p_{(i \mid j)} = 1$$

(Given that restaurant "j" was chosen last time, the sum of the conditional probabilities across all restaurants must equal 1).

If the current market shares (in terms of number of customers or food & beverage revenue) of the restaurants is known, then the first order Markov Model can be used to predict how market shares will change over time. Mathematically the new market shares can be represented as follows:

$$m_{j(t+1)} = \sum_{j=1}^{n} p_{(i \mid j)}\, m_{it}$$

Where,

$m_{j(t+1)}$ = Market share of restaurant "j" at time (t+1).
m_{it} = Market share of restaurant "i" at time "t", given i = 1 . . . n.
$p_{(i/j)}$ = Is described above.

In other words, the market share of restaurant "j" at time t + 1 equals the share of each restaurant at time t times the proportion of customers who switch from that restaurant to restaurant "j", summed over all the restaurants in the market.

Named after a turn of the century Russian mathematician, Markov Models are used to describe stochastic processes–that is, random processes that evolve over time. In the hospitality industry, Morkov Models have had limited applications. One paper (Uysal, Marsinko, and Barrett, 1995) examined the effect of switching trip type on market share using Markov Chain Models. In another paper (Wesley, Kalyan, and Murali, 1994) developed a general Markov Model that relates probabilistic aspects of recall to consumer and marketing mix variables. The authors demonstrate how the model can be used in numerical analyses to evaluate awareness-building strategies in industries such as the restaurant industry.

The most common application of Markov models are in health and it is used to characterize the possible prognoses experienced by a given group of patients. This entails modeling the progress over time of a notional group of patients through a finite number of health states. Patients are initially placed in one of the health states, and the probabilities of transition to the other states in the model are defined within a given time period, known as a Markov cycle (Sonnenberg, and Beck, 1993).

As with all models, Markov models are simplifications of the real world. An important limitation of Markov models is that they lack memory. This is termed the Markovian assumption–that is, the probability of moving from one state to another is independent of the history of the subject before arriving in that state. The extent to which the Markovian assumption is limiting is a matter of opinion, on which the researcher is not qualified to judge. Although the limitations of the Markovian assumption should be kept in mind, the adept modeler will often find ways around the assumption using a combination of time dependent transition probabilities and distinct states for subjects with different purchasing histories.

A SIMULATION APPLICATION

For the purpose of this research a simple (probably oversimplified!) market situation is considered only to demonstrate the application of the conceptual model. Many assumptions about the market are made that may not be necessarily true in the real world but will satisfy the needs of this research paper without unduly distracting from its primary purpose of showing how the model can be applied. There is definitely a need to validate the model by using real market data.

Consider the example of a restaurant market with five restaurants that are directly in competition with each other as shown in Table 1. In Table 1, PPA is the Per Person Average or the average sales per person (guest check average, as commonly known in the industry) patronizing the restaurant. STR-A refers to the Seat Turnover Ratio (Actual) and STR-P refers to the Seat Turnover Ratio (Potential) which is a proxy for the maximum number of covers that the restaurant can have during a day. For the sake of this simulation, it is assumed that each restaurant turns each table a maximum of four times per meal period (breakfast, lunch and dinner). Hence the maximum turnover ratio (assumed to be the same for all restaurants) is 4 turnovers × 3 meal periods = 12 STR-P.

TABLE 1. Restaurant Market

	Seats	PPA	STR-A	STR-P
Restaurant A	156	$12.50	9.5	12
Restaurant B	154	$13.25	7.6	12
Restaurant C	122	$16.50	10.5	12
Restaurant D	241	$11.25	9.4	12
Restaurant E	165	$10.95	8.4	12

Table 2 shows the initial market positions of the restaurants in terms of revenue and market share. The formulae for calculating all the variables described in this paper are shown in Table 3.

Figure 1 captures the initial market performance in terms of revenue yield index score and the market share index score. The matrix can be split into four quadrants–Market Leaders, Share Performers, Yield Performers, and Stragglers. Market Leaders are restaurants that are above market average in both revenue and market share. Stragglers on the other hand are restaurants that perform below market average in both revenue and market share. The Share Performers are restaurants that achieve above market average market share but below market average revenue. The Revenue Performers are restaurants that achieve above market average revenue but below market average market share. It should be noted that Stragglers and Revenue Performers are restaurants that are losing market share to Market Leaders and Share Performers.

In the market depicted in Figure 1, restaurants D and C are the clear Market Leaders while restaurants B and E are Stragglers. Restaurant A is a Revenue Performer in the market. The question is what would happen in the current market if a new entrant comes in to compete directly?

For the purpose of this simulation, it is assumed that the current players in the market (Restaurants A through E) are concerned about the potential new competition (Restaurant F) and are able to get information about the new restaurants from reliable sources such as the local Chamber of Commerce, or the potential owners themselves. For the current simulation Restaurant F is assumed to have 165 seats, a PPA of $10.50 and an actual Seat Turnover Ratio equal to the average of the market. Using this information, the current owners conduct a survey describing each of the current restaurants and the new restaurant across some common variables such as: the PPA, menu and price, type of cuisine, theme and decor, location, and other relevant amenities. To engage the respondents, the survey also includes sketches and photographs of all the restaurants in order to give the respondents a better idea about the layout and design of the estab-

TABLE 2. Market Revenue and Share Performance

	Actual Revenue	Potential Revenue	Revenue Yield	Revenue Yield IS	Market Share	Market Share IS
Restaurant A	$18,525.00	$23,400.00	79.17%	105	19.33%	97
Restaurant B	$15,507.80	$24,486.00	63.33%	84	16.18%	81
Restaurant C	$21,136.50	$24,156.00	87.50%	116	22.06%	110
Restaurant D	$25,485.75	$32,535.00	78.33%	104	26.59%	133
Restaurant E	$15,176.70	$21,681.00	70.00%	93	15.84%	79
TOTAL:	$95,831.75	$126,258.00	75.67%	100	20.00%	100

TABLE 3. Description of Variables

Variable	Formula	Description
PPA (Per Person Average)	Food and Beverage Revenue / Covers Sold	The average food and beverage sales per customer.
STR-A (Seat Turnover Ratio-Actual)	Actual Covers Sold / Number of Seats	The actual number of times the restaurant turns over each table in a day.
STR-P (Seat Turnover Ratio-Potential)	Maximum Covers that Can Be Sold / Number of Seats	The maximum number of times the restaurant can turn over each table in a day.
Actual Revenue	Seats × PPA × STR-A	The total food and beverage revenue in on day.
Potential Revenue	Seats × PPA × STR-P	The maximum food and beverage revenue the restaurant can generate in a day.
Revenue Yield	Actual Revenue / Potential Revenue × 100	The percentage of the maximum revenue that the restaurant is achieving.
Revenue Yield IS	Restaurant's Revenue Yield / Market's Average Revenue Yield × 100	The Revenue Yield Index Score. A value higher than 100 means the restaurant is better than average and vice versa.
Market Share	Restaurant Revenue / Total Market's Revenue × 100	The percentage of total market revenue that the restaurant generates.
Market Share IS	Restaurant Market Share / Average Market Share x 100	Market Share Index Score. A value higher than 100 means the restaurant is better than average and vice versa.

FIGURE 1. Initial Performance Matrix for the Market

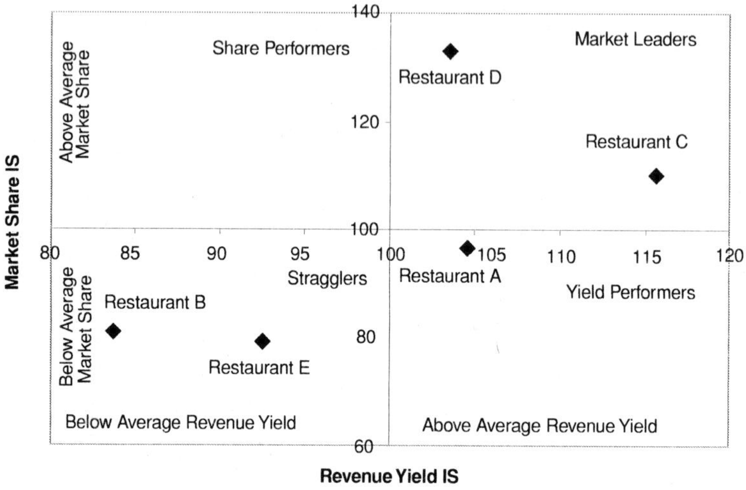

Share Performers
Market Leaders
Restaurant D
Above Average Market Share
Restaurant C
Market Share IS
140
120
100
80
60
80 85 90 95 100 105 110 115 120
Stragglers
Restaurant A
Yield Performers
Below Average Market Share
Restaurant B
Restaurant E
Below Average Revenue Yield
Above Average Revenue Yield
Revenue Yield IS

lishments. Figure 2 shows an example of the instrument that can be used for such a survey. At the end of the survey, the respondents in each restaurant are asked which of the restaurants described are they most likely to patronize on their next visit to the area based on the information provided. The respondents will be requested to choose only one of the restaurants described.

The survey should be conducted by all the current market players on a sample of their respective patrons. The result of such a survey would be a brand-switching matrix that gives the probability of a specific restaurant's patron coming back or going to a competition on the next visit. For the purpose of this simulation, Table 4 is assumed to show the brand-switching matrix of the current market.

Table 4 shows that the probability of an existing customer of Restaurant A will return on the next trip is 43.45%. The probability that a patron of Restaurant A will switch over to the new Restaurant F on the next trip is 12.53%. Similarly, the probability that a current patron of Restaurant B will switch over to the new Restaurant F is 11.78%. Similar percentages for Restaurants C, D and E are shown in column marked "F" in the table.

Brand switching data in the form of a square contingency table are often used to investigate competitive market structure. Brand-switching data provide a particularly compelling and intuitive behavioral measure of competi-

FIGURE 2. Example of the Instrument That Can Be Used for the Survey

<table>
<tr><th>No.</th><th colspan="2">Attribute</th><th>Description</th></tr>
<tr><td>1</td><td colspan="3">Design and Decor</td></tr>
<tr><td></td><td colspan="2"></td><td>Name: Restaurant B
Location: 123 Main Street, Anytown, USA
Features:
☐ 154 Seats
☐ Tablecloth restaurant
☐ Per Person Average (PPA)–$13.25
☐ Has been in business for 12 years
☐ Creative American theme</td></tr>
<tr><td>2</td><td colspan="3">The Main Course</td></tr>
<tr><td></td><td colspan="3">☐ Spring Vegetable and Chevre Raviolis with Sage and Pinenut Beurre
$15
☐ Prawn and Mussel Capellini with Spring Vegetables, Kalamatas and a Garlic Cream
$16
☐ Macaroni and White Vermont Cheddar Cheese with Prosciutto di Parma
$13
☐ Angus Beef Burger with Shoestring Potatoes and Garlic Aioli
$10
☐ Chicken Breast with Brie and a Roasted Garlic Cream Marinated in Dijon
$16
☐ The Ultimate Pot Roast Beef Chunks Braised to Perfection with Hearty Vegetables
$18
☐ Roasted Half Rossi Chicken with Garlic Mash Potatoes and a Citrus Glaze
$21
☐ Double Cut Pork Chop with Sweet Onion Jam and Currant Jus
$21
☐ Ahi Tuna Seared Rare with Shiitake Mushrooms and a Soy Ginger Glaze
$18
☐ Pacific Red Snapper with Blood Oranges, Asparagus and a Ruby Port Sauce
$23
☐ Grilled Ribeye Steak with Garlic Mash and a Cabernet Jus
$24
☐ Peking Duck Breast with Wild Rice and Cherry
$23
☐ Grilled Lamb T-Bones with Natural Jus, Couscous and Spring Vegetables
$25</td></tr>
<tr><td>3</td><td colspan="3">Restaurant Attributes</td></tr>
<tr><th colspan="2">Cuisine</th><th>Ambiance</th><th>Special Features</th></tr>
<tr><td colspan="2">Creative American, Californian, Continental, Eclectic, Healthy, Italian, Mediterranean, Nouveau American' Salads, Sandwiches, Seafood, Steaks, Vegetarian</td><td>Casual, Elegant, People Watching, Quiet, Romantic</td><td>Transmedia, Bar, Credit Cards, Daily Specials, Great Beers, Great Wines, Microbrews, Parking, Pre-Theater, Private Parties, Prix Fixe Menu, Wheelchair Access</td></tr>
<tr><td>4</td><td colspan="2">Based on the descriptions of all the restaurants, please check the restaurant you are most likely to visit on your next visit to the area. (Please check only one restaurant)</td><td>☐ Restaurant A
☐ Restaurant B
☐ Restaurant C
☐ Restaurant D
☐ Restaurant E
☐ Restaurant F</td></tr>
</table>

TABLE 4. Brand-Switching Matrix

	Restaurant	**Restaurant Purchase Occasion Two**						**TOTAL**
Purchase Occasion One		**A**	**B**	**C**	**D**	**E**	**F**	
	A	43.45%	6.41%	12.53%	21.73%	3.34%	12.53%	100%
	B	12.93%	48.56%	15.52%	6.61%	4.60%	11.78%	100%
	C	17.55%	14.26%	48.45%	4.39%	2.74%	12.61%	100%
	D	15.14%	16.48%	9.35%	44.10%	9.13%	5.79%	100%
	E	13.17%	18.05%	15.37%	3.66%	43.41%	6.34%	100%

tion, as one has a direct measure of movement among the restaurants. Applying the Markov Model described in this paper using the brand-switching probability data a contingency table of time t versus time t + 1 brand choice can be constructed. The model can be easily implemented on a spreadsheet and the brand choice for times t + 1, t + 2 or more can be generated to see how the market shares of the restaurant change with time. It should be noted that the initial market shares of the restaurants were obtained from Table 2.

Figure 3 shows the change in the market shares of the restaurants because of the entrance of the new Restaurant F.

From Figure 3 it can be seen that Restaurant F with zero market share initially, will stabilize its market share at 16% by round four. Since an average restaurant patron dines out about once a fortnight, the above chart shows the changes in the market shares of the restaurants in about a couple of months after the opening of the new Restaurant F. What is interesting from Figure 3 is that given the brand-switching probabilities of the restaurants, Restaurant D will be most affected by the entrance of Restaurant F since its market share is projected to drop from 27% to about 14%. It should be noted that while only 5.79% of the current patrons of Restaurant D say they would switch to Restaurant F on the next occasion, the expected new market share of Restaurant D is greatly affected because of the new dynamics in the market. The entrance of Restaurant F may have changed the relative market positioning of all the existing restaurants in the minds of the consumers.

In actuality, it is Restaurants A, B and C that are losing 12.53%, 11.78% and 12.61% of their customers respectively to the new Restaurant F. In other words, the potential customers of Restaurant F are perceiving Restaurant A, B

FIGURE 3. Change in Market Share Because of the New Entrant

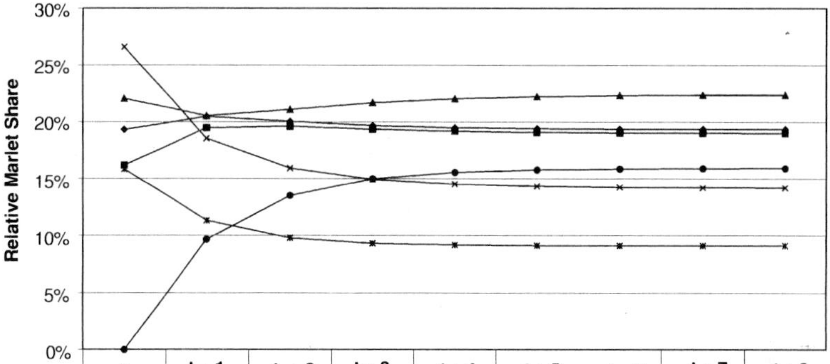

Relative Marlet Share

	t	t + 1	t + 2	t + 3	t + 4	t + 5	t + 6	t + 7	t + 8
A	19%	20%	20%	20%	20%	19%	19%	19%	19%
B	16%	19%	20%	19%	19%	19%	19%	19%	19%
C	22%	21%	21%	22%	22%	22%	22%	22%	22%
D	27%	19%	16%	15%	15%	14%	14%	14%	14%
E	16%	11%	10%	9%	9%	9%	9%	9%	9%
F	0%	10%	14%	15%	16%	16%	16%	16%	16%

Time Period

and C to be somewhat similar to Restaurant F. However, Restaurant D is the biggest loser and definitely needs to reevaluate its strategy.

Table 5 shows the forecasted market performance of all the restaurants after the new restaurant is opened. For the current simulation Restaurant F is assumed to have 165 seats, a PPA of \$10.50 and an actual Seat Turnover Ratio equal to the average of the market.

The forecasted revenue for Restaurant F was obtained by the following formula: Seats × PPA × Average Market STR-A, i.e., 165 × \$10.50 × 9.08 = \$15,731.10. Since Restaurant F is expected to have 15.91% of the market share at time t + 8, and all the market shares of the other restaurants are known, simple calculations can be used to determine the forecasted sales for all the restaurants. Restaurant F is also assumed to have a maximum Seat Turnover ratio of 12 like the other restaurants. Using this information, the potential revenue for Restaurant F can be determined. The rest of Table 5 follows the same calculations as shown in Table 3.

Figure 4 shows a snapshot of the market performance after Restaurant F enters the market.

TABLE 5. Forcasted Market Performance with the New Entrant

	Forecast Revenue	Potential Revenue	Revenue Yield	Revenue Yield IS	New Market Share	Market Share IS
Restaurant A	$19,158.59	$23,400.00	81.87%	120	19.38%	116
Restaurant B	$18,803.51	$24,486.00	76.79%	112	19.02%	114
Restaurant C	$22,130.95	$24,156.00	91.62%	134	22.39%	134
Restaurant D	$14,036.09	$32,535.00	43.14%	63	14.20%	85
Restaurant E	$9,004.36	$21,681.00	41.53%	61	9.11%	55
Restaurant F	$15,731.10	$20,790.00	75.67%	111	15.91%	95

Comparing Figure 1 with Figure 4 we notice the following changes in the restaurants' positions as shown in Table 6.

Table 6 shows that the entrance of Restaurant F in spite of grabbing a good share of customers will have little affect on Restaurants A, C, and E since they do not change their relative positions in the market. Actually, Restaurant B is helped with the entrance of Restaurant F. This may be because of the new dynamics in the market and the crossover effect of having more choices of similar type restaurants. The restaurant that is poorly affected by the entrance of restaurant F is Restaurant D. From being a Yield Performer, it changes position to a Straggler.

Strategically, Restaurant D has the most to worry with the entrance of Restaurant F. It may need to look at both its pricing structure and its relative competitive performance since it is a Straggler and is poor on both counts. While reducing the PPA may be the first impulse of the managers, better product positioning might be the best option for Restaurant D. They will have to develop long-term strategies to distinguish themselves from the other restaurants in order to overcome the competition.

CONCLUSIONS

While the methodology suggested seems easy to implement, it is not foolproof. The model makes many assumptions that may skew the results. For ex-

FIGURE 4. Post-Entrant Market Performance Matrix

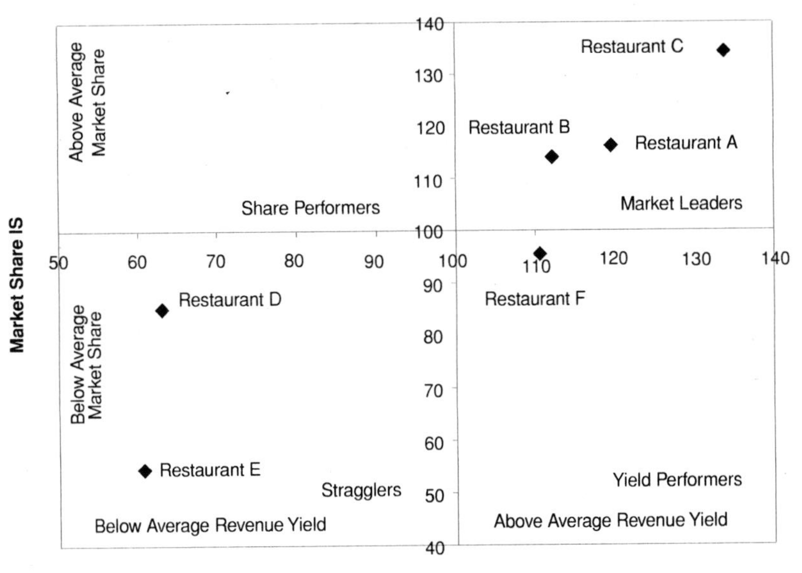

ample, the assumption that the brand-switching probabilities of the customers will remain constant with time may be unreasonable. However, a higher order of the model suggested can be used to account for the changing probabilities with time. Also, the assumption that customers will truthfully respond to the survey questions or really have enough information to make a purchase decision is a moot point. Numerous marketing research papers point out the difficulty of believing what consumers tell us, especially about purchase behavior.

There is a need to put the model suggested to a real-world test. Many questions about data validity, variables to be considered, and the assumptions made can only be answered after thoroughly testing the model with real world data. For example, are there better measures to evaluate the outcome? Or, will the model be more accurate if analyzed by each meal period? How do socio-demographic variables affect the brand-switching matrix? These are some of the questions that will need further research.

Managing new competition is an ongoing task of restaurant managers all over the world. It is always better for managers to make strategic decisions based on proven theory than making decisions in a vacuum. The methodology described in this paper offers restaurant managers a means to evaluate the af-

TABLE 6. Change in Restaurants' Positions

	Initial Position	New Position
Restaurant A	Yield Performer	Market Leader
Restaurant B	Straggler	Market Leader
Restaurant C	Market Leader	Market Leader
Restaurant D	Market Leader	Straggler
Restaurant E	Straggler	Straggler
Restaurant F	None	Yield Performer

fect of a new entrant in a competitive market. The methodology while easy enough to be implemented on a spreadsheet needs to be validated. If the management of a restaurant is able and willing to collect the data required for determining the probabilities for brand-switching, it should be a relatively easy task to find out how the new entrant will affect the market.

REFERENCES

Hamstra, M. (1998). Sandwich Leaders Tap Bolder Flavors in Shifts Away from Low Price Points. *Nation's Restaurant News*, 35, 108-112.

Koeppel, D. (1992). Fast food's new reality, *Adweek's Marketing Week*, 33,13, 22(2).

Lilien, L.G. (1993). *Marketing Management: Analytic Exercises for Spreadsheets*. San Francisco, CA: The Scientific Press.

Lilien, L.G., Kotler, P., and Moorthy, K.S. (1992). *Marketing Models*. Eaglewood Cliffs, NJ: Prentice Hall.

Siguaw, J.A., Mattila, A., and Austin, J.R. (1999). The brand-personality scale: an application for restaurants. *Cornell Hotel & Restaurant Administration Quarterly*, 40, 3, 48(8).

Sonnenberg, F.A., and Beck, J.R. (1993). Markov models in medical decision making: a practical guide. *Medical Decision Making*,13, 322-338.

Uysal, M., Marsinko, A., and Barrett, R.T. (1995). An Examination of Trip Type Switching and Market Share: A Markov Chain Model Application. *Journal of travel and Tourism Marketing*, 4 (Spring), 45-55.

Warner J. (1999). Dunkin' bows breakfast push. *ADWEEK New England Advertising Week*, 36, 9, 3(1).

Wesley, J. H., Kalyan, R., and Murali, K. M. (1994). Finding choice alternatives in memory: probability models of brand name recall. *Journal of Marketing Research*, 31, 4, 441(21).

Yesawich, P. (1997). The myth of the loyal guest. *Lodging Hospitality*, 53, 11, 8.

Study of the Changing Role of Multi-Unit Managers in Quick Service Restaurant Segment

W. Terry Umbreit

SUMMARY. Quick service restaurant (QSR) chains in the past decade expanded rapidly and continued to experience healthy increases in sales growth. The enhanced level of competition created by the growth brought about a number of structural and organizational changes. This study investigates the changing role of multi-unit managers as a result of organizational changes, as well as updating information on the position. Ten corporate QSR executives were interviewed. Results reveal the multi-unit manager position has grown in importance. Leadership and organization/time management skills are the biggest challenges faced by individuals during their transitions into the multi-unit manager role. *[Article copies available for a fee from The Haworth Document Delivery Service: 1-800-HAWORTH. E-mail address: <getinfo@haworthpressinc.com> Website: <http://www.HaworthPress.com> © 2001 by The Haworth Press, Inc. All rights reserved.]*

KEYWORDS. Multi-unit manager, quick service restaurants, change, leadership, time management, transition

In the new millennium, the quick service restaurant (QSR) segment will continue to be an important economic force in the United States. The QSR segment

---

W. Terry Umbreit, PhD, is Professor and Chair, Department of Hospitality Management, Washington State University Pullman.

[Haworth co-indexing entry note]: "Study of the Changing Role of Multi-Unit Managers in Quick Service Restaurant Segment." Umbreit, W. Terry. Co-published simultaneously in *Journal of Restaurant & Foodservice Marketing* (The Haworth Hospitality Press, an imprint of The Haworth Press, Inc.) Vol. 4, No. 4, 2001, pp. 225-238; and: *Quick Service Restaurants, Franchising and Multi-Unit Chain Management* (ed: H. G. Parsa and Francis A. Kwansa) The Haworth Hospitality Press, an imprint of The Haworth Press, Inc., 2001, pp. 225-238. Single or multiple copies of this article are available for a fee from The Haworth Document Delivery Service [1-800-HAWORTH, 9:00 a.m. - 5:00 p.m. (EST). E-mail address: getinfo@ haworthpressinc.com].

estimated sales for 1999 was approximately $116 billion, which represents more than 31% of the total foodservice industry volume (Technomic Inc., 1999). Sales growth was impressive at the close of the 1990s for quick service restaurants. Chain organizations are plotting growth strategies that assume their above-average sales surges can be sustained (Sheridan and Matsumoto, 2000). According to Susan Mills (1999) of the National Restaurant Association, foodservice industry growth is expected to soar from a projected $354 billion in 1999 to $577 billion by 2010. This would raise the current away-from-home average food consumption from 44% to 53%. Growth of this magnitude will have important managerial implications for restaurant organizations. One recent article in the trade press has sounded the alarm about the impending labor shortage in the foodservice industry (Bernstein, 2000). Reed Hayes, President and COO of the National Restaurant Association's Educational Foundation, estimates the industry will need more than one million well-trained unit general managers in the next several years. If the United States' low unemployment rate continues, it will only exacerbate the problem.

While restaurant unit managers are crucial to the success of any individual unit, chain restaurant organizations also use a multi-unit management structure to control and monitor a large number of geographically dispersed restaurants. Perhaps the most important role in the multi-unit management structure employed by QSR chains is the level immediately above unit manager. This first level of multi-unit management has a variety of titles including district manager, area supervisor, area coach or area manager. First-level multi-unit managers are critical to any QSR chain because these individuals are the closest to the unit managers and are responsible for enforcing standards; monitoring operations; developing unit teams; and creating a supportive operating culture. The role of multi-unit managers today in the QSR segment has become even more important because as chains expand, the employment pool is shrinking.

In the late 1980s and early 1990s, many QSR chains such as Taco Bell and KFC followed the strategies of other industries by embracing employee empowerment and other similar management philosophies for the purpose of enhancing operating results. This led to a significant amount of downsizing and corporate restructuring in the middle management ranks. The span of control for multi-unit managers increased from less than ten units to fifty in some chains. A few restaurant companies experimented with the concept of "leaderless" restaurants. These strategies were implemented in the name of efficiency, improved decision-making, employee commitment, and customer satisfaction. The predominant thinking was to push down the decision-making closer to the employee level, and reduce the layers of management.

During the past 5 years, the pendulum has begun to swing back to smaller spans of control, and the use of more management-support people (Brooks, 1996). Executives in the QSR segment began to realize that downsizing was a temporary solution and the experiment did not produce the desired results. Rob Doughty of Pizza Hut (1996) commented on the change: "Our service suffered and we were less focused on operations and restaurants." Unfortunately, the downsizing removed too many managers from running the restaurant, which negatively impacted operational and service performance. The greater spans of control meant that the multi-unit managers could not visit the restaurants as frequently to help solve operational and human resource issues. The spans were also reduced when the restaurant industry began to focus on managing existing units more effectively. This development was caused by the rise in the cost of real estate and construction, and the emergence of strong competitors in every market segment.

LITERATURE REVIEW

While academic and popular articles continue to discuss the growth and trends in the QRS segment, less is written about multi-unit management. Muller and Campbell (1995) acknowledged this when they stated, "the literature regarding multi-unit restaurant management is surprisingly sparse given the segment's economic magnitude." A recent study conducted by Gross-Turner (1999) on the United Kingdom's branded hospitality businesses from a human resource management perspective adds additional information about the position. He concluded that the role is predominantly one of implementor, not a creator of policy. The role continues to be one of policing and controlling. However, multi-unit managers in his study who used their own initiative obtained the best out of unit managers. In addition, results about multi-unit management in the United Kingdom revealed that the larger, mature companies were more control-oriented, while the smaller and entrepreneurial firms used a strategy of maintaining brand integrity through stressing the importance of values and culture. Gross-Turner's study encompassed more than QSR chains including an investigation of multi-unit managers in pubs and hotels.

The seminal work on multi-unit management in the United States was conducted by Umbreit and Smith (1991), Umbreit (1989), Mone and Umbreit (1989), and Umbreit and Tomlin (1986). Research in the latter study identified the five basic dimensions of a multi-unit manager's job as financial management; restaurant operations; marketing and promotions management; facilities and safety management; and human resource management. After analyzing responses from 164 multi-unit managers, Mone and Umbreit (1989) found that the job dimension recording the highest perceived value was restaurant operations, followed by human resource and financial management. In the same study multi-unit managers revealed that they were spending an increasing amount of time on human resource management, and wanted further training in this area. Human resource management was also identified as the dimension providing the greatest difficulty to individuals when making the transition from single-unit to multi-unit management.

Muller and Campbell (1995) studied the attitudes and perceptions of three levels of chain restaurant managers from a single, national, QSR company. Specifically, the study investigated the degree to which the views of three levels of management differed in regard to the skills required of a competent multi-unit manager. The three levels of management were store managers, area supervisors, and headquarters staff. Some of the research results dealing with area supervisors differed from the findings of Mone and Umbreit (1989). For example, area supervisors in this study reported they wanted further training in marketing, followed by finance. Human resources ranked fourth as a training need since many area supervisors appeared to believe they had sufficient human resource skills to succeed within the company. In this organization, all three management levels ranked human resources as the most important dimension. While this study was limited to one company, it did help further define the role of multi-unit manager as one of managing people more than one of managing restaurants.

PURPOSE OF STUDY

This study was designed to gain an understanding of the impact of recent organizational and market competitive changes on the role of the multi-unit manager in QSR chains and to update information on the position. Specifically, the investigation focused on obtaining information on the position including importance and span of control; recruitment, selection and training; compensation, evaluation and career perception; and transition issues and challenges. The study builds on previous information on the role of the multi-unit manager and assesses its importance to QSR organizations. It also identifies the strategies designed by QSR firms to prepare managers for the duties and responsibilities of the position.

METHODOLOGY

The study methodology is a qualitative design using a questionnaire to solicit information from selected QSR organizations. The questionnaire contained 30 questions derived from previous study results and new questions designed to investigate current organizational changes in the position. Questionnaire content was based on the Muller and Campbell (1995) and Mone and Umbreit (1989) studies. These studies formed the basis for the inquiry of transitional issues faced by managers moving from single-unit to multi-unit management. The questionnaire format included both structured and unstructured questions. There were several open-ended questions designed to obtain information on the position outside of the structured format. The questionnaire design also permitted interviews to take place in person or over the telephone.

A total of 10 QSR organizations participated in the study. They were selected to insure a diversity of company participation based on size, location, public versus private ownership, and chain versus franchise status. Six of the participating companies are national chains and four are regional. Two of the companies are considered franchisees. The participating QSR organizations operate more than 18,000 company and franchise units, and employ approximately 1,600 multi-unit managers. The smallest company operates 37 units and has 4 multi-unit managers. The largest company operates more than 7,000 units and has 1,000 multi-unit managers. Interviews were conducted with one company chairman, two presidents, three vice presidents, three regional vice presidents, and one senior human resource officer. Six of the ten interviews were conducted in person and four were over the telephone.

RESULTS

Position Importance and Span of Control

One of the interesting findings of the study was the change of position title for this level of multi-unit management. According to the executives interviewed, the titles for the position now include area coach, area supervisor, market leader, area or business consultant as well as district manager. The change from the previous managerial titles as district or area manager reflects a change in the role of the position from command and control to a position of influencer, coach, mentor, teacher, supporter, facilitator, and consultant. A president of a large QSR company commented that "today's multi-unit supervisor is a relationship job requiring the individual to work with store managers in helping them solve problems." The position today is that of a leader who helps crew members and managers in the restaurant units grow and succeed. Multi-unit managers now are required to build credibility with diverse employment groups including minorities, women, and with individuals of all ages and work experience backgrounds. The style of leadership is situational in which the managers adjust their style to the individuals to whom they are providing leadership. All of the executives interviewed indicated that the skill sets are different for multi-unit managers compared to 10 years ago. The focus in today's operating environment is leadership, not management.

Executives interviewed during the study uniformly agreed the first-level multi-unit manager is a critical position in their organizations. The multi-unit manager is the closest person to the restaurants with responsibility to set standards and develop a caring atmosphere for all employees. In the opinion of one executive, the multi-unit manager is crucial because he or she is the number one coach and the leverage point to touch each restaurant. Multi-unit managers build an effective team through frequent visits to the restaurants, and demonstrating care and concern for each employee. Multi-unit managers are communication conduits with responsibility for listening to suggestions and passing on information to their superiors. The position is important to all QSR chains because individuals in the role are responsible for the development of managers and employees. Managers and employees are the individuals that interact with the customers. QSR executives interviewed agreed that the role is entirely about the development of people. QSR chain growth cannot take place without the development of managers and staff that have acquired the skills to assume greater responsibilities. This is one of the key roles identified in the study. Multi-unit managers also are considered the experts in the field resolving issues dealing with technology, equipment, customer service, budgets, marketing and the introduction of new products. Multi-unit managers are in the position to help store managers solve problems and improve operating results.

A major aspect of the study was to investigate the current spans of control for multi-unit managers. The current spans of control reported by the ten QSR organizations varied from a low of 6 to a high of 30. Most of the companies had spans ranging from 6 to 8 restaurants. Two franchise organizations reported higher spans since they viewed the position as more of a consultant to franchisees or part owners. The study discovered entirely franchised organizations have higher spans, but the role is more of consultant with responsibility to assist the franchise owners in improving results through suggestions and recommendations. A non-franchise company located on the west coast recently established an associate position responsible for 6 to 8 restaurants with area managers in charge of 14 to 16 units. When asked if the span of control was increasing or decreasing, 7 out of 10 companies reported the span decreasing from previous levels. Two QSR firms have slightly increased their spans while one has remained the same. The seven executives reporting decreased spans indicated that the change took place to improve systems and procedures and enhance performance. Another reason is because the increased complexity of the business required more frequent unit visits to enhance communication. Several executives mentioned that the low unemployment rate necessitated more focus on the restaurant units because the lack of experienced managers required more training, coaching, and help with staff issues. An executive interviewed for the study acknowledged that "too much was left on the table" for the spans to remain at higher levels. The spans were basically reduced to increase sales, lower costs through tighter controls, and improve customer service. Spans of control identified in this study were comparable to those reported in a

1989 study (Umbreit) where 92.5% of the multi-unit managers were responsible for 8 or fewer restaurants.

Recruitment, Selection, and Training

The recruitment, selection, and training of multi-unit managers varied by each organization. The significant growth of new restaurants and the difficult labor market have forced QSR firms to recruit multi-unit managers more and more from other organizations and related businesses. Five of the organizations interviewed reported a hiring mix consisting of 50% within their company and 50% outside.

Two companies hire 80% of their multi-unit managers from within the organization and 20% outside. All of the QSR organizations interviewed have an internal system for identifying potential multi-unit candidates through annual performance reviews and career planning systems. Only three of the organizations require a college education for the position, while the remaining considered it less important. Recruitment of potential multi-unit candidates from inside QSR organizations is based on past performance, which includes possessing technical and tactical knowledge of managing a restaurant. Internal candidates must demonstrate the ability to achieve results such as building sales, controlling costs, and maximizing profit.

While experience in running a restaurant successfully and achieving sales and cost control numbers were important in the selection process, QSR executives were unanimous in their belief that candidates for multi-unit positions must possess certain soft skills. Soft skills are defined as leadership capabilities for developing teams, training, communicating, and motivating others to higher levels of achievement. According to one executive interviewed, candidates must have the genuine desire to see other people succeed. Other personnel qualities for the position include ambitious, self-motivated, self-confident, decision-maker, good planner and organizer, and willingness to serve other people. Some of the companies interviewed have developed a benchmarking process where they profile potential candidates against successful multi-unit managers. External candidates for the position were required to have previous multi-unit experience and were screened carefully for possession of the soft skills. Outside candidates also were tested and went through multiple interviews. Several of the executives participating in the study emphasized the importance of a cultural match; in other words, the understanding of the company's core values and operating procedures when hiring outsiders for their multi-unit positions. Most of the executives mentioned that their company had a unique culture, which required a defined set of skills to function effectively in the organization.

In regard to the training of multi-unit managers, the QSR organizations utilize a number of different approaches. Before they are promoted to a multi-unit position, internal candidates are given opportunities to develop their skills through special training programs designed by the company. Some of the internal candidates are sent to leadership classes outside of the organization. Other internal candidates while still restaurant managers are given special assignments or placed on task forces dealing with human resources, marketing, or purchasing. Outside hires for multi-unit manager positions are placed in company training programs to learn restaurant operations and management procedures.

Several companies attempt to develop the skills of multi-unit manager candidates by assigning them more than one restaurant to operate before the actual promotion. Five of the QSR organizations interviewed have developed specially designed training programs for multi-unit managers. These training programs last from 8 to 26 weeks. The content of these training programs dealt with leadership skills and learning special company systems and operating procedures. Specialized training for multi-unit managers included gaining a better understanding of legal issues, finances, new product introductions, computer applications, and communication techniques. Most of the companies interviewed provided new multi-unit managers with mentors to help them adjust to their changed role and get guidance and direction. Several of the companies utilize on on-the-job training with a reduced number of initial units supervised. Job shadowing was mentioned as an important approach in training multi-unit managers.

Communication, Compensation, and Evaluation. The results revealed valuable information on the multi-unit manager concerning, communication, compensation, turnover, and evaluation. The communication process utilized by QSR chains with their multi-unit managers involved a structure of weekly or monthly meetings together with scheduled quarterly and annual sessions. Companies utilized a variety of communication approaches but usually a meeting with multi-unit managers took place several times a month. These meetings focused on a discussion of operational issues faced by the market area or region. Meeting content included a review of restaurant unit operating performance, human resource challenges such as turnover and retention, and sales and marketing opportunities. Multi-unit managers at these meetings were given an opportunity to share ideas and successes. Eight of the QSR companies held quarterly meetings encompassing a larger group of multi-unit managers. These quarterly meetings were more strategic in content with financial updates provided along with a discussion of new initiatives. Some companies utilized the quarterly meetings for employee recognition purposes. Every executive mentioned the use of annual meetings for training, education, and recognition. Annual conventions held by QSR organizations provide a platform for a review of past performance and the announcement of new strategies with special programs provided to enhance leadership skills and motivate managers.

Multi-unit managers are well compensated for their performance. Executives participating in the study revealed a base compensation for the position ranging from $45,000 to $85,000. Multi-unit managers were eligible for a bonus of $4,000 to $30,000 annually, or from 10% to 20% of base salary. Bonuses were based on profits, sales, cost controls, and the overall performance of the assigned restaurants. Individuals in the position also received a car allowance and some received stock options.

Five of the companies evaluated their multi-unit managers formally on an annual basis and five twice a year. Immediate supervisors usually conducted evaluations of multi-unit managers. On occasion, divisional vice presidents would participate in the process. In two companies multi-unit managers were required to complete self-evaluations. Evaluations contained two parts. One was devoted to a discussion of each restaurant's operating performance involving sales, controls, profitability, and future projections. The evaluation of the restaurants contained a discussion of individual restaurant management staffing issues and employee retention strategies. The second part focused on developmental needs and how multi-unit managers can improve their performance. Some executives indicated an interest in working toward semi-annual evaluations to increase communications, and provide an opportunity for an additional emphasis on personal development. Quarterly reviews were conducted for the purpose of evaluating assigned restaurant profitability and achievement of sales goals.

Executives interviewed during the study were asked to provide information on managerial turnover for the position both voluntary and involuntary. Turnover for multi-unit managers ranged from 5% to 20% annually. Of the 10 companies participating in the research study, 2 stated the turnover was increasing, 2 reported no change and 6 reported a trend of decreasing turnover for multi-unit managers. Executives mentioned the decrease was a result of better job design, selection, training, and field support. About half of the turnover was involuntary for performance-based reasons in which the multi-unit managers were not achieving the numbers or were not fitting into the culture of the new organization. Voluntary reasons given included a preference for a life style change that included less hours and travel. Of the individuals who left voluntarily about 50% stayed in the restaurant industry and 50% left the segment.

According to the interviewed QSR executives, the position of multi-unit manager is a very desirable position in the organization, and is looked upon as a significant promotion from restaurant manager. The unit managers aspire to the position for several reasons. First, individuals promoted from restaurant manager leave day-to-day restaurant operations, and enjoy a much greater degree of independence and freedom of decision-making. Second, individuals in multi-unit management positions receive an attractive benefit package including a company car, stock options, and the potential for larger bonuses. The study found at least 50% or more of multi-unit managers considered their job as a career position. As many as 80% of multi-unit managers in several QSR companies stayed in the position for a substantial period of time. Individuals working for QSR organizations also understand that the multi-unit position can lead to other promotional assignments such as field training manager, franchise consultant or marketing manager.

TRANSITION ISSUES AND CHALLENGES

An additional focus of the investigation was to identify the issues faced by QSR organizations when their managers were promoted to multi-unit level responsibilities. The two major transitional issues identified by the QSR executives surveyed include those involving leadership and organization/time management. Executives viewed the possession of leadership skills as paramount to the success of managing multiple units. New multi-unit managers making the transition have difficulty giving up control. They have problems moving away from the focus on operational issues to those dealing with people. The position now requires multi-unit managers to become supportive individuals who lead by influencing others. A vice president of a large QSR company stated the problem succinctly when he declared, "this job is about diagnostics and writing prescriptions." New multi-unit managers have difficulties stepping back and working through others to accomplish the objectives set by the organization. The most effective multi-unit manager today is a coach, trainer, motivator, and leader. All of the executives felt strongly about the importance of multi-unit managers developing their subordinates for promotion to support new store openings.

Management of multiple units places a premium on the individual to build credibility and trust among the restaurant management staff and employees. Newly appointed multi-unit managers also have problems according to the study, working with various personalities. Multi-unit management requires managers to adjust to the style of subordinates since they are a diverse group with age, ethnic, and cultural differences. Multi-unit management requires situational leadership and the ability to understand the different capabilities of individuals working in restaurants.

The second major transition issue identified in the study was organization/time management. All of the executives interviewed pointed out that this problem occurs during the transition process. Based on executive observations, newly appointed multi-unit managers have difficulty prioritizing their time and organizing their schedules. The problem centered on multi-unit managers learning how to maximize their time in the restaurants during visits and conducting the appropriate follow-ups. New multi-unit managers have a tendency to get engrossed in operational details rather than delegating the responsibility of correcting problems to unit managers. Newly assigned multi-unit managers are overwhelmed initially with the volume of information generated when supervising 8 to 10 restaurants. Managers new to the position need to learn how to handle the volume of communications, while still conducting store visits and completing other assignments.

Transition issues identified in this study were similar to those originally reported by Mone and Umbreit (1989). In that study the issues centered around the areas of human resources and restaurant operations. It appears that the position has not changed during the past ten years in regard to the requisite skills and abilities needed by individuals to succeed in multi-unit management. In the 1980s multi-unit managers were interested in learning leadership skills, while leadership is now the key to the success of managing multiple restaurants. The low unemployment rate and restaurant chain growth has greatly increased the focus on effective multi-store staffing and recruitment. Multi-unit managers today are forced to be better leaders to insure the successful operation of restaurants. Multi-unit managers have been assisted in their management of restaurant units through the use of improved communications systems as a result of technology. For example, multi-unit managers can communicate more quickly with their unit managers using cell phones, e-mail, pagers, and computers.

Out of 10 executives, 6 rated their companies effective in managing the transition process. Four of the companies were working on improving their transition strategies. The reasons for the improved effectiveness consisted of better training programs, selection and identification of highly qualified candidates for multi-unit positions, and field support such as the use of training and marketing specialists. All executives in the study recognized the importance of improving the transition process as a way of achieving operational effectiveness and remaining competitive in the market place.

SUMMARY

The study revealed a number of role changes for multi-unit managers in the quick service restaurants as a result of the segment's growth and expansion during the past ten years. The increased level of competition and the recent labor shortage have also impacted the position. For example, competition forced QSR chains to reduce the spans of control for multi-unit managers so they could spend more time in the units to increase restaurant operational effectiveness and profitability. Multi-unit managers today also have to spend additional time in restaurant units because of the difficulty of finding qualified managers and employees. The managers and employees hired are less skilled than previously and are from more diverse backgrounds. These factors have combined to elevate the status of the position and change its role. A summary of the multi-unit manager role changes appear in Table 1.

Multi-unit managers must now be able to create an effective operating environment for diverse employment groups by developing a supportive atmosphere. The new role of multi-unit manager requires the individual to be a coach, influencer, mentor, and developer of managers and employees. He or she must be a disseminator of information, problem-solver and expert. The role is one of human resource developer who constantly is finding, hiring and training restaurant managers and assistants. The role also requires a multi-unit manager to be an effective communicator using today's modern technology.

In response to the changing role of multi-unit managers, QSR organizations have placed much more emphasis on the position. Corporate selection, development and training programs have been greatly improved in recent years. Multi-unit managerial candidates are hired on the basis of leadership and possession of other soft skills. Compensation packages have been improved. Companies are focusing on improving the transition process from single to multi-unit by enhancing

TABLE 1. Summary of Changing Role of Multi-Unit Managers

Multi-Unit Managers– PAST	Multi-Unit Managers– PRESENT
1. Acts as a manager for several units.	**1.** Functions as a coach, mentor, influencer, and teacher for management teams.
2. Expects to manage a homogeneous group of managers with similar culture, values, expectations, socio-ethnic background.	**2.** Expected to manage diverse groups of managers with differing value systems, cultural and socio-ethnic-demographic backgrounds.
3. Considers him/herself as a manager of managers.	**3.** Expected role includes development of managers (not management of situations).
4. Disseminator of information.	**4.** Sharing of information and exchange of information to resolve problems.
5. Perceives oneself as an expert and tries to solve all unit problems.	**5.** Perceives oneself as a leader of group that solves problems by group thinking.
6. Supervises several restaurants.	**6.** Supervises managers that manage restaurants (fewer in number).
7. Often promoted from within from the ranks of low level supervisors.	**7.** Recruited from outside to foster new ideas and new perspective.
8. Primarily compensated by salary.	**8.** Primarily compensated by performance incentives and people development.
9. Occupies a management position in the organization.	**9.** Occupies a step in career ladder.
10. Management philosophy is more focused on production goals.	**10.** Management philosophy takes more business approach.
11. Primarily responsible for production targets.	**11.** Primarily responsible for financial, marketing and human resource objectives.

leadership development and organizational/time management skills. The multi-unit management position enjoys a great deal of status in quick service firms and is considered a career opportunity.

IMPLICATIONS AND RECOMMENDATIONS

Study results on the changing role of multi-unit managers have significant implications for industry practitioners and educators. The information obtained in this study on the changing role of an important position in the QSR segment is valuable in the development of future executive training programs and hospitality curriculums. In particular, the study pointed out the importance of leadership skills in the management of today's diversified workforce. The preparation of future leaders in this segment or the hospitality industry in general should incorporate training in the soft skills including communication, motivation, team building, human resource development and situational leadership. In addition, future leaders need to learn time management and organizational skills. The use of technology as a communication tool is also beneficial today and in the future.

Researchers can build on the results of this investigation by conducting further research on multi-unit management. Suggestions include surveying a representative sample of multi-unit managers to identify their training needs and requirements. Additionally, it would be of value for researchers to examine the current content of multi-unit management training programs to assess the techniques and tools utilized in developing leadership skills. A more focused study on transitional issues would be valuable in terms of identifying specific problem areas and how these can be overcome through training and development programs.

REFERENCES

Bernstein, Charles (2000), "Industry of 'Choice'? Tough Road Looms," *Weekly Restaurant Connections*, 3 (14), 2.

Brooks, Steve (1996), "Downsizing's Comeuppance," *Restaurant Business*, 95(7), 80-85.

Bradach, Jeffrey L. (1998). *Franchise Organizations*. Boston: Harvard Business School Press.

Doughty, Rob (1996), as quoted in "Downsizing's Comeuppance," *Restaurant Business*, 95 (7), 84.

Gross-Turner, Steven (1999), "The Role of the Multi-Unit Manager in Branded Hospitality Chains," *Human Resource Management Journal*, 9 (4), 39-57.

Hayes, Reed (2000), as quoted in "Industry of 'Choice'? Tough Road Looms," *Weekly Restaurant Connections*, 3 (14), 2.

Lombardi, Dennis (1996), "Trends and Directions in the Chain-Restaurant Industry," *Cornell Hotel & Restaurant Administration Quarterly*, 37 (3), 67-83.

Mills, Susan (1999), National Restaurant Association.
Mone, Mark A., and W. Terry Umbreit (1989), "Making the Transition from Single Unit to Multi-Unit Fast-Service Management: What are the Requisite Skills and Educational Needs?" *Hospitality Education and Research Journal*, 13 (3), 319-331.
Muller, Christopher C., and Douglas F. Campbell (1995), "The Attributes and Attitudes of Multi-Unit Managers in a National Quick-Service Restaurant Firm," *Hospitality Research Journal*, 19 (2) 1-18.
Peacock, M. (1995) "A Job Well Done: Hospitality Managers and Success," *International Journal of Contemporary Hospitality Management*, 7 (2/3), 48-51.
Sheridan, Margaret, and Janice Matsumoto (2000), "A Road Map for 2000," *Restaurants & Institutions*, 110 (1) 66-74.
Technomic Inc. (1999), 1999 U.S. Foodservice Industry Segments.
Umbreit, W. Terry, and Donald I. Smith (1991), "A Study of the Opinions and Practices Of Successful Multi-Unit Fast Service Restaurant Managers," *The Hospitality Research Journal*, 14 (2) 451-458.
Umbreit, W. Terry (1989), "Multi-Unit Management: Managing at a Distance," *The Cornell Hotel and Restaurant Administration Quarterly*, 30 (1), 53-59.
Umbreit, W. Terry, and Warren J. Tomlin (1986), "Identifying and Validating the Job Dimensions and Task Activities of Multi-Unit Food Service Managers," *Proceedings Of the 40th Annual Conference of the Council on Hotel, Restaurant, and Institutional Education*, 66-72.

An Exploratory Study of Determinant Factors for the Success of Chinese Quick Service Restaurant Chains

Sandy Changfeng Chen
John T. Bowen

SUMMARY. The Chinese restaurant segment can be lucrative, but the blending of eastern and western cultures required for American Chinese restaurants produces a unique set of characteristics necessary for success. This study identified ten characteristics of successful Chinese quick service restaurant chains. Large Western restaurant companies that lacked these characteristics have lost millions of dollars in their ventures into the Chinese restaurant segment. This study analyzed successful Chinese quick service restaurant chains. Market challenges and opportunities for Chinese quick service restaurants and implications for both the industry and future research in this area are provided at the end of the study.

KEYWORDS. Chinese restaurants, quick service restaurants, foodservice, Chinese restaurant industry, qualitative research

---

Sandy Changfeng Chen, MS, is a PhD Candidate and John T. Bowen, PhD, is Professor, both at College of Hospitality Management, University of Nevada Las Vegas, Las Vegas, NV.

[Haworth co-indexing entry note]: "An Exploratory Study of Determinant Factors for the Success of Chinese Quick Service Restaurant Chains." Chen, Sandy Changfeng, and John T. Bowen. Co-published simultaneously in *Journal of Restaurant & Foodservice Marketing* (The Haworth Hospitality Press, an imprint of The Haworth Press, Inc.) Vol. 4, No. 4, 2001, pp. 239-262; and: *Quick Service Restaurants, Franchising and Multi-Unit Chain Management* (ed: H. G. Parsa and Francis A. Kwansa) The Haworth Hospitality Press, an imprint of The Haworth Press, Inc., 2001, pp. 239-262. Single or multiple copies of this article are available for a fee from The Haworth Document Delivery Service [1-800-HAWORTH, 9:00 a.m. - 5:00 p.m. (EST). E-mail address: getinfo@haworthpressinc.com].

INTRODUCTION

As the flagship of Asian cuisine, Chinese food has maintained its popularity as one of America's favorite ethnic foods. In 1989, a national survey of 635 adults found the majority of U.S. consumers considered Chinese, Italian (excluding pizza), and Mexican cuisine to be their ethnic favorites cuisines. Seventy-eight percent of the respondents had tried Chinese food in the past. In 1992, Research Advantage surveyed one thousand consumers and found 84% had dined on Italian food, followed by 82% for Chinese food, 75% for Mexican food, 43% for French food, and 42% for Japanese food. Eighty percent of the quick service restaurant (QSR) chains polled by Research Advantage served Chinese food. In the 1999 study, Research Advantage polled restaurant operators and again found Chinese food to be one of the most frequently carried ethnic foods (Nation's Restaurant News, April 26, 1999).

Chinese food is gaining renewed popularity with the surge in popularity of ethnic foods. The National Restaurant Association's 1998 survey found that popular foods included ethnic cuisine, international flavors, and imported ingredients (Nation's Restaurant Association, 1999). There are two ways restaurants provided ethnic items on their menus. One approach is to add ethnic items to the mainstream menus, such as restaurants adding one or two Chinese dishes to their restaurant's menu. For example, a recent study conducted by Strategic Foodservice Solutions of Tampa, Florida revealed that more than two-thirds of all foodservice operations offered at least one ethnic item on their menus (Nation's Restaurant News, April 26, 1999). The other approach is to establish restaurant concepts based on ethnic cuisine. The focus of this study is on the second approach, the Chinese quick service restaurant (QSR) segment.

Despite the fact that Chinese food has maintained its popularity for years, the Chinese restaurant sector has long been an under-developed market, dominated by single-unit independent Chinese restaurants. For decades, Chinese food was served in family-style restaurants, in Chinatowns catering to Chinese immigrants. San Francisco and New York City were the home to the two largest and oldest Chinatowns. In 1970 these two Chinatowns had more than 50% of all the Chinese restaurants in the United States (Lao, 1975).

There are many barriers for growth and development of the Chinese QSR segment. In the 1970s, almost all of the attempts by large companies, such as General Mills and Pillsbury failed when they tried to implement a national Chinese restaurant chain. However, these failures did not discourage other attempts to enter this elusive market. Gradually, successful restaurant chains emerged, such as Panda Express, Manchu Wok, Mark Pi, Leeann Chin, Pick-Up-Stix, Ho-Lee-Chow, Chop-Chop, Magic Dragon, Chin Chin, and P.F. Chang's China Bistro. The majority of these chains are quick service operations. The two leading national chains, Panda Express and Manchu Wok, are quick service chains with a total of 498 restaurants (see Table 1).

The Chinese QSR segment has proved to be a lucrative one for those companies who have developed successful delivery systems. These chains developed and implemented successful formulas for strategically tapping the Chinese QSR chain sector. These formulas are unique and different from those applied in the western restaurant management. However, most of the information on Chinese multi-unit restaurant chains is anecdotal. There is a lack of published research. This study will help fill a void in the literature.

The objectives of this study were to describe the development of Chinese restaurants in the last three decades, document the evolution of Chinese QSR chains, and to provide insights to those factors, which differentiate successful Chinese QSR chains from unsuccessful ones. It is hoped this study will provide restaurateurs or franchisers with useful information and inspire other researchers to investigate this restaurant segment.

HISTORICAL PERSPECTIVES ON CHINESE RESTAURANTS

Chinese restaurants have become an ubiquitous feature of American urban and suburban life (Levenstein, 1993). While the total number of Chinese restaurant units in the United States was only 2,829 in 1971 (Lao, 1975) that number increased to 17,859 units in 1987, which accounted for approximately 5% of total U.S. restaurants (Kochak, 1988). By 1992 the number of Chinese restaurants in 1992 had grown to 30,000, accounting for nearly a third of all "ethnic restaurants", with revenues of $9 billion (Karnow, 1994; Lu & Fine, 1995). Presently, there are 35,769 Chinese restaurant units in the United States (SBS, 1999). This is impressive considering the number of Americans of Chinese ancestry represents less than one percent of the total population (Lu & Fine, 1995).

History of the Development of Chinese Restaurants

The first Chinese restaurant was built in San Francisco in 1849. Most of Chinese restaurants of the nineteenth century were modest affairs, primarily catering to working-class diners, and spread through the western United States along the rail lines (Pillsbury, 1990; Karnow, 1994; and Lu and Fine, 1995). The growth and cultural legitimation of Chinese restaurants began in the 1890s as Chinatowns were transformed from vice districts to tourist attractions (Light, 1974; and Lu and Fine, 1995). In the coming decades, Chinese restaurants gradually expanded out of the Chinatowns. Other driving forces that led to this change were the wave of new Chinese immigrants and the growth of the second generation of old Chinese immigrants. Unlike many older immigrants who lived in an urban Chinatown, the new Chinese immigrants and the new generation did not form a closed ethnic community. They did not rely on mutual help and support from other Chinese for their survival and success. Thus, social ties between modern Chinese immigrants were looser than those found in a traditional Chinatown (Lu and Fine, 1995). To reach higher economic goals, as well as, avoid competition with traditional restaurants located in Chinatowns, the new immigrants and new generation built restaurants outside of Chinatowns. These restaurants were forced to adapt their food to suit American tastes. By the late 1970s, Chinese restaurants became a familiar sight throughout America, even expanding into small towns and rural areas as the food quickly became popular with Americans (Epstein, 1993; and Lu and Fine, 1995).

Typology of Chinese Restaurants

Exploring the Chinese food and restaurant market in the last three decades, this study traced great breakthroughs and some significant changes. These changes symbolize a new era and the revolution of the Chinese food and restaurant industry in the United States. Chinese restaurants finally expand beyond the "wall" of Chinatowns and transform from "Mom-and-Pop" businesses into multi-unit operations. Chinese food has become a popular alternative product for all restaurateurs rather than the ethnic product of Chinese restaurateurs only.

With these changes, three types of Chinese restaurants evolved; restaurants in Chinatowns, restaurants outside of Chinatowns, and multi-unit operations. The restaurants in Chinatowns have authentic cuisine, employees are usually members of the same family, and the restaurants are usually single-unit independent operations. This is the first type of Chinese restaurant in America and is still popular with Chinese-Americans and tourists.

The second type of Chinese restaurants, single-unit independent ones outside of Chinatown, provide customers with both classical and regional Chinese foods that mainly caters to American customers. This type of Chinese restaurant started to grow in the 1890s and still serves meals family-style.

In the late 1970s, the third type of Chinese restaurant, multi-unit restaurant operations appeared. Some of these noticeable establishments include Canada-based Manchu Wok, South Pasadena, California-based Panda Express, Minneapolis-based Leeann Chin, Columbus, Ohio-based Mark Pi, Phoenix Arizona-based P.F. Chang's China Bistro, California-based Pick-Up-Stix, Seattle-based Magic Dragon, and home-delivery concepts Ho-Lee-Chow and Chop-Chop. Figure 1 provides a summary of the characteristics of these three different types of Chinese restaurants.

TABLE 1. Seven Well-Recognized Chinese Restaurant Chains in the USA

Year*	Company	Headquarter	Ave. Annual Revenue *	Business Method	Current Services	Total Units by Feb. of 2000	Operating Environments// Demographics
1972	Mark Pi's Express	Columbus, Ohio	N/A	Franchising	Quick service	50	Locations which have both high traffic and potential for quick turnover business //in 6 states.
1980	Leeann Chin, Inc.	Bloomington, Minnesota	65	Privately owned	Buffet, quick service, catering	57	Grocery stores, food courts, other high traffic area//Minneapolis/St. Paul Metropolitan region, Michigan and the Kansas City area.
1980	Manchu Wok	Markham, Ontario, Canada	N/A	Franchising	Quick service	194 (both U.S. & Canada)	Food courts in shopping malls, airports, office buildings and universities .
1983	Panda Management Company, Inc., (including Panda Express, Panda-Panda, & Panda Inn)	South Pasadena, California	300	Privately owned	Quick service, casual dining and fine dining	370 (304 for Panda Express, the quick service concept in 34 states)	For Panda Express: shopping mall food courts, leading supermarkets and retail chains, neighborhood shopping centers and Key Intersections, university and college campuses, airports, Casinos, and Sports Arenas// in 35 states.
1983	Chin Chin International	Santa Monica, California	N/A	Privately owned	Café	8	High traffic areas/places//in 2 states.
1989	Pick-Up-Stix, Inc. (Pick-Up-Stix & STIX)	California	N/A	Privately owned	Quick service, casual dining	47	Affluent areas with median household income of $60,000 and population of 50,000 within three miles//South California and San Diego.
1993	P.F. Chang's China Bistro	Scottsdale, Arizona	138.5	Franchising	Full-service	39	In front of malls, near theaters, in area with household incomes north of $70,000//in 18 states.

*Note:

- Year: the year in which the first unit was established
- Revenue: in US$ and million (company estimates)

Source: All the information was obtained from the companies web sites, fact sheets, and different media reports.

EVOLUTION OF CHINESE RESTAURANT CHAINS

Difficulties in the Development of Chinese Restaurant Chains in the 1970s

One of the enigmas to the Chinese restaurant sector was that even though thousands of Chinese restaurants had been built throughout the country, as late as 1980, there were no national Chinese restaurant chains. Independent restaurants, well known as family businesses, continued to dominate the Chinese restaurant sector. One of the few research studies focussing on Chinese restaurants

FIGURE 1. A Typology of Chinese Restaurants: A Summary of the Characteristics of Three Different Types of Chinese Restaurants in the Current Chinese Restaurant Segment

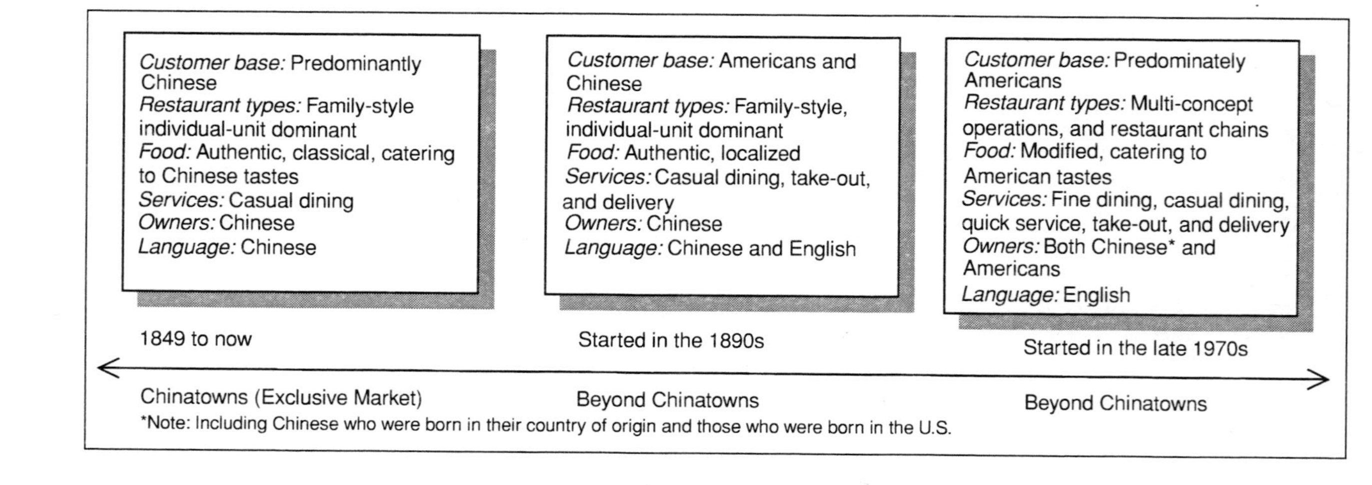

investigated the reasons why family restaurants dominated this segment. Lao (1975) cites the following barriers for the development of restaurant chains:

1. *Cultural. Most owners were not American born and had saved their money to open their own businesses. The majority were uneducated and distrusted "corporate methods of management," and were also apprehensive about their ability to handle additional problems and details entailed with expansion and incorporation.*
2. *Financial. Since most Chinese restaurants were typically family operations and had been for some time, they were slow to develop, and ceased to grow when some optimum size was reached which supported the then present family.*
3. *Psychological. The American public had been "educated" to associate Chinese restaurants with extensive menus. The concept of a restaurant chain had traditionally had a limited menu and high turnover, but the Chinese restaurant image, with quite the opposite tradition, was at odds with this concept of modern restaurant operation and management.*
4. *Social. For Chinese restaurants, expansion required a supply of key personnel. Unfortunately, in recent years there had been a scarcity of competent Chinese Chefs. Considering the normally extensive menu in a Chinese restaurant, plus the fact that every dish was cooked to order, an experienced chef was a necessity if the operation was to run smoothly. As absent chef spelled disaster. This was quite contrary to Western-style operations like steak houses and family-style restaurants where cooking skills could be taught to non-skilled workers. These operations normally utilized food items high on the raw-to-ready scale as a rule.*

The Driving Forces and Marketing Appeals for the Emergence of Chinese QSR Chains in the 1980s and 1990s

The literature review revealed four main reasons as the driving forces for the emergence of Chinese restaurant chains in the 1980s and 1990s. The first three reasons made Chinese food more popular than ever before, which included: (1) a changing demographic makeup of the customer population–more immigrants from different countries came to the United States, and the American population was more diverse than before; (2) a changing customer profile. People started to pay more attention to their diets and perceived Chinese food as low fat and healthy; and (3) the more advanced transportation and communication, which made it much easier to obtain ingredients needed for cooking Chinese food. The last reason, which made the Chinese restaurant market a great appeal for the QSR chain sector, was the intense competition at the QSR chain segment.

As large restaurant chains expanded, many restaurant market segments entered the mature stage of the industry life cycle (Parsa and Khan, 1993; and West, 1988). For example, the QSR industry was characterized as having very intense competition and it became increasingly difficult to find good locations for new QSR units. While large restaurant chains were taking advantage of their high brand-awareness to diffuse into different market segments such as institutional foodservice, contract feeding, and international markets, some restaurant practitioners found a domestic market niche, the ethnic food segment (Parsa and Khan, 1993).

The Chinese restaurant sector, with a surging demand and few multi-unit operations, was regarded by many restaurateurs as "the last large untapped chain frontier" (Brumback, 1995). It became an attraction to ambitious restaurateurs. In the late 1970s, a lot of small local chains emerged. Fast food concepts, including dine-in, take-out, and delivery, became hot start-ups for these small chains due to the low financial risk and the established chain management systems for the fast-food industry. Burger King was one of the first American food service companies to enter the market when it bought a two-unit Texas-based Chinese fast food concept called Quick-Wok. Burger King had visions of making Quick-Wok into a national chain. However, the vision did not become a reality. Burger King sold the chain to Pillsbury Company, which developed it into a thirteen-unit regional chain. But the regional chain was not profitable. Another American company, General Mills, bought Leeann Chin, a casual-dining Chinese concept in 1985, but had to sell it back to its original Chinese founder after three years (Kochak, 1988).

The entry of western restaurant giants, such as Burger King, General Mills, and Pillsbury into the Chinese QSR sector defined a different tone for this market. Also, it drew great attention to this niche market from different media. These American chains failed in their attempts to build a Chinese restaurant chain. As Kochak (1989) stated, " . . . there are a few regional Chinese restaurant chains, but all attempts to create a national Oriental-concept chain have failed." Although most attempts failed, many restaurateurs were not disenchanted with the segment. The 1980s and 1990s witnessed the emergence of some national Chinese restaurant chains, whose rapid growth and success forever changed the landscape of Chinese restaurant market.

Manchu Wok, Canada's largest Chinese fast food restaurant chain, crossed the border and built 245 units in 34 states from 1980 to 1990. Manchu Wok became the first nationwide Chinese restaurant chain in the United States. The company's public relations department kept a very low profile. Manchu Wok did not receive much press coverage and the name of the chain received little recognition outside of the trading area of its restaurants. In the early 1990s, another Chinese fast food chain, Panda Express, was established, and expanded at an average annual rate of 25% (*Restaurant USA*, November, 1992). Panda Express quickly exceeded Manchu Wok and became the largest Chinese QSR chain in the United States. Meanwhile, two former Domino's executives, Senior VP Robert Cotman and Thomas Burnham, established two Chinese food take-out and home-delivery concepts: Chop Chop and Ho-Lee-Chow. The casual dining segment continued to grow with Mark Pi and Leeann Chin as the main players, who also expanded into the quick service sector. The P.F. Chang's China Bistro quickly dominated the middle-upscale segment. This Phoenix-based chain, was established in 1993 with average annual sales approximately $4 million per unit (Carlino, 1996). Table 1 provides some facts and figures about the seven well-recognized Chinese restaurant chains in the current market.

Major Barriers to Chinese Restaurant Chains

The proliferation of Chinese restaurants, and the rapid growth of nationwide and regional chains in the last three decades, did not result in a profit for all players. Some restaurateurs started to have a "taste of success," while other restaurateurs continued to experience the "bitterness of failure." American food service companies' entrance into the Chinese restaurant market often resulted in failure. For example, General Mills' China Coast went out of business in 1995. The five-year effort which resulted in 52 restaurants did not provide a successful outcome. China Coast was the second attempt by General Mills (the parent of Darden Corp.) to break into the Chinese food industry. The closure of the nationwide chain resulted in a $20 million loss during fiscal 1995 (Carlino, 1995). Another American attempt was Mr. Ching, a Chinese fast food and home-delivery chain, established in the late 1980s in Atlanta, Mr. Ching failed after five years. Nevertheless, the two founders, Mr. Ed Sear and Mr. Frank Lane, insisted that there was a $1.2 billion market for the home-delivery of Chinese food in the U.S. (Hayes, 1991).

The failures by successful American companies made restaurateurs realize the Chinese restaurant segment was unique. Techniques that made other chains successful would not work in the Chinese segment.

Major Barriers to Chinese QSR Chains

By examining the successful Chinese QSR chain cases, which consist of the majority of Chinese restaurant chains, this study found that internal management factors rather than external-marketing forces were the major causes of the failures. The problems due to internal factors were related to cultural differences, a lack of familiarity with Chinese food, the inability of western managers to manage Chinese chefs, and the inefficient delivery of Chinese food to American customers.

Many of the western restaurateurs aggressively transferred the western fast food preparation and delivery systems to Chinese food but ignored cultural differences. Some major barriers that challenge Chinese QSRs are: (1) culinary authenticity; (2) consistency of food quality; (3) menu variety; (4) the preparation and cooking systems; (5) cost control; (6) employee management; and (7) competition from independent restaurants.

These barriers created problems not only for non-Chinese owners but also for Chinese owners. Although "authenticity" is a perception, "all diners, whatever their culture, wanted food that was suitable to their tastes" (Lu & Fine, 1995). For many diners, the perception of "authentic food" is largely based on the presence of a Chinese chef. It was unlikely that a Chinese chef would be found for each unit of a restaurant chain. The consistency of food quality largely depended on the cooking skill of the chefs. Cooking Chinese food required technique and skill. It is an art rather than a science. It is claimed to be "heart smart" or "the art of eyeballs" is needed to measure the amount of different condiments needed for a dish. A good chef usually needs a long period of professional training and practice. The food is regarded as an extension of a chef's personality. Even with standardized recipes, different chefs created different tastes. This makes it extremely difficult for a restaurant chain to control the food quality and consistency when it expanded from one or two units to more units.

Other factors that affect the consistency of food quality included unique characteristics of Chinese food. Chinese food has long been famous for being fresh and healthy. The freshness and flavor of some hot dishes could be lost in a very short time after food is cooked. This is a big challenge, especially for fast food operations where the nature of steam-table service and a quick-service format makes the delivery of truly traditional Chinese cuisine hard to accomplish. Without authenticity and a consistent quality of food, a Chinese restaurant chain lacks the necessary advantage to compete with independent restaurants that were well known for quality food, low prices, and originality.

Menu variety is another problem. Chinese cuisine is famous for its extensive menu. Usually, in a full-service independent Chinese restaurant at least sixty food items are found on the menu. However, a multi-unit concept requires a limited menu. This made the menu selection with the right food items particularly important. For example, Mr. Ching's menu was changed three times during five years. Menu items first were increased to 64 items from 40, but finally shrunk to 9 items!

The amount of manual work needed in a Chinese fast food kitchen is approximately five or six times that of a Wendy's or a McDonald's (Kochak,

1989). For example, 70% of the raw materials are vegetables that need to be prepared fresh daily. In Kochak's (1989) report, a QSR owner stated, "two or three workers would cut vegetables everyday and do nothing else." The complex cooking process of Chinese food requires labor-intensive work. This results in high labor costs. Family-style independent Chinese restaurants are able to control labor costs because of the participation of family members, who were paid little or not paid at all.

To Chinese chefs, kitchen management is an art rather than a science. Chinese chefs are very temperamental but they are essential to the success of independent Chinese restaurants. The employment of Chinese servers is another management controversy discussed by many chain restaurant owners. Americans have long perceived that a Chinese worker should serve Chinese food in an Asian ambience. Many Chinese restaurant owners felt that Chinese-descent employees were necessary for successful operations. They argued that non-Chinese employees need extensive training to be familiar with Chinese food and do not understand the Chinese service culture. Chinese employees also reinforce the customers' perception of "the authenticity of the food." However, some restaurateurs believed that western restaurant skills could be applied and "standards" could be established for a Chinese restaurant chain. They believed Chinese cooks and servers were not necessarily important components of a successful Chinese restaurant chain.

METHODOLOGY

The most important goal of this research was to obtain insights and an understanding of the unique characteristics of management expertise applied in those successful Chinese restaurant chains. These characteristics were different from those applied in the western restaurant management and service and were the determinant factors that led to the success of those Chinese restaurant chains. In order to achieve this goal, this study examined the Chinese QSR chain sector. The main reason for observing the QSR chain sector is that the majority of the existing Chinese restaurant chains are in the quick service business. Currently, there are only two well-recognized regional restaurant chains, P.F. Chang's Bistro and Chin Chin International, in the middle-upscale sector, and two regional chains, Mark Pi and Leeann Chin, in the casual dining sector. The QSR chains include Panda Express, Manchu Wok, Pick-up Stix, Magic Dragon, Ho-Lee-Chou, Chop Chop, and many small local chains. Two full-service operators, Mark Pi and Leeann Chin, also entered the quick service segment. The largest national Chinese chains, Panda Express and Manchu Wok, both are in the quick service sector.

A qualitative research methodology was used for this exploratory study. There were two reasons for this decision: a limited number of Chinese QSR chains and the lack of previous research and study. The major participants in this study were executives of the leading Chinese QSR chains. Two qualitative research methods, in-depth interview and mail panel, were used in the study.

The data collection included three major steps. First, secondary data were collected from newspapers, trade journals, the Internet, and restaurant companies. Concurrently with this procedure, in-depth interviews were conducted with six single-unit restaurant managers, four of whom were from the Las Vegas, Nevada area. These managers provided insights into what makes a Chinese restaurant successful. Two former managers of China Coast, a failed chain, were interviewed via the telephone.

The first interview was completely unstructured. The interviewer asked some questions obtained through partial literature review, participant observations, and some informal conversations. After gaining insight about the major concerns of a restaurant owner/manager, a semi-structured list of questions was developed and used to guide the rest of the interviews. Changes and adjustments were made to this list as more interviews progressed. All of the interviews with the six single-unit managers were tape-recorded with the permission of the managers. The conversations were transcribed, categorized, and analyzed.

As a result of these efforts, a structured questionnaire was developed. This questionnaire contained 59 variables both suggested by these managers and obtained from the literature review that may greatly contribute to the success or failure of a Chinese QSR chain. Also, it included nine open-ended questions to obtain further participant information and eleven demographic questions.

The second step involved contacting two founders and four executives from five well-recognized Chinese QSR chains. Three of them accepted telephone interviews, while the other three were willing to answer a mail questionnaire. The interviews with the two founders were conducted via telephone, with each interview lasting approximately one hour. Notes were taken during the interviews and transcribed right away thereafter. One face-to-face interview with a chain general manager was conducted in one of the company's restaurants located in Las Vegas. All the interviews were guided by the questionnaire developed at the first step. The other three answered the mail survey.

In order to establish positive interaction between the interviewer and respondents, the interviewer followed guidelines prescribed by Merriam (1998): the interviewer was respectful, non-judgmental, and non-threatening during interviews; the conversations were started by greetings and talks about the participants' company facts and figures and gradually led to the pre-designed questions which the interviewer was seeking answers. The interviewees were very cooperative and enthusiastic about sharing their expertise, thoughts, feelings, and opinions. The responses via first-class mail were received on time, while at the same time the interviewer and informants were communicating via e-mails. After all the interviews and mail surveys were completed, the participants' opinions were coded and listed. Seven categories with a total of 65 variables identified. The seven categories were: (1) the most important attributes in determining the success of a Chinese quick service restaurant chain in terms of operations and management; (2) the most serious challenges facing those QSR chains; (3) strategies used by the investigated companies to keep them competitive and successful; (4) trends in the future Chinese restaurant market in the United States; (5) reasons that make the growth of Chinese restaurant chains a threat to individual Chinese restaurants; (6) the top competitors for a Chinese QSR chain; and (7) personality traits that have made the participants successful.

The last step was the mail panel. The list of seven categories with 65 attributes and a thank-you letter were mailed back to the same participants, the two founders and four executives who participated in the second step interview and survey, to look for confirmation and agreement. They were asked either to confirm or disconfirm the importance of the factors identified by at least one of them at the second step. This method was used by Hambrick (1983). At the end of each category, extra space was given to encourage the panel members to add any comments. After all the responses were received, the results were displayed on a single spreadsheet. Comparisons were made to look for agreements among these panel members.

RESULTS

Ten of the 26 most important factors in the first category were agreed upon by all the six of the participants (see Table 2). These attributes include a good concept, consistency of food quality, standardized preparation and cooking processes, employee training, cost control, western management expertise, marketing, store brand awareness, customer satisfaction, and location.

A Good Concept

"Chinese food" is what those chains sell to customers. However, Chinese food includes such a broad array of items that it is easy for many restaurateurs to lose their focus. A good concept is considered one of the most important factors in establishing a successful business. As stated by one panel member, "A good business always has a strong concept. The owner knows what he is sell-

TABLE 2. Occurrences of Agreement Based on Six Panel Members: The Most Important Attributes in Determining the Success of a Chinese QSR Chain

Attributes	Occurrences of Agreement Based on Six Panel Members
Consistency of food quality	6
Western management expertise	6
Cost control	6
Standardized preparation and cooking processes	6
Marketing (promotion, advertising, research, intelligence, WOM)	6
Store brand name awareness	6
Customer satisfaction	6
A good concept	6
Standardized training programs	6
Location	6
Development of new food items	5
Fixed menu	5
Fresh ingredients	5
Sanitation	5
Chinese food with American touch	5
Good service	5
A commissary of sauces, vegetables, and meat	4
Good relationship with community	4
Sufficient capital	3
Healthful food	2
Exhibition kitchen	2
Trustworthy employees	2
Extra value	2
Early entry of the market	1
Ambience blending Chinese and American style	1
Franchising strategy	0

ing and sticks to that concept." In another panel member's words, "A good concept is a good brand, a brand that has a strong value statement and distinct features."

Consistency of Food Quality

Consistency of food quality is called the "soul" of a Chinese restaurant chain. Chinese food is famous for its rich tastes, but cooking Chinese food is a complex process. Too many factors can affect the quality of a dish. An owner explains, "I'll give the same recipe to ten cooks, and ten cooks can make ten different dishes using the same ingredients." Keeping the consistency of food quality is a tough task for all the QSR chains, but is more difficult for Chinese restaurants because of the sauces. Rather than leave sauces up to cooks at each restaurant, some chains use a central commissary with trained chefs overseeing the production of their sauces. The sauces are then transported to the individual restaurants. Conversely, some chains use unit managers to vigorously supervise the daily making of sauces. Employee training systems are very important in this type of restaurant. All the chains have printed formal workbooks teaching cooks the procedures of product handling. In addition, all the cooks undergo training before they formally start their work in a restaurant kitchen.

Standardized Food Preparation and Cooking Processes

Another method utilized by all the investigated chains is the standardized food preparation and cooking processes. The preparation process and cooking methods in a Chinese restaurant are time-consuming and labor-intensive. The standardization of the preparation processes and cooking methods not only helps keep the consistency of food quality, but also lessens the need for Chinese Chefs at each of the restaurants. Anyone who has cooking experience can be trained to cook "Chinese food." Another way to utilize local cooks is the Americanization of the Chinese cooking method. Although Chinese wok cooking is the major cooking method practiced by the chains, some restaurants utilize American cooking methods to prepare some ingredients. For example, one chain cooks the beef strips for their beef and broccoli dish on a broiler. The strips are added to steamed broccoli and a sauce from the commissary to create the final product. Another chain uses a grill in a majority of locations to prepare Teriyaki chicken. This makes it easier to train and use local cooks.

Cost Control

The standardization of food preparation processes and cooking methods also makes efficient food cost control possible and shortens the time needed to prepare and cook a dish so as to reduce labor work and cost. Individual restaurants maximize their profits by the participation of family members who are usually unpaid or paid very little. These family members are usually skilled in Chinese food preparation methods. For QSR chains, the recruitment of low-wage labor, standardization of the food preparation and cooking processes, and a commissary production of some food products, are critical factors that help reduce labor cost. The respondents also mentioned good purchasing techniques as a way to control the food cost. Other methods common to good restaurant management were also mentioned; inventory control, employee theft control, food waste control, and dish portion size control. The portion size is critical as Chinese dishes are often plated from a line. There are not pre-set portions common to other types of QSRs.

Training Programs

All of the chains have standardized training programs designed to provide quick and comprehensive training for chefs, servers, and unit managers. These programs may run from a couple of days to several months. Workbooks with picture illustrations are used. Training programs are analogous to the "software" of a business. They enable and reinforce the standards and rules. Tests or quizzes are followed to evaluate the progresses of both trainees and old employees.

Western Management and Eastern Expertise

The appropriate blend of western management and eastern expertise is another critical determinant for the success of Chinese QSRs. All of the chains have leaders who have extensive experience with restaurant management. Some of them used to work with large western restaurant chains, such as Red Lobster, Domino's, Wendy's, and Chi-Chi's. These leaders are the driving forces for the growth of their Chinese chains. They are both visionary and passionate about applying their rich management and marketing experience to the Chinese restaurant sector. Meanwhile, successful operations almost always have a Chinese personality masterminding the enterprise. This person commonly held the position of chief executive and was responsible for food quality. These restaurants apply western management systems, while ensuring food quality, by having a Chinese chef manage food production. Another perfect blending is the use of American employees, who are well trained about Chinese food and have no language barrier, to serve Chinese food to American customers. This creates a rather different experience to customers. Tastes of the food are adapted to satisfy the desires of the American market. Menus containing Chinese food items that have been well recognized and accepted by the American people for decades are used because customers are already familiar with them. For example, some favorite food items are Beef and Broccoli, Orange Flavored Chicken, Chicken with Mushrooms, Mongolia Beef, Black Pepper Chicken, and Sweet and Sour Chicken.

Marketing

Traditional Chinese restaurants seldom believe in advertising. In contrast, the successful QSR chains are actively engaged with not only local media, but also regional and national media. These factors practice sophisticated marketing plans. Their marketing intelligence makes the chains flexible and can quickly respond to fluctuating market trends.

In addition to the traditional promotion and advertising methods such as the yellow pages, local tourist brochures, coupons, newspapers, radio and TV, innovative distributing channels are utilized by these chains. All the chains have a nice presence on the World Wide Web. Some chains use exhibition kitchens as promotional tools. One participant says, "an exhibition kitchen can give customers a sense of the freshness and healthiness of Chinese food." For example, Panda Express has bins of fresh vegetables stored where they can be viewed by the customer. Another participant feels an exhibition kitchen can gain customers' trust by visualizing the whole preparation and cooking process. To be effective, exhibition cooking requires a lot of extra work to keep the workstation clean and neat because cooking Chinese food is usually a "messy process."

Other marketing practices valued by these chains include good relationships with local communities, word-of-mouth (WOM), and marketing intelligence. All the chains are involved in local charities and are active sponsors for public activities. Discount or complimentary coupons are distributed around the neighborhood whenever there is a new unit open. Marketing information is collected on a daily basis by marketing and sales department personnel. Marketing research such as focus groups and customer satisfaction surveys are conducted when necessary.

Customer Satisfaction

Customer satisfaction is regarded as the highest mission by the chains. One manager put it this way, "No customers, no business. All our efforts are to create customer satisfaction," says a participant. In addition to quality food, training and procurement assure customer service in these chains programs. Also, attention is paid to their customers' likes and dislikes. Based on customers' preferences, some menu items are frequently modified, and weekly or monthly specialties are tested to obtain customers' feedback.

Brand Name Awareness

All the chains have established wide brand name awareness either on a regional basis or on a national basis. The majority of the QSR chains were born out of their original single-unit full-service restaurant that was well recognized in the local area. The full service concept was then adapted to make it suitable for a quick service operation. As quoted from a participant, "Not only was the new concept's kitchen modified for a quick service environment, but also modified were the concept's cooking styles, preparation techniques, core entrée recipes and service approach; many of the new concept's management team members came out of the old concept." Thus, the effect of brand awareness and the experience learned from the old concept are considered very important success factors.

Location

Location is an old topic but with new content. Chinese QSR chains are entering various non-traditional venues. These venues include shopping mall food courts, leading supermarkets, retail chains, neighborhood centers, key intersections, university and college campuses, airports, casinos, and sports arenas. Panda Express has been very successful locating units within a supermarket. Chinese food is often purchased as a take out item. By locating in supermarkets the restaurants serve as a convenient way for shoppers to provide a meal for their families. Panda Express has also developed units to fit food courts at shopping malls.

Other Factors

Other variables were partially agreed upon by panel members, reflecting different degree of importance of these variables to the panel members' businesses. The last variable, franchising strategy, did not obtain any votes from the panel members. Although two of the companies in the study are franchising, the panel members did not consider that franchising strategy greatly contributed to their business success. Franchising can contribute to brand awareness by expanding the name of the restaurant chain.

OTHER ISSUES RELEVANT TO CHINESE QSRs

In addition to finding out what the restaurant executives felt were the most important attributes in determining success of a Chinese QSR, the surveys and interviews also asked them about other issues that affected the success of a restaurant. The responses to these questions are summarized below.

Challenges Facing Chinese QSR Chains

The panel members agreed on that employee recruitment is the most serious challenge facing Chinese QSR chains (Table 3). Employee retention is another serious challenge agreed upon by five panel members. The employee recruitment has two folds: the recruitment of experienced service employees and

TABLE 3. Occurrences of Agreement Based on Six Panel Members: The Most Serious Challenges Facing the Investigated QSR Chains

Attributes	Occurrences of Agreement Based on Six Panel Members
Employee recruitment	6
Employee retention	5
Securing great sites	4
Rising wage rates	3
Labor cost control	2
Competition from other quick service restaurants	2

qualified management level personnel. The latter is more serious because of its irreplaceable nature and the recruitment and retention of talented people will be critical for a QSR chain to keep competitive in the future market. Other factors mentioned by at least one panel member include securing great sites, labor cost control, rising wage rates, and competition from other quick service restaurants.

Trends in the Chinese QSR Sector

The Chinese market is growing and its potential is big. However, it will become more competitive as more successful multi-unit operations develop (see Table 4). Additionally, Americans' general knowledge about Chinese food is limits and it has been influenced by the "Mom-and-Pop" style of Chinese restaurants. Four members of the panel agree that the successful presence of their chains has shown the American people a new image of Chinese food and restaurants. These restaurants are thus differentiating themselves from individual "Mom-and-Pop" style restaurants.

Strategies for a QSR Chain to Keep Competitive and Successful

Despite some challenges, the future for Chinese QSR chains is optimistic. The new services planned by some chains include adding delivery and drive-through services. However, to keep their companies competitive and successful in the future, the panel members agreed on that the following strategies are very important (Table 5): establishment of store brand name recognition, continuous development of new food items, and consistency of food quality. Other strategies partially agreed upon by panel members include a better Chinese restaurant management system, a right position of the company, and extra value. Extra comments given by some panel members included that improving people's knowledge of Chinese food should be a

TABLE 4. Occurrences of Agreement Based on Six Panel Members: Trends in the Chinese QSR Sector

Attributes	Occurrences of Agreement Based on Six Panel Members
The market is growing	6
The potential is big	6
It is competitive	6
The image is changing from Mom-and-Pop stereotype to professional and upscale businesses	4
More education about Chinese food is needed for the public	3

TABLE 5. Occurrences of Agreement Based on Six Panel Members: Strategies for a QSR Chain to Keep Competitive and Successful

Attributes	Occurrences of Agreement Based on Six Panel Members
Store brand recognition	6
Continuous development of new food items	6
Consistency of food quality	6
A better restaurant management system	4
A right position of the company	3
Extra value	2

long-term goal for all the Chinese restaurateurs and that successful employee training and retention would be the guarantee of a successful business.

Top Competitors for a QSR Chain

The panel members regarded all the other QSRs in the same area as top competitors. Four of them agreed that all the restaurants in the same area are top competitors, while only one considered full-service Chinese restaurants in the area as top competitors. One panel member commented, restaurants in your competitive set that have good business should be studied to see what lesson you can learn from them (see Table 6).

Advantages of a Chinese Restaurant Chain

In the interviews and surveys, the respondents did not consider single-unit independent full-service Chinese restaurants as competitors of chain restaurants. The panel members in the third step confirmed this view. As stated in the last paragraph, only one panel members thought that full-service restaurants in the same area as top competitors. This was an unexpected finding. Perhaps the

TABLE 6. Occurrences of Agreement Based on Six Panel Members: Top Competitors for a QSR Chain

Attributes	Occurrences of Agreement Based on Six Panel Members
All the fast/QSRs in an area	6
All the Chinese fast food/QSRs in an area	6
All the restaurants in an area	4
Full-service Chinese restaurants in the area	1

respondents felt the advantages of a chain removed them from competition with single unit restaurants. These advantages include: a chain has more buying power for media and wider brand recognition, a chain is more professional, its food and service are more consistent, and a chain has more buying power for products (see Table 7).

The Personality Traits That Have Made the Participants Successful

The study also attempted to investigate the personality traits shared by the six panel members that led to their successful restaurant career. These attributes included passion, perseverance, hard work, vision, self-motivation, ability to inspire employees, ability to develop people, confidence, good plans, be sensitive to marketing information, sufficient knowledge about Chinese food, and sufficient knowledge about American culture. These attributes were agreed upon by all the panel members.

IMPLICATIONS

This study has implications for both the industry and academic research. The attributes listed in Tables 2 through 8 can serve as useful checklists for restaurateurs who are intent to capitalizing on Chinese or ethnic food segment.

Implications for Practitioners

The study identified ten determinant factors for the success of the leading Chinese QSR chains, the major issues they face and personality traits shared by all the panel members. Although this study is based on a small sample, the respondents represent the leading Chinese companies QSRS. These restaurateurs have been successful in a market that has proved to be very difficult for chain operators. Thus, those thinking of entering the market should heed their advice.

The entire QSR chain segment has a brilliant future. Crecca (1997) predicts QSR chains in the United States will reach $116 billion in annual sales. This

TABLE 7. Occurrences of Agreement Based on Six Panel Members: Advantages of a Chinese Restaurant Chain Over Individual-Unit Restaurants

Attributes	Occurrences of Agreement Based on Six Panel Members
A chain has more buying power for media	6
A chain has wider brand recognition	6
A chain is more professional, and its service and food are more trustworthy	5
The food in a chain restaurant is more consistent	5
A chain has more buying power for products	3

TABLE 8. Occurrences of Agreement Based on Six Panel Members: The Personality Traits That Have Made the Participants Successful

Attributes	Occurrences of Agreement Based on Six Panel Members
Passion	6
Perseverance	6
Hard work	6
Vision	6
Self-motivation	6
Ability to inspire employees	6
Ability to develop people	6
Confidence	6
Good plans	6
Be sensitive to marketing information	6
Sufficient knowledge about Chinese food	6
Sufficient knowledge about American culture	6

will be half of the total consumer food expenditures. Chinese QSR chains are positioned to gain a healthy share of this market. As the Chinese QSR market matures it is likely to be dominated by large companies such as Panda Express, Manchu Wok, Mark Pi, and Leeann Chin.

Implications for Research

There is little research on Chinese QSRs. It is hoped that this study will inspire research on this segment. Researchers can study success factors in more detail. The Chinese full service segment can be studied to see if the ten success factors discovered in this study hold in the full service segment. Additionally, there are other opportunities that were not covered by this study, that require research. The first opportunity is the breakfast segment; the second is the kids' segment; and the third is other Asian cuisine segment. Often touted as the "most important meal of the day," breakfast is becoming more important for quick-service restaurant operators who want to tap into the eat-on-the-run market and combat coffee specialty competitors (Strauss, 1999). Another void is the kids' menu. The literature review reveals that only a few Chinese QSR chains provide a kids' menu. Can McDonald's and Burger King's success in targeting kids be replicated in the Chinese sector? Another opportunity is to add some popular Asian food items to the current Chinese menu. The Chinese segment is emerging as an important multi-unit segment. It is still in the growth stage and provides many opportunities for researchers.

CONCLUSIONS

This study is one of the first to provide a comprehensive description of the evolution Chinese QSR chains in the United States. The study systematically investigates and identifies success factors for Chinese QSR chains. Ten factors are identified as the most important ones contributing to the success of five well-recognized Chinese QSR chains, which include a good concept, consistency of food quality, standardized preparation and cooking processes, employee training, cost control, western management expertise, marketing, store brand awareness, customer satisfaction, and location. The implications of these factors indicate the unique characteristics of Chinese food being utilized as a popular alternative product for all the restaurateurs. The insights from the six panel members have presented valuable tips for successfully establishing a Chinese QSR chain and overcoming different barriers. The Chinese market is still lucrative one, and the findings of this study will aid in restaurants successfully entering this market.

REFERENCES

Brumback, N. (1995), "Asian unchained." *Restaurant Business*, 94 (18), 100+.

Carlino, B. (1995), "Darden gives up on China coast, shutters units." *Nation's Restaurant News*, September 4, 1.

Carlino, B. (1996), "Hot concepts! P.F. Chang's: far east meets west at Chinese Bistro." *Nation's Restaurant News*, May 20, 80.

Crecca, D. H. (1997), "Models for success in foodservice 2005." *Restaurant and Institutions*, May 1, 117-118+.

Epstein, J. (1993), "A taste of success." *New Yorker*, April 19, 50-56.

Hambrick, D. C. (1983), "An empirical typology of mature industrial-product environment." *Academy of Management Journal*, June 26, 213-230.

Hayes, J. (1988), "Chinese delivery on a roll." *Nation's Restaurant News*, May 30, 22(22), 1+.

Hayes, J. (1991), "Mr. Ching: same concept, different city." *Nation's Restaurant News*, July 29, 25(29), 50.

Karnow, S. (1994), "Year in, year out, those eateries keep eggrolling along." *Smithsonian 24*, January, 86-94.

Kochak, J. (1988), "Market segment report: oriental." *Restaurant News*, March 1, 87(4), 177-192.

Kochak, J. (1989), "Market segment report: oriental." *Restaurant Business*, 88 (3), 141+.

Konvitz, J. (1975), "Identity, self-expression and the American cook." *Centennial Review*, Spring, 85-95.

Lao, C.K. (1975), "Chinese restaurant industry in the United States: Its history, development and future." *A Monograph Presented to the Faculty of the Graduate School of Cornell University for the Degree of Master of Professional Studies in Hotel Administration.*

Levenstein, H. (1993), *Paradox of Plenty: a social history of eating in modern America*. New York: Oxford University Press.

Lu, S., and Fine, G. A. (1995), "The presentation of ethnic authenticity: Chinese food as a social accomplishment." *The Sociological Quarterly*, 36 (3), 535-553.

Marchetti, M. (1996), "Creating Panda-monium." *Sales and Marketing Management*, 148(1), 14+

Merriam, S. B. (1998), *Qualitative Research and Case Study Applications in Education*. (2nd ed.). CA. San Francisco: Jossey-Bass. 85-85.

National Restaurant Association (1999), "A caring culture: 1999 restaurant industry forecast." *Restaurants USA*, December 1998, F27-28.

Nation's Restaurant News (1999) *Operator survey*. April 26, 3 (1), 17-24.

Parsa, H. G., & Khan, M. A. (1993), "Quick-service restaurants of the 21st century: an analytical review of macro factors." *Hospitality Research Journal*, 17 (1), 161-173.

Restaurant USA (1992), "Popularity of ethnic foods likely to grow." November, 10 (2), 38-39.

SBS (1999), *Name List of Chinese Restaurants in the United States*. Web site: http://www.chineserestaurantnews.com/. Los Angeles: Author.

Strauss, K. (1999) "QSR operators wake up to burgeoning breakfast opportunities." *Nation's Restaurant News*. November, 33 (44), 103-106.

Tasoulas, M. (1999), "Hotter than Wasabi." *Restaurant Business*, September 1, 43-48.

West, J.J. (1988), "Environmental Scanning, Strategy Formulation and Their Effect Upon Firm Performance." *Unpublished doctoral dissertation*, VA. Blacksburg: Virginia Polytechnic Institute and State University.

Integrating "My Fair Lady" with Foodservice Management: A Theoretical Framework and Research Agenda

Dennis Reynolds

SUMMARY. This conceptual paper explores the potential consanguinity between supervisory expectations and self-efficacy. Enhanced performance resulting from the positive expectations of others–commonly referred to as the Pygmalion effect–has been studied in a managerial context with the goal of increasing organizational effectiveness. The Golem effect is the result of less-than-desirable performance that results from expressed negative expectations. Unlike the Pygmalion effect, however, the Golem effect has received little empirical attention. Pygmalion and Golem serve as attingent vectors with each resulting from the corresponding supervisory expectations; both portend considerable application to foodservice management particularly in the multiunit foodservice arena wherein the role of manager as motivator is key. Similar to this social-expectancy construct, self-efficacy theory is relevant to many aspects of human-resource management–particularly managerial motivation–and organizational behavior. The construct offers utility given its application as an antecedent and potential predictor of performance. While these constructs have been explored, the relationship between supervisory expectations and subordinates' self-efficacy has been largely ignored in terms of empirical work

Dennis Reynolds, PhD, is affiliated with the School of Hotel Administration, Statler Hall, Cornell University.

[Haworth co-indexing entry note]: "Integrating 'My Fair Lady' with Foodservice Management: A Theoretical Framework and Research Agenda." Reynolds, Dennis. Co-published simultaneously in *Journal of Restaurant & Foodservice Marketing* (The Haworth Hospitality Press, an imprint of The Haworth Press, Inc.) Vol. 4, No. 4, 2001, pp. 263-301; and: *Quick Service Restaurants, Franchising and Multi-Unit Chain Management* (ed: H. G. Parsa and Francis A. Kwansa) The Haworth Hospitality Press, an imprint of The Haworth Press, Inc., 2001, pp. 263-301. Single or multiple copies of this article are available for a fee from The Haworth Document Delivery Service [1-800-HAWORTH, 9:00 a.m. - 5:00 p.m. (EST). E-mail address: getinfo@haworthpressinc.com].

that explores more than just their mediating effects. Moreover, recent research indicates this relationship may be moderated by variables related to the subordinate. Following a literature review, this paper draws a theoretical link between supervisory expectations and self-efficacy. It concludes with a research agenda that advocates the exploration of potential psychometrics related to the constructs as well as a rigorous research design.

KEYWORDS. Motivation, Pygmalion, Golem, expectations, self-efficacy, human resources

In the workplace, the Pygmalion effect describes a situation in which a subordinate's performance is improved as a result of a manager's positive expectations, discernible in his or her behavior, pertaining to that performance. The theoretical foundation for the phenomenon is embedded in Merton's (1948) self-fulfilling-prophecy theory. In essence, the theory suggests that a person's set of beliefs regarding a certain outcome or event lead to representative behaviors. These behaviors lead to the realization of the respective outcome, thereby fulfilling the original prophecy.

The manifestation dates back much earlier than Merton's theory, however. In Greek mythology, Pygmalion was a celibate prince of Cyprus who sculpted an ivory statue of the ideal woman. Pygmalion's sculpture was so perfect that he fell in love with his masterpiece. Seeing the prince's deep love, the goddess Aphrodite brought the statue to life, naming her Galatea, and the two lived happily ever after (Seton-Williams, 1993). In 1913, George Bernard Shaw developed a play titled "Pygmalion" that reflected a similar theme, that one person can transform another person by projecting expectations. The film of the same name, as well as the later rendition (titled "My Fair Lady"), delivered the concept to the mainstream.

The Pygmalion effect offers considerable utility for foodservice management. For example, a number of studies have recorded the effect of positive expectations on work-related behavior in such settings as manufacturing (King, 1974), the military (Crawford, Thomas, & Fink, 1980; Eden & Ravid, 1982; Eden & Shani, 1982), retail business (Sutton & Woodman, 1989), and quick-service restaurants (Shimko, 1990). In addition, the theoretical construct suggests that the foodservice industry's long-standing "my way or the highway" approach to management may be less than optimal. This is underscored by research on negative Pygmalion effects, also known as Golem effects, which suggests that a supervisor's low expectations, made apparent through his behavior, can negatively impact subordinates' performance (e.g., Babad, Inbar, & Rosenthal, 1982; Oz & Eden, 1994).

Despite the provocative findings, two gaps exist in the Pygmalion-research domain. First is the model formulation. While Pygmalion research is provocative and suggestive of widespread application, the conceptual models used to understand the phenomenon are potentially lacking. Early models focused exclusively on defining a typology of supervisory behaviors (e.g., Rosenthal, 1971, 1973). Later models, such as those developed by Sutton and Woodman (1989) and Eden (1984, 1990a), identified subordinate performance and included the theoretically mediating variable of self-expectations (stemming from research on the Galatea effect).

The inclusion of self-expectations as the primary mediator between supervisory behavior and subordinate performance may not be correct, however. Self-expectations are not task specific and pertain more to judgments of self-worth. Global constructs such as self-expectations may provide some explanatory power in predicting behavior but are not as good in terms of predictive utility as other, more directly linked determinants of performance (Pajares & Kranzlewr, 1995; Pajares & Miller, 1994).

Some researchers, such as Eden and Zuk (1995) and Gist (1987), have suggested that a more appropriate mediating variable preceding performance is self-efficacy. Self-efficacy refers to one's belief in the capability to perform a specific task (Bandura, 1977). Specifically, self-efficacy has been defined as *an individual's judgments of his or her capabilities to organize and execute courses of action required to attain designated types of performances* (Bandura, 1986). The theory has been applied as a useful approach to understanding human behavior (e.g., Gecas, 1989). Moreover, it is relevant to many aspects of organizational behavior (King, 1974) and human-resource management (Gist, 1989; Locke, Frederick, Lee, & Bobko, 1984), with specific application to managerial motivation (Bandura, 1997).

Self-efficacy theory offers particular utility in studying the effect of behavior related to supervisory expectations on subordinates, however, given its application as an antecedent and potential predictor of performance. For example, it has been linked to sales performance in the insurance industry (Barling & Beattie, 1983) and has demonstrated a strong predictive relationship with academic performance (Wood & Locke, 1987). Other researchers have found the same relationship in a variety of settings and industries (e.g., Eden & Aviram, 1993; Hill, Smith, & Mann, 1987; Frayne & Latham, 1987; Gist, 1989).

Of topical importance, self-efficacy has been empirically found to be a better predictor of subsequent performance than past behavior (e.g., Bandura, 1977; Eden & Zuk, 1995; Gist, Schwoerer, & Rosen, 1989). Its role in the relationship with supervisor's behavior oriented toward expectations is therefore deserving of empirical attention.

The related problem with the recent theoretical models in the self-fulfilling-prophecy research domain is that none recognizes the potentially moderating effect of variables associated with subordinates. That is, while the independent-dependent variable relationship underlying Pygmalion effects is evident, researchers have seldom asked if conditional variables affect either the strength or the direction of this causal relationship. As discussed by Cooper and Hazelrigg (1988; see also Hall and Briton, 1993) moderators in Pygmalion studies have been limited and have focused largely only on such popular themes as gender (e.g., Compton, 1970; Dvir, Eden, & Banjo, 1995), and to a lesser extent, personality (e.g., Hazelrigg, Cooper, & Strathman, 1991).

In general, these studies have demonstrated that Pygmalion effects are moderated. However, the ubiquity of the phenomenon has seemingly kept researchers from studying other moderators. This may be, as Baron and Kenny (1986) suggest, because interest in moderators is usually secondary to interest in the independent variable (in this case, supervisory behavior associated with expectations). However, the limits of generalizability pertaining to the independent-dependent variable causal chain become evident as researchers begin to explore moderators and mediators, with moderators sometimes elucidating a mediating process (Taylor, 1993).

The second major gap in the research surrounding Pygmalion pertains to the Golem effect. To date, most conclusions have been drawn from anecdotal evidence. McNatt's (2000) meta-analysis of Pygmalion-related research conducted within management contexts included 17 studies. Of these, only two investigated Golem effects. The likely reason pertains to the ethical implications involved in experimentally affecting an individual's performance by artificially lowering the supervisor's expectations (Eden, 1990a).

Two exceptions to the anecdotal evidence exist (Oz & Eden, 1994; Davidson & Eden, 2000). These studies, described later in this paper, provided adequate rigor but did not directly assess the Golem effect. That is, the researchers did not introduce a treatment truly reflecting negative expectations in contrast to a control group. This shortcoming limits the findings' utility.

The purpose of this paper, then, is (1) to identify a new model designed to better understand the phenomenon and (2) to outline a research agenda aimed at empirically testing one of the primary causal linkages within the model. Specifically, this paper reviews past research surrounding interpersonal expectations with particular focus on research conducted in the workplace.

Models used to explain the phenomenon are also described and critiqued. Then, a new model is presented, which notably includes self-efficacy as an antecedent of behavior. The model includes a moderating variable targeted at explaining why some subordinates respond differently to managers' behavior associated with expectations. Finally, a research agenda is detailed, which is aimed at exploring the supervisory behavior-self-efficacy link (including both positive and negative expectations) and the role of moderators in the causal chain; an approach to assess the effects of behavior associated with both positive and negative expectations is also discussed.

BACKGROUND

The Pygmalion Effect

Rosenthal and Jacobson (1968) provided the first empirical evidence of the Pygmalion effect. In a field experiment using an educational setting, the researchers confirmed their hypothesis that students for whom teachers held higher expectations would perform better, despite the random assignment of students to what teachers were told was the overachieving group.

Rosenthal and Jacobson's seminal work made such an impact in both academic and lay circles that it rated front-page coverage in the *New York Times,* as noted by Postman and Weingartner (1969). However, the pioneering study also received considerable criticism, including "biased reporting, magnification of selected findings, and over-dramatization of conclusion" (Braun, 1976, p. 188). Specifically, methodological concerns and generalizability of the findings have been the subject of wide debate (e.g., Elashoff, Dixon, & Snow, 1971; Mendels & Flanders, 1973; Snow, 1969; Thorndike, 1968). For example, the researchers relied entirely on norms established for the "Tests of General Ability," which are considered inadequate for the age and socio-economic status of the students involved in the experiment (Braun, 1976). Furthermore, a reanalysis of the original data by Elashoff, Dixon, and Snow (1971) revealed no treatment effect in two of the groups, although Rosenthal and Jacobson reported support for their hypothesis both for these and the other groupings of students.

Nonetheless, the study did establish a novel approach to understanding the Pygmalion phenomenon. Later studies, including Babad (1993, 1995), Dusek, Hall, and Meyer (1985), and Rosenthal and Rubin's (1978) meta-analysis of interpersonal expectancy research, as well as Rosenthal's (1991) update on the state of Pygmalion research in classrooms and Harris and Rosenthal's (1985) summary, firmly established the existence of positive-expectations effects and demonstrated the magnitude of the phenomenon in education.

Livingston (1969) is credited with first recognizing the application of the Pygmalion effect to management contexts. In his article, which was targeted toward practitioners, he theorized that raising supervisory expectations regarding workplace productivity would serve as a determinant of subordinate performance. While he offered no empirical evidence, the paper did serve as a catalyst for research beyond educational settings.

In the 31 years since Livingston's Pygmalion-in-management article, hundreds of studies have been conducted in non-business settings (Rosenthal, 1994). However, as demonstrated in McNatt's (2000) meta-analysis of Pygmalion studies in management-related contexts, the number of studies within the management domain is small. Underscoring the importance of the phenomenon in the management domain, McNatt reported an overall corrected estimate of the average effect of $d = 1.13$ (58 effect sizes, n = 2,784). Similarly, Kierein and Gold (in press) reported an overall effect size of $d = 0.81$ (13 effect sizes, n = 2,853) in the 9 studies included in their meta-analysis.

King (1970, 1971) and Mase (1970) were the first to respond to Livingston's call for empirical research in a business setting. King (1970) conducted a controlled field experiment employing the methodology adopted by Rosenthal and Jacobson (1968) using a training program for disadvantaged persons. The dissertation involved five studies of which three were fully analyzed. (Data pertaining to the two omitted groups were incomplete.) King (1971) reported the overall results by aggregating the findings of the three studies. While the sample size was small (n = 14 for the treatment group), the results were impressive in that both perceptual and cognitively based outcomes were affected. The trainees, consisting of industrial pressers, welders, and mechanics, who had been designated to their trainers as possessing greater aptitude, received significantly higher scores on objective achievement tests, higher supervisory and peer ratings, and shorter learning times. They also had lower dropout rates.

Mase (1970) focused on learning achievement in an industrial setting as part of his dissertation research. Mase's sample consisted of 130 National City Bank of New York trainees. Divided into six different training classes, a total of 30 trainees were randomly designated as high achievers; no information was offered to instructors about the remaining trainees, who served as the control group. Performance was assessed on the basis of (1) scores in each of the five subject areas tested at the conclusion of the training sessions, (2) gain on standardized achievement tests, (3) number of days absent during training session, number of times late, and successful completion of the program, (4) number of days absent or late during first two months on the job and number of people remaining with the company, and (5) supervisory ratings of job performance after 60 working days on the job. Maze reported support for the hypothesized Pygmalion effects, such as a significant difference in terms of performance measures of time with the company following the program and number of absences during the training period between the high achievers and the control group. Possibly owing to the execution of the research design in which some trainers "indicated a suspicious attitude towards the study" (p. 52), however, results pertaining to the other measures of performance were inconclusive.

Rosenthal (1974) reanalyzed the original data from King's (1970) dissertation in order to explore potential differences in the smaller groups that were not discussed in King's (1971) published report. The analysis confirmed King's findings but also uncovered a difference based on the gender of trainees in terms the effect of the supervisors' expectations; Rosenthal reported that the women involved in the study were less subject to Pygmalion effects than the men.

In 1974, King tested the effect on a somewhat different sample wherein plant managers' expectations regarding productivity were artificially raised. Each plant had 10, six-person crews and 30 support personnel. In the two treatment plants, the managers were given artificial reports about previous findings pertaining to job-enlargement and job-rotation programs and were told that the programs had been demonstrated to substantially raise productivity in other plants; the same programs were instituted in the control plants but plant managers were told that the programs were only maintenance measures and that "increased output was not expected to occur" (p. 222). Productivity improved in all four plants but significant productivity improvements in terms of plant output were found in those plants where managers' expectations were artificially elevated versus the control plants.

Lundberg (1975) replicated King's study of plant productivity in an experimental-laboratory setting using 108 college students. The intent of the experiment was to test King's findings under the more strict conditions of the laboratory and to test the impact of job scope in concert with supervisory expectations. Thirty-six groups of students, each consisting of three undergraduate students, were randomly assigned to either treatment or control conditions. The positive-expectations treatment was manifest in three forms: First, the confederate manager expressed enthusiasm regarding the task at hand; second, the manager smiled after checking output and quality; and finally, he responded to a question posed by the experiment after 12 minutes that the employees were doing well and that he expected high production output. The control treatment did not include any of these. The author reported that positive expectations had a modest effect ($p < .10$) on output as compared to the control group.

Shimko (1990) demonstrated the Pygmalion effect's applicability to the quick-service-restaurant segment in mitigating turnover through strengthening supervisory expectations of line employees. In her qualitative study designed to identify issues related to turnover as well as training needs in the quick-service segment, the researcher interviewed 38 fast-food restaurant managers. Her objective was to categorize the findings into causal groupings. Issues related to Pygmalion factors represented 48 percent of the total. It is important to note, however, that Shimko's approach was not thoroughly reported and data were not fully described in the article.

Still others have tested Pygmalion in the workplace. Sutton and Woodman (1989; see also Sutton, 1986) tested the Pygmalion effect in two department stores. However, their findings were not conclusive most likely owing to experimental-design flaws (Eden, 1990a). The researchers noted in their summary, too, that one possible reason for the inconclusive findings was that supervisors did not have enough time to interact differently with treatment employees than with their control counterparts (Sutton & Woodman, 1989).

Markin and Lillis (1975) suggested a methodology utilizing Pygmalion principles to enhance performance of salespersons. Chowdhury (1997) responded by providing empirical evidence demonstrating that sales employees' performance was enhanced through positive supervisory expectations in his study involving 193 sales associates in two retail stores. Correlations between expectations and performance ratings based on three different employee-related outcomes were .42 or above.

Other researchers have applied Pygmalion in an organization-specific context using the military as a test environment and found significant effects. Eden and Shani (1979, 1982) successfully tested the Pygmalion effect with soldiers in a combat-training setting. Similarly, Eden and Ravid (1981, 1982) showed strong Pygmalion effects in military personnel enrolled in a clerical-training course. Eden and Ravid also found that women were less susceptible to supervisory expectations than their male counterparts. Crawford, Thomas, and Fink (1980) instilled artificially high expectations for select supervisors outside the normal chain of command of eight individuals previously identified as poor performers. The significant results, as well as post-experimental interview data, indicated strong support for the Pygmalion effect based on the comparison of the small treatment group with the control group (n = 20).

Dating back to Rosenthal's (1969) early work, the common approach has been to randomly designate some group members as high performers and to plant such an idea as a seed in the supervisor or teacher. Researchers then compare output from these perceived high performers with the others who are treated as controls. Eden (1990b) suggested that such an approach does not control for interpersonal contrast effects that result from subordinates' improvement assessed when the referent is the control that is part of the same group of individuals under study. The researcher studied 29 platoons in the Israel Defense Forces (IDF) by randomly assigning entire platoons to either Pygmalion or control groups. Using objective measures of performance such as multiple-choice exams oriented toward the respondents' areas of expertise, members of the Pygmalion platoons achieved significantly higher scores. Analyses suggested that nearly a fifth of the variance in mean platoon performance was determined by the experimental treatment. The article underscored the importance of controlling for interpersonal contrast effects with the implicit message that previous studies failed to integrate such a control. However, earlier studies, such as King (1974), did control for interpersonal effects but did not discuss this facet of the study when describing the research design.

Dvir, Eden, and Banjo (1995) also used the IDF but focused on the moderating effects of gender. Conducting two experiments, the researchers confirmed the Pygmalion hypothesis among men led by a man and among women led by a man but not among women led by a woman. The authors concluded that the Pygmalion effect could be produced for both genders but suggested the results may be less discrete when the supervisor is female. Davidson and Eden (2000) did find effects for female leaders of female subordinates in their two-part study of IDF basic training in the Women's Corps. (This study is discussed at length in the following section.)

In summary, there is considerable empirical evidence supporting the existence of the Pygmalion effect in a variety of settings including management contexts. As demonstrated, the number of studies exploring Pygmalion effects in the workplace is relatively small given the potential impact on productivity. This is likely due to the inherent difficulties associated with instilling "treatment" expectations. Another reason may be that this aspect of self-fulfilling prophecy is so ubiquitous and is so embedded in the social fabric of the workplace that it transcends everyday scrutiny, as noted by Eden (1990a) and Weick (1979).

The Golem Effect

The Golem effect is the negative or dark version of Pygmalion: behavior reflecting low or negative supervisory expectations generates negative results in subordinates' performance. Babad et al. (1982) first used the term *Golem* to describe the negative version of Pygmalion effects drawing from Hebrew slang, where the word means *oaf* or *fool*. The term originates from a contraption of Jewish legend wherein a creature was created and brought to life to eradicate evil but ultimately became a monster owing to its increasingly strong, corrupting power (Collins & Pinch, 1998). Another reference to Golem (in this case spelled "Gollum") can be found in mid-twentieth-century American literature. Tolkien (1954) used the name for a fictional character who was corrupted by using magical powers and who in turn sought to inflict harm to others. Consonant with how Golem is referenced in a managerial context, Tolkien showed how Gollum's iniquitous behavior ultimately led to his ruin.

The outcome of the Golem effect can manifest either as net declines in subordinates' performance or simply lower than otherwise attainable levels of performance. As noted earlier, the majority of recent evidence referenced in the literature is anecdotal, drawing on extrapolations from Pygmalion experiments. The reason is likely the ethical implications involved in experimentally affecting an individual's performance by artificially lowering the supervisor's expectations. Even Babad et al. (1982) opted to not apply a treatment of artificially lowered teacher expectations toward students; rather, the researchers experimentally raised teachers' expectations toward some students and compared the performance results with students for whom teachers had either naturally occurring high or low expectations. Regrettably, the teachers' familiarity with the low-expectations students likely confounded the results.

In reviewing the psychology literature, only a handful of studies were found that pertain loosely to the Golem effect. Beez (1971) randomly assigned 60 college-student tutors toward 60 Head Start kindergarten-age students to either high or low expectation conditions and then asked for evaluations of the students following ten-minute sessions between tutor and student. Each tutor was asked to teach a series of words to the child and to present clues pertaining to the jigsaw-puzzle task. A third party evaluated the students on the basis of the number of words learned.

The expectation manipulation was communicated through a fictitious psychological report given to the tutors for each student. The wording in each report was identical with the exception of name, age, and gender of the particular child. In addition, the folder for "low ability" children reported lower scores on two intelligence tests. These scores were construed in the summary section more negatively, stressing negative aspects of cultural deprivation, and predicted that school-related tasks might be difficult for the child. Conversely, for "high-ability" children, higher scores were reported; information was also included suggesting the students, while also culturally deprived, would do well on the tasks used in the experiment. Results indicated the effect of the treatments produced significantly different outcomes.

Feldman and Prohaska (1979) conducted an experiment in which confederates acting as students emitted either positive or negative expectations toward 40 subjects acting as their teachers. Results indicated there was a general, significant impact on the subjects' behavior resulting from the treatments. Feldman and Theiss' (1982) study of the joint effects of teachers' expectations about students and students' expectations about teachers on the performance and attitudes of both groups produced similar results.

Subjects (n = 60) were designated as teachers and were led to expect either high or low performance from randomly assigned students whereas students (n = 60) were independently led to expect a teacher with either a high-or low-level of competence. When the lesson was finished, the researchers collected data about the subjects' attitudes pertaining to the lesson, their partner, and themselves, as well as student performance on a test of the lesson content. The results of the randomly paired subjects confirmed the hypotheses that student performance (as measured by the lesson-content test) was a function of low or high expectations and that student attitudes were affected by their expectations about the teacher.

Within the management-literature domain, there are two studies that go beyond offering anecdotal evidence of the Golem effect. The first is the research completed by Oz and Eden (1994). The researchers randomly led treatment-assigned squad leaders (n = 17) in a military unit to believe that low scores on physical fitness tests were not indicative of subordinates' ineptitude while control squad leaders (n = 17) were not told how to interpret test scores. Tests indicated that low-scoring individuals in the experimental squads improved more than those in the control squads. While the researchers employed a respectable research design and were cautious to abide by ethical standards, the sample was small (given the level of analysis). Of greater relevance, the researchers failed to truly lower supervisors' expectations and compare the results with a control group in a manner consonant with the majority of work designed to test the Pygmalion effect (e.g., Rosenthal & Jacobson, 1968). Therefore, it could easily be argued that the researchers failed to directly assess the Golem effect. The study is important, however, because it was the first to attempt identification of Golem effects in a management context.

The second exception is the recently published two-part study conducted by Davidson and Eden (2000). The researchers replicated the Oz and Eden approach by using a treatment designed to prevent low expectations from forming on the part of instructors of trainees of disadvantaged women in the IDF. For the first part of the experiment, the researchers used 360 female training recruits, separated into 30 squads participating in a remedial training program. Soldiers were special recruits who had scored just above the minimal cutoff required for admission to the military. The IDF places such recruits into remedial training because it considers them at risk of failure if engaged in the standard, more-rigorous training program. As the authors note, "The labeling and the special program render this fertile soil for low, Golemizing expectations to take root" (p. 388).

As mentioned earlier, the goal of the research was to test whether de-Golemization could be effected. To this end, the researchers randomly instructed leaders from half of the platoons that the recruits possessed substantially higher than usual abilities for special recruits. Results indicated the Golem effect was evident in the lower performance achieved by the control squads.

In the second part of the study, the researchers targeted trainees who possessed even lower qualifications. Using 20 squads consisting of a total of 225 individuals, the researchers replicated the same de-Golemizing results found in the first part of their experiment. It is important to note, however, that in both the Oz and Eden (1994) and Davidson and Eden (2000) studies, the researchers did not introduce a treatment consisting of lowering supervisory expectations toward the subordinates. Hence, the findings, while provocative, fail to explicitly test the Golem effect.

While not central to the management domain, Vrugt (1990) adopted a more thorough design and conducted an experiment to investigate whether artificially induced negative expectations of therapists might be conveyed by nonverbal behavior toward clients. Beginning male psychotherapists (n = 18) and male psychology students (n = 18) were randomly assigned to either treatment or control conditions. The significant findings confirmed the hypothesis that the therapists' negative expectations did affect clients in a negative manner.

Hence, indications exist that the Golem effect is as pervasive and relevant to workplace behavior as Pygmalion-related outcomes. The studies discussed above underscore the application to the workplace; the implications are similarly clear for foodservice industry. Lacking still, however, is empirical evidence that tests the Golem effect in a manner consistent with recent Pygmalion studies.

MODEL EVOLUTION AND DEVELOPMENT

A number of models have been presented in the literature designed to explain the phenomenological underpinnings of the Pygmalion effect in the workplace. Most of these involve a similar process: (1) the supervisor or source formulates expectations about the subordinate or target; (2) the supervisor behaves toward the target in accordance with the expectations; and (3) the subordinate responds to the supervisor's behavior (Darley & Fazio, 1980, Harris, 1993; Jussim, 1986). In general, the models vary with respect to the detail they offer pertaining to what is involved between each of these steps.

Rosenthal (1971, 1973, 1974, 1981) developed an early theoretical four-factor model of how expectations are transmitted; the factors include climate, feedback, input, and output.[1] Climate, also referred to as socioemotional climate, refers to nonverbal messages from the manager. These can range from nonverbal communication tactics, such as extended eye contact or body language, to overt gestures. Feedback refers to expressed verbal feedback pertaining to performance. Input consists of the amount of information given to an employee. According to Rosenthal, an employee for whom a manager holds high expectations is likely to receive more input in the form of training or related resources, thereby allowing the subordinate greater opportunity to perform well. Finally, the amount of output that a manager encourages from the employee serves as the fourth factor. High-achieving employees are generally encouraged to offer opinions and are given more assistance in finding solutions to work-related problems.

The model served as a good basis for later work but was lacking in three distinct ways. First, the division between socioemotional climate and feedback is often unclear, particularly when non-verbal communication is paired with verbal expression. Second, the variable that Rosenthal titled *output* is closely related to the exchange resulting from feedback. In other words, the role of the receiver can be equally important in the communication exchange. Of greater importance, this factor pertains largely to the subordinate, although the factors were originally intended to relate directly to the supervisor (in the form of mediators). Finally, no explicit mention is made regarding the subordinate's behavior, which is the dependent variable of interest in understanding Pygmalion effects.

Rosenthal (1989, 1993) simplified and refined the model considerably, which he called the affect/effort model. Using an educational setting as an example, this model portends that a change in the level of expectations held by a teacher for the performance of a student is first translated into a change in the affect expressed by the teacher toward the student. This is followed by a change in the degree of effort exerted by the teacher in instructing that student. Accordingly, greater levels of expectations should lead to more affect shown toward the student; this, in turn, should lead to great effort expended on behalf of the student. While a large improvement over the earlier model, this model again ignores the vital role of outcome variables, among others.

Rosenthal (1981; see also Harris & Rosenthal, 1985) also proposed a 10-arrow model for the study of interpersonal-expectancy effects. As shown in Figure 1, this model begins to account for resulting outcomes and moderating and mediating variables. The moderating variables, such as the supervisor's affinity for the subordinate, are preexisting variables that influence the size of the Pygmalion effects on the part of the supervisor. Mediating variables, such as verbal or written feedback regarding performance, refer to the behaviors through which expectations are communicated. The predictor, mediating, and

FIGURE 1. Ten-Arrow Model

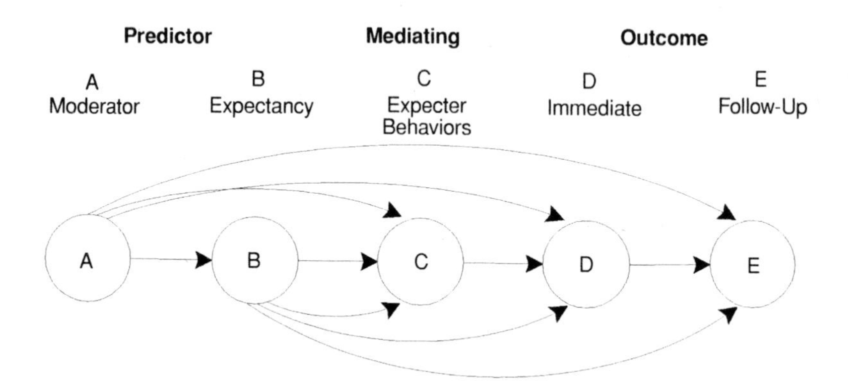

outcome variables align with those described in the early paragraphs of this section. That is, they describe the basic elements of the Pygmalion phenomenon. While the distinction between immediate and follow-up outcomes is interesting, the differentiation is less important than other

Sutton and Woodman (1989) developed what they called an *integrated model* (see Figure 2) that incorporates Rosenthal's (1973) four factors and some aspects of his 10-arrow model (1981) as well as construct components from others such as Braun (1976), Brophy and Good (1970, 1974), Cooper (1979), Darley and Fazio, (1980), and Eden (1984). Again using four factors, Sutton and Woodman proposed that the Pygmalion effect occurs when a perceiver's expectations (factor one) cause the perceiver to behave differently (factor two) toward the subordinate. That behavior affects the subordinate's self-expectations (factor three), which leads to changes in the subordinate's behavior (factor four).

This model represents a considerable improvement in that it ties cognitive attributes of the subordinate directly to the subordinate's behavior. Moreover, it eliminates the focus on potential mediators preceding the supervisor (as included in Rosenthal's models) and focuses more on the expectations, the manifestation of such expectations, their cognitive effect on the subordinate, and the resulting behavior.

A similar model is presented by Eden (1984, 1990a). As depicted in Figure 3, the Pygmalion effect begins with a manager or supervisor's expectations. On the basis of these expectations, the manager "allocates better leadership"

FIGURE 2. Integrated Model of the Pygmalion Effect

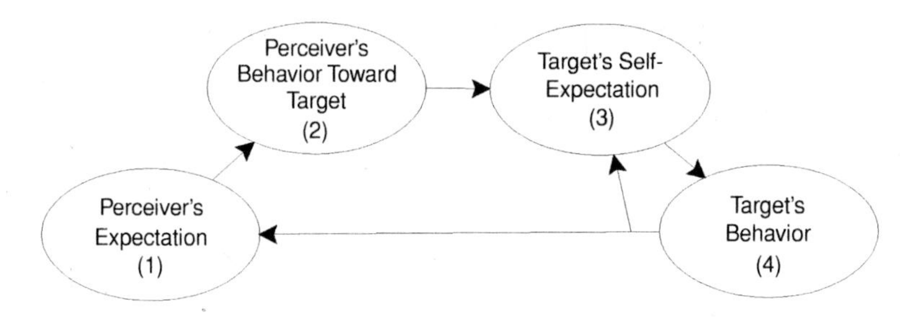

FIGURE 3. Model of Self-Fulfilling Prophecy at Work

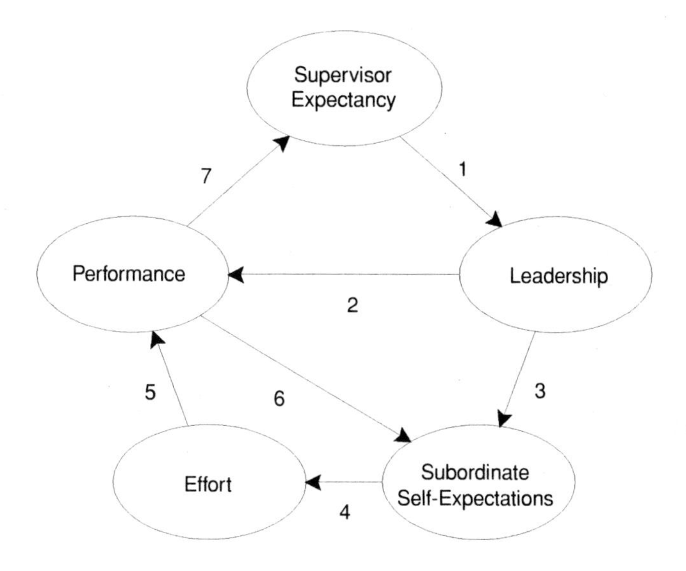

(Eden, 1990a, p. 69) that entails communication of these expectations (arrow 1). In addition, these leadership actions, such as providing training resources, directly facilitate enhanced performance by ensuring that the subordinate has the necessary tools or information (arrow 2). Pertaining specifically to the subordinate, the subordinate's self-expectations are enhanced as a result of the allocation of leadership-oriented actions (arrow 3), leading to increased motivation for the subordinate to exert more effort toward the task (arrow 4). Investing greater effort produces enhanced performance resulting in increased achievement or performance (arrow 5).

This achievement, then, affects both the manager and the subordinate. The performance feeds the subordinate's self-expectations regarding future performance (arrow 6). The manager evaluates the subordinate's achievement (arrow 7); high achievement confirms and justifies the manager's high initial expectations, reinforces them, and prompts continued similar behavior. The self-fulfilling-prophecy cycle is thereby sustained.

The model reflects the inclusion of cognitive processes mediating the supervisor's behavior and that of the subordinate. This is critical owing to the research showing such mediating processes are critical to understanding Pygmalion effects (Taylor, 1993). Eden (1984) also makes the case that positive expectations are represented in all aspects of leadership. He proposes that such actions as goal setting (as defined by Locke, Saari, Shaw, & Latham, 1981) and management by objectives (outlined by Odiorne, 1969) are direct ways of communicating high-performance expectations. However, as noted by House (1977), while the communication of high expectations of subordinate performance is an important component of charismatic leadership, it is not the only aspect of leadership. Eden's move to differentiate supervisor's behavior is therefore inaccurate.

Eden's (1990) inclusion of a variable titled "effort," which he argues serves as a mediator of performance, is unsatisfactory. Effort is but one of a host of factors that govern the level and quality of performance (Bandura, 1997). Individuals tend to integrate salient factors such as the knowledge, skills, and strategies they possess or can obtain rather than solely how much effort they can exert. As Wood and Bandura (1989) note, performance such as that found in the workplace often calls for ingenuity and resourcefulness; such output is far more dependent on adroit use of skills, specialized knowledge, and analytic approaches than on simple application of effort.

As mentioned earlier, these models share much of the same processes. Two key components, however, are missing from each. First and foremost is the role of subordinate's self-efficacy. Second are the moderating effects of variables associated with the subordinate (of which one category is demographics). Before explaining why self-efficacy is integral to a Pygmalion model, however, it is necessary to discuss the surrounding theory. The critical role of moderators is explained later in the paper. At this point in the discussion, it is sufficient to report Baron and Kenny's (1986) assertion that in such instances "moderators may hold theoretical as well as pragmatic significance" (p. 1175).

Self-Efficacy

Nested in social cognitive theory, the construct of self-efficacy has been applied in a variety of settings with the intention of understanding motivation and performance-related behaviors in much the same way that Merton's (1948) theory of self-fulfilling prophecy has been effectively used to understand a broad range of behavioral phenomena. Self-efficacy refers to one's belief in the capability to perform a specific task (Bandura, 1977). Specifically, self-efficacy has been defined as an individual's judgments of his or her capabilities to organize and execute courses of action required to attain designated types of performances (Bandura, 1986).

Self-efficacy has been conceptualized and assessed along three dimensions: magnitude, strength, and generality (Bandura, 1977; Gist, 1987; Stajkovic & Luthans, 1998b). Magnitude pertains to the level of task difficulty and complexity that a manager believes is attainable. Strength refers to the level of conviction associated with the magnitude; it can vary from weak to strong. Generality is the degree to which the personal efficaciousness is generalized across similar domains, which vary on modalities and external factors.[2] Gist and Mitchell (1992) offered one of the few variants to this conceptual framework by formulating a three-dimensional model encompassing the distinctions of externality, stability, and controllability provided by Weiner (1985). This model, while arguably providing for the malleability of the construct, remains grounded in Bandura's conceptualization.

Related to the dimensions of magnitude, strength, and generality are the four determinants of Bandura's (1977, 1997) self-efficacy model: enactive-mastery experiences, vicarious learning, psychological and emotional arousal, and verbal persuasion. Enactive-mastery experiences represent repeated accomplishments experienced in the past. Considered to provide the strongest effect on changing efficacy beliefs, enactive mastery is the only antecedent that provides directly related performance information (Bandura, 1982). Two components of this determinant are situational factors, which relate to the environment, and conception of ability, which integrates the individual's perceptions of ability (Stajkovic and Luthans, 1998b).

Vicarious learning or modeling occurs when one observes competent individuals performing a similar task and, as a result, gains a reinforced message from the observation (Bandura, 1977). The greater the perceived similarity between the individual and observed person, the greater the influence of the vicarious experience. Interestingly, this effect is stronger when the observed individual succeeds after overcoming considerable difficulty (Bandura, Adams, Hardy, and Howells, 1980).

Another major determinant of self-efficacy is psychological arousal. An individual may equate his or her perceptions of psychological or emotional level of arousal as debilitating or as an increasingly potential vulnerability (Gist, 1987). Furthermore, since high levels of stress are likely to debilitate performance, managers may be more inclined toward efficaciousness when not suffering from symptoms related to stress or burnout (Reynolds and Tabacchi, 1993; Stajkovic and Luthans, 1998b). Psychological arousal does not pertain to optimal stress levels, such as those described by McHugh and Brennan (1992) in which the level of stress leads to high levels of effectiveness; rather, the stress or arousal in the context of self-efficacy is so extreme to negatively affect performance.

Verbal persuasion by someone the person (i.e., subordinate) trusts and perceives as competent is the final source of efficacy information. Bandura (1982) asserted that verbal persuasion is an important source of efficacy beliefs and is likely more consequential to individuals than psychological arousal. Conversely, it may not be as effective as enactive mastery experiences or modeling. However, drawing from the transformational-leadership literature, it has been demonstrated that feelings of trust are indeed associated with behaviors related to verbal persuasion such as articulation of the vision for an organization, which can in turn be paramount in effecting changes in subordinate's behavior (e.g., Bass, 1999; House, 1977).

Researchers have traditionally treated verbal persuasion not as a means to increase skill level, as in the case of a training seminar, but rather as a means of focusing an individual on the likelihood of success. Chambliss and Murray (1979) demonstrated the value of verbal persuasion and demonstrated that its effects are greatest when the target individual is receptive to the input. In a managerial context, Baron (1988) reported that, given similar performance levels, disparaging criticism lowers efficacy whereas constructive feedback aids in maintaining or increasing a sense of personal efficacy. Wilson, Wallston, and King (1990) demonstrated that a persuasive message affects self-efficacy; its import, however, is predicated on a minimum level of self-efficacy.

In summary, self-efficacy entails the belief that one can successfully execute behaviors needed to produce desired outcomes. The construct offers theoretical support that individuals can be led by others to believe they can accomplish specific tasks. As noted by Harris and Rosenthal (1989), the "provision of proper tools and condition for potential success can further contribute to feelings of self-efficacy" (p. 944). Such assertions suggest a natural link between supervisory expectations and subordinates' self-efficacy.

Before discussing the proposed model that integrates self-efficacy, it is important to make the distinction between self-efficacy and self-expectations.

Self-expectations are not necessarily task specific and pertain more to judgments of self-worth. Such self-appraisal has been analyzed as self-concept (e.g., Rogers, 1959; Wylie, 1974). This self-concept is a composite view, which corresponds more to individuals' attitudes toward themselves and pertains more to self-image. Global concepts, like self-expectations, may provide some explanatory power in predicting behavior but are not as good in terms of predictive utility as self-efficacy (Pajares & Kranzlewr, 1995; Pajares & Miller, 1994). Moreover, as Bandura (1997) notes, "Self-concept loses most, if not all, of its predictiveness when the influence of perceived efficacy is factored out" (p. 11).

Additionally, research on self-efficacy (e.g., Barling & Beatie, 1983; Gist, 1989; Hill, Smith, & Mann 1987; Peake & Cervone, 1989; Stajkovic & Luthans, 1998a; Telch et al., 1982) as well as attribution theory (Miller & Turnbull, 1986, Weiner, 1985) underscore the link between supervisors' expectations and subordinates' behavior. As articulated by Harris and Rosenthal (1989), if the subordinate believes that the supervisors' behavior is unrelated to the subordinate's potential or behavior, the subordinate's self-efficacy is likely to go unchanged. The researchers provide the example of a supervisor who indiscriminately gives words of encouragement to several subordinates; these subordinates' efficaciousness is likely to go unchanged. Conversely, if a supervisor identified select individuals and offered them encouragement, the selected individuals would likely enjoy an increase in self-efficacy regarding the commended task.

A possible explanation as to why self-efficacy has not been formally integrated into previous Pygmalion models is that self-efficacy is often incorrectly operationalized. Eden (1988; see also Davidson & Eden, 2000) postulated that self-efficacy should be conceptualized as a continuum. Woodruff and Cashman (1993) suggested that domain self-efficacy–self-efficacy specific to the domain and, when appropriate, the tasks predominate within that domain–is a conceptually accurate method for assessing self-efficacy. Regrettably, researchers have been too general in their assessment of the construct (as discussed earlier) or too specific.

A cogent model that offers utility in explaining the Pygmalion phenomenon, while also achieving the goal of providing a foundation for empirical investigation, is depicted in Figure 4. The model includes four variables–of which the most notable is self-efficacy–that form a causal path. It also integrates subordinate-specific moderating variables.

As with previous models, supervisory expectations lead to behavior consistent with these expectations (arrow 1). Rosenthal's (1971, 1973, 1974, 1981) four factors discussed above relate directly to the linkage denoted by arrow 1. Using an academic setting, for example, Kester and Letchworth (1972) dem-

FIGURE 4. Proposed Pygmalion-Effect Model

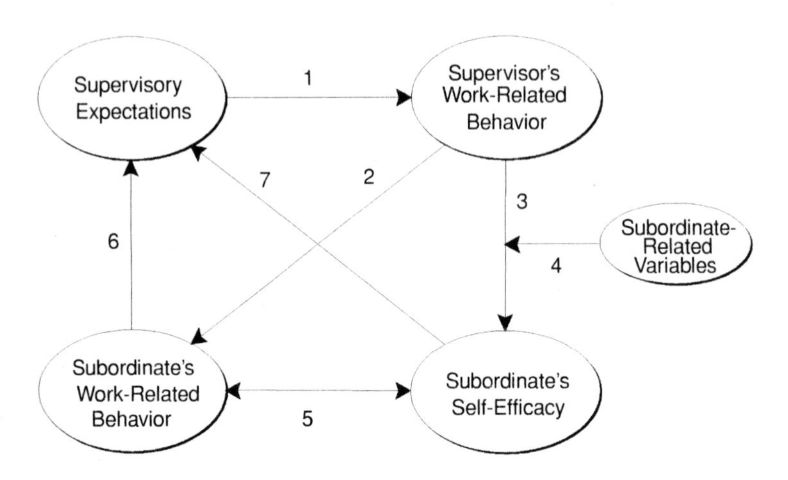

onstrated that teachers provide a warmer climate for interaction with students they consider to be bright. Brophy and Good (1970) and Good, Cooper, and Blakely (1980) found teachers offer more praise while Cornbleth, Davis, and Button (1974) found teachers provide more elaborate and informative feedback to students whom they deem to be overachievers. Finally, research indicates teachers give students they perceive as high performers greater attention (Kester & Letchworth, 1972; Rothbart, Dalfen, & Barrett, 1971) and more opportunities to contribute to class discussion (Rubovits & Maehr, 1973). Of particular interest to this paper, Likert (1961) stressed the importance of communication by supervisors to subordinates in order to convey expectations as a vital facet of supervisory behavior. This step is paramount for studies of Pygmalion effects owing to the fundamental role of supervisory expectations in the model. Further evidence of the importance and empirical significance is provided by Eden and Ravid (1982), who showed that supervisors demonstrate their expectations by means of differential treatment of their subordinates.

This linkage is the same as that found in Sutton and Woodman's (1989) integrated model, and is also similar to that found in Rosenthal's (1981) 10-arrow model. It is different, however, from Eden's (1984, 1990) model in that Eden linked supervisory expectations with leadership. As discussed earlier, constraining the supervisor's behavior to leadership-related behavior is insufficient. The critical differentiation is that expectations are important if they are communicated by way of any behavior.

As with Eden's (1984, 1990a) model, this behavior may directly lead to performance owing to the allocation of resources facilitating accordant outcomes (arrow 2). Supervisors' work-related behaviors might take the form of providing training or making opportunities available to subordinates for whom they have positive expectations. Such resources could be withheld, or actions might be taken to omit subordinates from certain activities in the presence of negative expectations. Bowers and Seashore (1966) also describe such management behaviors as "work facilitation" behaviors. They cite examples such as when a manager shows a subordinate how to improve performance, or offers ideas for solving problems on the job. Again, these behaviors can be, but do not have to be, cosseted in leadership behaviors.[3]

The proposed model markedly differs from its predecessors, however, as depicted in the linkage between supervisor's behavior and subordinate's self-efficacy (arrow 3) and, in turn, the relationship between self-efficacy and performance (arrow 5). Building on the earlier discussion on self-efficacy theory, self-efficacy offers particular value in studying the effect of supervisory expectations on subordinates given its application as an antecedent of performance (Gist, 1987; Gist & Mitchell, 1992). To this end, self-efficacy research has produced several consistent findings pertaining to work-related performance in such areas as sales attainment in the insurance industry (Barling & Beattie, 1983), academic performance (Wood & Locke, 1987), adaptability to new technology (Hill, Smith, & Mann, 1987), cognitive modeling (Gist, 1989), training (Gist, Schwoerer, & Rosen, 1989; Eden & Aviram, 1993), goal choice and associated task performance (Locke, Frederick, Lee, & Bobko, 1984), self-management (Frayne & Latham, 1987), and performance of endurance tasks (Garland, Weinberg, Bruya, & Jackson, 1988; Lerner & Locke, 1995).

Self-efficacy's inclusion in the model is fundamental given that it has been empirically found to be a better predictor of subsequent performance than past behavior (e.g., Bandura, 1977; Gist, Schwoerer, & Rosen, 1989). This is underscored by the findings of Stajkovic and Luthans (1998a) in their meta-analysis (114 studies, n = 21,616) that demonstrated a significant weighted-average correlation between self-efficacy and performance.[4]

Furthermore, Eden and Zuk's (1995) study, which integrated the self-fulfilling-prophecy paradigm, provides empirical evidence for the linkage. Naval Cadets (n = 25) in the IDF were randomly assigned to either a control or treatment condition. The treatment involved brief interviews between a psychologist and each cadet. During the interviews, the psychologist provided spurious information to each cadet in the treatment group; this information was allegedly drawn from a survey the cadets had completed five days earlier. The psychologist stated, "Based on your questionnaire responses, as well as other information, you have the qualifications to overcome seasickness very well and to outperform others at sea" (p. 630). The treatment resulted in significantly better self-efficacy scores for the treatment group.

As noted earlier, the proposed model also uniquely includes the potentially moderating role of subordinate–related variables (arrow 4). This is a potentially critical addition since little is known about how expectation-related effects might be moderated. This is particularly important given that researchers have reported differences in how some subjects respond to expectations (e.g., Eden & Ravid, 1982; King, 1971; Sutton, 1986).

The literature surrounding these effects has indicated moderating variables do exist and deserve future research. The handful of researchers who have studied moderators associated with the subordinate or student have found empirical evidence suggesting that self-fulfilling-prophecy theory cannot be adequately described without the inclusion of receiver-related moderators (Cooper & Hazelrigg, 1988; Eden & Ravid, 1982). To date, however, empirical investigation of moderators in Pygmalion studies has been rather limited and has focused largely only on such popular themes as gender (e.g., Briton & Hall, 1992; Compton, 1970; Dvir, Eden, & Banjo, 1995) and, to a lesser extent, personality (e.g., Hazelrigg, Cooper, & Strathman, 1991).

Moderators, also frequently referred to as *specifiers* or *conditional variables,* are of interest due to the expectations-outcome relationship that, at times, suggests a certain lack of veridicality (Locke et al., 1984). That is, some findings have been somewhat inconclusive in which instances the authors frequently cite the likely moderating effects of untested variables (e.g., Chowdhury, 1997; Sutton & Woodman, 1989). Given the lack of empirical research in this area, information on any category of moderators would be useful, whether the findings support the proposition or not.

Consonant with earlier models, the concluding link (arrow 6) closes the chain. The manager evaluates the subordinate's performance against the benchmark of his own expectations. In the presence of positive expectations, high performance confirms and perpetuates the supervisor's expectations; similarly, poor performance may confirm negative expectations.

Finally, the linkage depicted by arrow 7 suggests there is a relationship between self-efficacy and a supervisor's expectations. In other words, a subordinate's task-specific self-confidence (self-efficacy) may influence a supervisor's expectations until disproved. In support of this, Orpen (1989) found a direct correlation between subordinates' self-confidence and their supervisors' appraisals of their work. While this could be a reciprocal relationship, Zalesny (1990) provided empirical evidence that supervisor's evaluations were linked to the confidence of the subordinate regarding key job tasks only when the two had limited time together or when the supervisor was new to the position, suggesting subordinates' self-efficacy has an impact on their supervisors' attitudes until disproved.

RESEARCH AGENDA

The primary linkage of empirical interest depicted in the new model is (1) the relationship between supervisory behavior and subordinate self-efficacy and (2) the role of subordinate-related variables that may moderate this part of the causal chain. Hence, the first area of inquiry builds on the theoretical foundation of self-efficacy and its role in the new model in the Pygmalion phenomenon. The propositions supporting this line of inquiry, then, are:

P_{1a}: Supervisory behavior associated with positive supervisory expectations positively influence subordinates' self-efficacy.

P_{1b}: Supervisory behavior associated with negative supervisory expectations negatively influence subordinates' self-efficacy.

P_{1c}: Absence of supervisory expectations produces no significant change in subordinates' self-efficacy.

These hypotheses serve to test the linkage indicated by arrow 3 in figure 4. If such a relationship exists between supervisors' behavior and subordinates' self-efficacy, then it is of interest to test the moderating role of subordinates' demographics (arrow 4). As mentioned earlier, there are possibly a myriad of subordinate-related variables that may moderate the linkage between supervisory behavior and subordinate's self-efficacy. Research in related areas has indicated that demographic differences affect workplace performance and perceptions of subordinates by supervisors. These findings pertain to differences based on subordinates' ethnicity (Igbaria & Wormley, 1995; Stumpf & London, 1981), age (Eden & Shani, 1982; Ferris & King, 1992; Saks & Waldman, 1998), gender (Dion & Banting, 2000; Eden & Ravid, 1981; Fletcher, 1999; Saunders & Saunders, 1999; Shore & Thornton, 1986; Sutton, 1986), hierarchical level or title (DiPadova & Faerman, 1993), and work experience (Quinones, Ford, & Teachout, 1995).

Demographic variables are of particular interest to organizations, especially in the multi-unit restaurant industry, because of their utility in such processes as career development, performance management, and organizational change involving diversity. In the event a certain demographic variable is integral to enhancing the effect of positive supervisory expectations on subordinate performance, for example, such knowledge could be instrumental in teaching superiors of subordinates who possess that characteristic to maximize the appropriate technique of communicating expectations. In the performance-appraisal process, a better understanding of how demographics moderate subordinate behavior could aid in the delivery of feedback. Information pertaining to ethnicity might help organizations maximize the communication efforts geared toward different groups of employees. Still other uses for this information extend to team-member allocation. The added benefit of such a moderator is that demographic information is readily available and is, in some cases, readily modified (such as level in the organization).[5] Finally, demographics are critical to Pygmalion research specifically, and workplace-behavior research in general, owing to the rapid rate of change in workforce demography and the concurrent emphasis on maximizing human capital, which is particularly salient to foodservice operators (Golembiewski & Osuna, 1995; Woolridge, 1994).

Specifically, demographic variables that would represent a good starting point for research intended to test moderating effects in the Pygmalion model include: (P_2) level of management, (P_3) gender, (P_4) age, (P_5) ethnicity, (P_6) level of education, (P_7) number of years in the industry, and (P_8) number of years in the position. Each of these is discussed below. The reason for including these specific demographic variables is: they are easily identifiable by organizations and researchers, and they offer the greatest promise based on recent theoretical and empirical work.

Level of management, also referred to as organizational rank, is of interest because of the contradictory findings to date. Researchers have reported that lower levels of management are relatively similar in terms of motivating needs and stimuli (Chusmir, 1986; McClelland and Boyatzis, 1982; Stahl and Harrell, 1982). However, more recent work (e.g., Reynolds, 1999; Brownell & Reynolds, 2000) indicates that motivational characteristics vary with managerial rank. Similarly, researchers (e.g., Van Velser, 1988; Roach, 1986) have reported differences in preferred styles of management and associated communication on the basis of organizational rank.

There are also differences in the way managers respond to expectations given the associated differences in power and influence that are linked to the organizational hierarchy. For example, Goldhaber (1993) suggests that power may affect any number of organizational transactions owing to the differences in perceived power held by each individual involved. These levels of power typically are related to hierarchical positioning in the organization. Furthermore, the various types of interpersonal and group power suggest that verbal messages are likely affected by level of power and influence (c.f. Carli, 1999; Colwill, 1987; Friedkin, 1993).

Cornelius and Lane (1984) demonstrated that individuals with a higher need for power migrate toward positions of higher organizational rank and may respond differently to management-related stimuli. Most notably, Harris (1993) suggests that the unequal power involved in supervisor-subordinate relationships directly impacts situations involving expectations. This is underscored by early reviews of Pygmalion research that suggest stronger effects are obtained when the person in the supervisory role is of higher status (Rosenthal, 1969).

By extension, then, power afforded by one's legitimate position may moderate the relationship between expectations and self-efficacy given that managers at various hierarchical levels possess and exert varying degrees of both power and influence. This is supported by studies that suggest individuals with relatively less formal power and status also form expectations concerning a higher power and status target, such as a supervisor (Hollander & Offermann, 1990; Miller & Jablin, 1991; Mowday, Porter, & Steers, 1982). Thus, the corresponding proposition is:

> P_2: Subordinates' organizational rank will moderate the effects of supervisory expectations on subordinates' self-efficacy such that the higher the rank, the greater the effect of the expectation.

Early work on Pygmalion effects indicated that women are more influenced than men by the expectations expressed by others (Lenney, 1977; O'Leary, 1977). Similarly, Reynolds (1999) and Smits, McLean, and Tanner (1993) offered empirical evidence demonstrating differing motivational characteristics between men and women.

Four other studies indicate that self-fulfilling-prophecy effects–at least in terms of positive expectations–may be extremely weak among women. King's (1970) findings, when reanalyzed and reported by Rosenthal (1974), indicate that women are less responsive to positive expectations. Eden and Ravid (1981) were unable to replicate the effect of positive expectations when using female military trainees although the findings were strong for their male sample. Sutton (1986) failed to find related effects in her sample of sales associates. She suggested that the largely female sample used in her study explained her failure to find evidence of Pygmalion effects in retail sales personnel. Finally, Dvir, Eden, and Banjo (1995) conducted two experiments testing the Pygmalion effect among women. They found positive effects were produced only when the supervisor of the women was male.

Hall and Briton (1993), however, make a persuasive theoretical case for predicting larger effects in male supervisor-female subordinate dyads than male supervisor-male subordinate dyads. The researchers suggest, "Men's communications are more closely monitored than women's by virtue of men's higher status in society. Thus, any cues sent by men may be influential, and if those cues happen to be biasing cues, then men will bias the responses of [subordinates] more than women will" (p. 285). Similarly, they assert that women will be more influenced by the male supervisor. Another view is found in the assertions of Eagly and Johnson (1990), who concluded in their meta-analysis of 162 studies on leadership and gender that women and men simply lead and respond to leadership in different ways.

Confounding the issue of gender as a moderator of Pygmalion effects, none of the aforementioned studies included tests pertaining to negative expectations. This is disappointing, since empirical evidence indicating that Pygmalion and Golem effects are increased for one gender would help researchers better understand both the phenomenon and the differences between genders in subordinate roles.

The associated proposition for such inquiry is:

P_3: Subordinates' gender will moderate the effects of supervisory expectations on subordinates' self-efficacy such that the effect will be greater for women than men.

Some of the earliest research on the Pygmalion effect indicated age was a moderator. For example, Rosenthal and Jacobson (1968) found that positive effects were strongest in the youngest children in their sample. Livingston (1969) applied the same concept by arguing that the Pygmalion effect should be strongest for young managers just entering the managerial ranks. Eden's (1990a) review of Pygmalion-at-work studies also suggested that managers who were just beginning their careers were particularly susceptible to positive expectations. This was underscored in Ferris and King's (1992) study. Saks and Waldman (1998) also predicted and found a negative correlation between age and perceptions of performance by supervisors in their study of entry-level employees in public accounting firms.

However, as Eden (1990a) notes, assertions that Pygmalion effects are strongest may be due to the ease that researchers have in gaining access to lower-level managerial ranks. Conventional wisdom suggests older managers would be more resistant to views by others, including their supervisors. However, interpersonal-expectancy theory (Rosenthal, 1971, 1973) suggests that age should not be a moderator. Indeed, knowledge of whether age is a factor in either Pygmalion or Golem effects is of practical value since such information might help organizations anticipate career phases when supervisory expectations have greater salience.

In reference to age as a moderator, the proposition is:

P_4: Subordinates' age will moderate the effects of supervisory expectations on subordinates' self-efficacy such that subordinate's age inversely affects the effect.

Studies of differences in motivation characteristics on the basis of ethnicity also have gained attention in the last few decades. Watson and Barone (1976), for example, examined African-American and Caucasian managers and concluded that the motivating needs of the two groups were very similar. They also posited that differences that did initially exist between the two groups would likely disappear after the sampled African-Americans had gained more managerial experience. Chowdhury (1997), on the other hand, found differences among Hispanic managers as compared to their Caucasian, African-American, and Asian counterparts in terms of positive effects.

Increasing workforce diversity in foodservice management would also suggest the need for research designed to better understand the characteristics that may moderate Pygmalion effects. This is emphasized by the trade literature that details a trend in the chain-restaurant industry to include more women and people of color (King, 2000).

Other researchers have also suggested that differences may exist in how people respond to managerial settings based on ethnicity or minority status. For example, Stumpf and London (1981) reported performance, as perceived by the supervisor, is often biased on the basis of a subordinate's ethnicity; that is, minority employees are viewed negatively. Similarly, Igbaria and Wormley (1995) reported that, compared to the black employees, white employees "were rated higher on job performance ratings, [and] were more likely to have their performance attributed to internal factors" (p. 92).

This situation may in itself create a self-fulfilling prophecy, although the evidence to support it is anecdotal. Minority managers may become acculturated to lower expectations and therefore become less susceptible to manifest expectations pertaining to specific tasks. In other words, they may simply develop a tolerance to stated expectations, operating under the assumption that less is always expected of them. The proposition targeted at testing the moderating role of minority status is:

P_5: Subordinates' minority status will moderate the effects of supervisory expectations on subordinates' self-efficacy such that minority status lessens the effects.

Level of education applies to the exploration of Pygmalion effects owing to the role this variable may play in the relationship between supervisor and subordinate. Knotts (1975) noted that managers with increased levels of education respond differently to different workplace stimuli. The work of McClelland and Boyatzis (1982) suggests that leadership patterns and long-term success are directly linked to level of education. While there have been no studies that explored level of education as a moderating variable, these aforementioned references, as well as related work in the field of leadership and leader-member exchanges, indicates this variable may play a role in moderating Pygmalion and Golem effects (e.g., Dansereau, Graen, & Haga, 1975; Hollander & Offermann, 1990; Liden, Wayne, & Stilwell, 1993).

Furthermore, it seems likely that level of education would moderate this relationship in much the same way as age. That is, subordinate managers who possess higher levels of educations would be less affected by expectations since their baseline level of self-efficacy might preclude any dramatic change. The relevant proposition, therefore, is:

> P_6: Subordinates' level of education will moderate the effects of supervisory expectations on subordinates' self-efficacy such that more education lessens the effects.

The number of years in the industry, and within a position, are of interest because these variables are quickly identifiable to organizations using such sources as employment records and intuitively may serve as moderators in the Pygmalion-self-efficacy relationship in much the same way as do related variables such as age and education. As noted earlier, Livingston (1969) asserted that susceptibility to Pygmalion effects is especially high when people first enter the workforce. Moreover, changes that manifest into midlife and midcareer crises, as reported by Levinson (1978), may influence Pygmalion effects. McClelland and Boyatzis (1982) and Stahl and Harrell (1982) indicated that motivational characteristics of managers change with tenure, particularly years of experience in management. Furthermore, Taylor, Audia, and Gupta (1996) explored years in the position, which they referred to as job tenure, and noted differences on the basis of organizational commitment and turnover.

These variables, then, may serve to moderate Pygmalion effects and deserve investigation. The corresponding sets of propositions are:

> P_7: Subordinates' number of years in the industry will moderate the effects of supervisory expectations on subordinates' self-efficacy such that increases in industry tenure lessen the effects.

> P_8: Subordinates' number of years in the position will moderate the effects of supervisory expectations on subordinates' self-efficacy such that increases in job tenure lessen the effects.

These propositions represent the first work of its kind at understanding how demographic variables moderate Pygmalion effects, including both positive and negative expectations. From the earlier section of the paper, it is clear that Pygmalion and Golem effects can be produced in the workplace. The theoretical foundations of each construct underscore this. Moreover, self-efficacy serves as an adequate dependent variable, particularly given the evidence from recent research. It is also appropriate given its likely role as an antecedent of performance. Finally, these proposition and the future research that explores them directly answer Eden's (1990a) statement pertaining to the need to explore the generalizability of self-fulfilling prophecy theory: "We need more research on self-fulfilling-prophecy effects among workers at different career stages, different levels in the organization, and different ages to know what limitations should govern generalizability" (p. 68).

FUTURE RESEARCH

Without question, this agenda is challenging, owing to its complexity and ethical issues. One design that would serve the purpose of empirically testing these linkages is a field experiment involving a three-stage process. First, a baseline level of self-efficacy pertaining to subordinates' specific, job-related tasks would need to be established. Three treatments, consisting of verbally expressed positive, negative, and no apparent expectations, respectively, and aimed at the same tasks, could then be applied by random assignment to subordinates from a given sample. Such treatments would need to be administered to separate groups (i.e., group one receives treatment one, etc.). Finally, subordinates' self-efficacy could then be reassessed in a manner consistent with the initial measurement. The unit of analysis would be at the group level, using net gain/decline in self-efficacy as the dependent measure.

This general design conforms to the methodology used in early Pygmalion research while also serving to directly assess Golem effects. Moreover, the design affords complete control of contrast effects that have caused concern in past studies, as noted by Eden (1990b). This elaboration of the Solomon (1949) Four-Group Design also controls most major threats to internal validity. Furthermore, such an experimental design "deservedly has higher prestige and represents the explicit consideration of external validity factors" (Campbell & Stanley, 1963, p. 24).

Several issues need to be addressed to apply such an approach. First, one or more organizations within the multi-unit foodservice segment needs to be identified that is interested and would grant access to multiple levels of its management ranks. Second, an adequate operationalization of self-efficacy must be formulated that integrates tasks relevant to the identified levels of management within the target organization. Third, treatments need to be designed, which can be used to express expectations pertaining to these tasks. Finally, the application of the field-experiment format needs to be properly executed. While such designs are somewhat unusual and are not often conducted in the social sciences owing to the substantial investment of time and resources required, this design affords the researcher increased strength in terms of the inferences drawn (Sproul, 1995). This type of empirical fieldwork satisfies Yukl's (2000) call for studies that supplant the all-too-common survey approach with research that attempts to truly capture the phenomenon of interest.

CONCLUSION

Theory and research on Pygmalion effects and the role of self-efficacy in the workplace have progressed considerably since Rosenthal's early work in the classroom during the late 1960s and Bandura's first paper on self-efficacy in 1977. However, the streams have been parallel and have been treated as largely independent areas of inquiry. The conceptualization presented here represents a step toward blending these streams by demonstrating that supervisory expectations may affect subordinate self-efficacy.

What is needed next is not more conjecture but empirical work that tests the proposed model, or at least facets of it, which will, in turn, direct researchers toward a path of greater understanding of the Pygmalion effect. Human resource executives in the restaurant industry have lauded the practice of appropriately encouraging subordinates and avoiding the all-too-common custom of berating employees in the ongoing quest to maximize productivity (c.f. Reynolds, 2000). Empirical work within the industry is now required; this includes tests of both Pygmalion and Golem effects. Furthermore, subordinates' demographic variables as moderators need to be tested in such studies in order to provide the foundation for more inclusive models representing this phenomenon.

The effects of supervisory expectations and their apparent link to subordinate behavior attenuate the appropriateness for further study geared to enhancing managerial and, in turn, organizational performance. Ultimately, more robust models that explain this valuable phenomenon will help us to better appreciate the application of "My Fair Lady" tactics in foodservice management, and maximize the resulting outcomes in the workplace.

NOTES

1. This four-factor model is often referred to, particularly by Rosenthal (1973, 1981), as a four-factor theory. However, as noted by Harris and Rosenthal (1985), it is not a formal theory. The researchers suggest, "... it is merely a convenient grouping of behaviors into molar categories which are easy to understand, but which are inductively derived and are not based on any prior theory of interpersonal behavior and influence" (p. 365).

2. It is important to note that while generality is one of the dimensions of self-efficacy, the construct cannot be applied as a general measure of social or physical self-efficacy. While some researchers have attempted to formulate such measures (e.g., Ryckman, Robbins, Thornton, & Cantrell, 1982; Sherer, Maddux, Mercandante, Prentice-Dunn, Jacobs, & Rogers, 1982), such tools violate the basic assumption of the multidimensionality of self-efficacy as noted by Bandura (1997). Whereas domain-specific measures of self-efficacy are good predictors of performance, global measures do not offer much in terms of predictive utility. Bandura (1997) provides a good example. He suggests that if efficacy beliefs were measured by assessing someone's general beliefs about academic performance, without specifying tasks or domains, such a measure would likely be a weak predictor of attainments in a particular scholastic domain such as mathematics.

3. Other behaviors might include planning, organizing, decision making, and providing leadership (Schoorman, Schechter, Moeller, & Schneider, 1988).

4. It should be noted that the arrow connecting subordinate's self-efficacy is bi-directional. Subordinates' self-efficacy is continually modified owing to the role of past performance as an antecedent of self-efficacy.

5. In no way is this discussion intended to suggest such information should be used for decisions–in any way–regarding hiring or promoting certain categories of individuals.

REFERENCES

Babad, E. (1993). Pygmalion–25 years after interpersonal expectations. In P. D. Blanck (Ed.), *Interpersonal expectations* Cambridge, England: Cambridge University Press,125-153.

Babad. E. (1995). The "teacher's pet" phenomenon: Students' perceptions of teachers' differential behavior and students' morale. *Journal of Educational Psychology*, 87, 361-368.

Babad, E., Inbar, J., & Rosenthal, R. (1982). Pygmalion, Galatea, and the Golem: Investigations of biased and unbiased teachers. *Journal of Education Psychology*, 74, 459-474.

Bandura, A. (1977). Self-efficacy: Toward a unifying theory of behavioral change. *Psychological Review*, 84, 191-215.

Bandura, A. (1982). Self-efficacy mechanism in human agency. *American Psychologist*, 37, 122-147.

Bandura, A. (1986). *Social foundations of thought and action: A social-cognitive view*. Englewood Cliffs, NJ: Prentice-Hall.

Bandura, A. (1997). *Self-efficacy: The exercise of control*. New York: W. H. Freeman.

Bandura, A., Adams, N. E., Hardy, A. B., & Howells, G. N. (1980). Tests of the generality of self-efficacy theory. *Cognitive Therapy and Research*, 4, 39-66.

Barling, J., & Beattie, R. (1983). Self-efficacy beliefs and sales performance. *Journal of Organizational Behavior Management*, 5, 41-51.

Baron, R. A. (1988). Negative effects of destructive criticism: Impact on conflict, self-efficacy, and task performance. *Journal of Applied Psychology*, 73, 199-207.

Baron, R. M., & Kenny, D. A. (1986). The moderator-mediator variable distinction in social psychology research: Conceptual strategies and statistical considerations. *Journal of Personality and Social Psychology*, 51. 1173-1182.

Bass, B. M. (1999). Ethics, character, and authentic transformational leadership. *Leadership Quarterly*, 10(2), 181-219.

Beez, W. V. (1971). *Influence of biased psychological reports on "teacher" behavior and pupil performance*. Unpublished doctoral dissertations, Indiana University.

Bowers, D. G., & Seashore, S. E. (1966). Predicting organizational effectiveness with a four-factor theory of leadership, *Administrative Science Quarterly*, 11, 238-263.

Braun, C. (1976). Teacher expectation: Socio-psychological dynamics. *Review of Educational Research* 46(2), 185-213.

Brophy, J. E., & Good, T. (1970). Teachers' communication of differential expectations for children's classroom performances: Some behavioral data. *Journal of Educational Psychology*, 61, 367-374.

Brophy, J. E., & Good, T. (1974). *Teacher-student relationships: Causes and consequences*. New York: Holt, Rinehart & Winston.

Brownell, J., & Reynolds, D. (2000). Implications of MBTI type: A study of on-site foodservice managers. *FIU Review*, 18(1), 1-18.

Campbell, D. T., & Stanley, J. C. (1963). *Experimental and quasi-experimental designs for research*. Chicago: Rand McNally College Publishing Company.

Carli, L. L. (1999). Gender, interpersonal power, social influence. *Journal of Social Issues*, 55(1), 81-99.

Chambliss, C. A., & Murray, E. J. (1979). Efficacy attribution, locus of control, and weight loss. *Cognitive Therapy and Research*, 3, 349-354.

Chowdhury, M. S. (1997). *Motivating the salesforce: The Pygmalion effect*. Unpublished doctoral dissertation, School of Business and Entrepreneurship, Nova Southeastern University.

Chusmir, L. (1986). How fulfilling are health care jobs? *Health Care Management Review*, 11, 27-32.

Collins, H., & Pinch, T. (1998). *The Golem: What you should know about science* (3rd ed.). Cambridge, England: Cambridge University Press.

Colwill, N. L. (1987). Men, women, power, and organizations. *Business Quarterly*, 51(4), 10-13.

Compton, J. W. (1988). Experimenter bias: Reaction time and types of expectancy information. *Perceptual and Motor Skills*, 31, 159-168.

Cooper, H. (1979). Pygmalion grows up: A model for teacher expectations, communications, and performance influence. *Review of Education Research*, 49, 389-410.

Cooper, H. M., & Hazelrigg, P. (1988). Personality moderators of interpersonal expectancy effects: An integrative research review. *Journal of Personality and Social Psychology*, 55, 937-949.

Cornbleth, C., Davis, O. L., Jr., & Button, C. (1974). Expectations for pupil achievement and teacher-pupil interactions. *Social Education*, 38, 54-58.
Cornelius, E., & Lane, F. (1984). The power motive and managerial success in a professionally oriented service industry organization. *Journal of Applied Psychology*, 69, 32-39.
Crawford, K. S., Thomas, E. E., & Fink, J. J. (1980). Pygmalion at sea: Improving the work effectiveness of low performers. *The Journal of Applied Behavioral Science*, 16, 482-503.
Dansereau, F., Graen, G., & Haga, W. (1975). A vertical dyad approach to leadership within formal organizations. *Organizational Behavior and Human Performance*, 13, 46-78.
Darley, J. M., & Fazio, R. H. (1980). Expectancy confirmation processes arising in the social interaction sequence. *American Psychologist*, 35, 867-881.
Davidson, O., & Eden, D. (2000). Remedial self-fulfilling prophecy: Two field experiments to prevent Golem effects among disadvantaged women. *Journal of Applied Psychology*, 85(3), 386-398.
Dion, P. A., & Banting, P. M. (2000). Comparisons of alternative perceptions of sales performance. *Industrial Marketing Management*, 29(3), 263-270.
DiPadova, L. N., & Faerman, S. R. (1993). Using the competing values framework to facilitate managerial understanding across levels of organizational hierarchy. *Human Resource Management*, 32(1), 143-171.
Dusek, J. B., Hall, V. C., & Meyer, W. J. (1985). *Teacher expectations*. Hillsdale, NJ: Erlbaum.
Dvir, T., Eden, D., & Banjo, M. L. (1995). Self-fulfilling prophecy and gender: Can women be Pygmalion and Galatea? *Journal of Applied Psychology*, 80(2), 253-270.
Eagly, A. H., & Johnson, B. T. (1990). Gender and leadership style: A meta-analysis. *Psychological Bulletin*, 108, 233-256.
Eden, D. (1984). Self-fulfilling prophecy as a management tool: Harnessing Pygmalion. *Academy of Management Review* 9(1), 64-73.
Eden, D. (1988). Pygmalion, goal setting, and expectancy: Compatible ways to boost productivity. *Academy of Management Review* 13(4), 639-652.
Eden, D. (1990a). *Pygmalion in management: Productivity as a self-fulfilling prophecy*. Lexington, MA: D.C. Heath and Company.
Eden, D. (1990b). Pygmalion without interpersonal contrast effects: Whole groups gain from raising manager expectations. *Journal of Applied Psychology*, 75(4), 394-398.
Eden, D., & Aviram, A. (1993). Self-efficacy training to speed reemployment: Helping people to help themselves. *Journal of Applied Psychology*, 78(3), 352-360.
Eden, D., & Ravid, G. (1981). *Effects of expectancy on performance among male and female trainees*. Paper presented at the International Interdisciplinary Congress on Women, Haifa, Israel.
Eden, D., & Ravid, G. (1982). Pygmalion vs. self-expectancy: Effects of instructor-and self-expectancy on trainee performance. *Organizational Behavior and Human Performance*, 30, 351-364.
Eden, D., & Shani, A. (1979). *Pygmalion goes to boot camp: Expectancy, leadership, and trainee performance*. Paper presented at the meeting of the American Psychological Association, New York.

Eden, D., & Shani, A. (1982). Pygmalion goes to boot camp: Expectancy, leadership, and trainee performance. *Journal of Applied Psychology*, 67, 194-199.

Eden, D., & Zuk, Y. (1995). Seasickness as a self-fulfilling prophecy: Raising self-efficacy to boost performance at sea. *Journal of Applied Psychology*, 80(5), 628-635.

Elashoff, E., Dixon, J., & Snow, R. (1971). *Pygmalion reconsidered: A case study in statistical inference: Reconsideration of the Rosenthal-Jacobson data on teacher expectancy*. Belmont, CA: Wadsworth Publishing Company.

Feldman, R. S., & Prohaska, T. (1979). The student as Pygmalion: Effect of student expectation on the teacher. *Journal of Educational Psychology*, 71(4), 485-493.

Feldman, R. S., & Theiss, A. J. (1982). The teacher and student as Pygmalions: Joint effects of teacher and student expectations. *Journal of Educational Psychology*, 74(2), 217-223.

Ferris, G. R., & King, T. R. (1992). The politics of age discrimination in organizations. *Journal of Business Ethics*, 11(5), 341-349.

Fletcher, C. (1999). The implication of research on gender differences in self-assessment and 360-degree appraisal. *Human Resource Management Journal*, 9(1), 39-46.

Frayne, C. A., & Latham, G. P. (1987) Application of social learning theory to employee self-management of attendance. *Journal of Applied Psychology*, 72, 387-392.

Friedkin, N. E. (1993). Structural bases of interpersonal influence in groups: A longitudinal case study. *American Sociological Review*, 58(6), 861-865.

Garland, H., Weinberg, R., Bruya, L., & Jackson, A. (1988). Self-efficacy and endurance performance: A longitudinal field test of cognitive mediation theory. *Applied Psychology: An International Review*, 37, 381-394.

Gecas, V. (1989). The social psychology of self-efficacy. *Annual Review of Sociology*, 15, 291-316.

Gist, M. E. (1987). Self-efficacy: Implications for organizational behavior and human resource management. *Academy of Management Review*, 12, 472-485.

Gist, M. E. (1989). The influence and training method on self-efficacy and idea generation among managers. *Personnel Psychology*, 42, 787-805.

Gist, M. E., & Mitchell, T. R. (1992). Self-efficacy: A theoretical analysis of its determinants and malleability. *Academy of Management Review*, 17, 183-211.

Gist, M. E., Schwoerer, C., & Rosen, B. (1989). Effects of alternative training methods on self-efficacy and performance in computer software training. *Journal of Applied Psychology*, 74, 884-891.

Goldhaber, G. M. (1993). *Organizational communication* (6th ed.). Madison, WI: Brown & Benchmark Publishers.

Golembiewski, R. T., & Osuna, W. (1995). Diversity and structure: Exploring a neglected but critical interface. *International Journal of Public Administration*, 18(8), 1161-1208.

Good, T. L., Cooper, H. M., & Blakely, S. L. (1980). Classroom interaction as a function of teacher expectations, student sex, and time of year. *Journal of Educational Psychology*, 72, 378-385.

Hall, J. A., & Briton, N. J. (1993). Gender, nonverbal behavior, and expectations. In P. D. Blanck (Ed.), *Interpersonal expectations*. Cambridge, England: Cambridge University Press. 276-295.

Harris, M. J. (1993). Issues in studying the mediation of expectancy effects: A taxonomy of expectancy situations. In P. D. Blanck (Ed.), *Interpersonal expectations*, Cambridge, England: Cambridge University Press, 350-378.

Harris, M. J., & Rosenthal, R. (1985). Mediation of interpersonal expectancy effects: 31 meta-analyses. *Psychological Bulletin*, 97, 363-386.

Hazelrigg, P. J., Cooper, H., & Strathman, A. J. (1991). Personality moderators of the experimenter expectancy effect: A reexamination of five hypotheses. *Personality and Social Psychology Bulletin*, 97, 363-386.

Hill, T., Smith, N. D., & Mann, M. F. (1987). Role of efficacy expectations in predicting the decision to use advanced technologies. *Journal of Applied Psychology*, 72, 307-314.

Hollander, E. P., & Offermann, L. R. (1990). Power and leadership in organizations: Relationships in transition. *American Psychologist*, 45, 179-189.

House, R. J. (1977). A 1976 theory of charismatic leadership. In J. G. Hunt & L. L. Larson (Eds.), *Leadership: The cutting edge*. Carbondale, IL: Southern Illinois University Press.

Igbaria, M., & Wormley, W. M. (1995). Race differences in job performance and career success. *Association for Computing Machinery*, 38(3), 82-98.

Jussim, L. (1986). Self-fulfilling prophecies: A theoretical and integrative review. *Psychological Review*, 93, 429-445.

Kester, S. W., & Letchworth, G. A. (1972). Communication of teacher expectations and their effects on achievement and attitudes of secondary school students. *Journal of Educational Research*, 66, 51-55.

Kierein, N., & Gold, M. A. (in press). Pygmalion in work organizations: A meta-analysis. *Journal of Organizational Behavior*.

King, A. S. (1970). *Managerial relations with disadvantaged work groups: Supervisory expectations of the underprivileged worker*. Unpublished doctoral dissertation, Texas Tech University.

King, A. S. (1971). Self-fulfilling prophecies in training the hard-core: Supervisors' expectations and the underprivileged workers' performance. *Social Science Quarterly*, 52, 369-378.

King, A. S. (1974). Expectation effects in organizational change. *Administrative Science Quarterly*, 19, 221-230.

King, P. (2000, May 15). Industry execs face a Catch-22 in hiring minorities. *Nation's Restaurant News*, 34(20), 18.

Knotts, R. (1975). Manifest needs of professional female workers in business-related occupations. *Journal of Business Research*, 3, 267-274.

Lenney, L. (1977). Women's self-confidence in achievement settings. *Psychological Bulletin*, 84, 1-12.

Lerner, B. S., & Locke, E. A. (1995). The effects of goal setting, self-efficacy, competition, and personal traits on the performance of an endurance task. *Journal of Sport & Exercise Psychology*, 17(2), 138-145.

Liden, R. C., Wayne, S. J., & Stilwell, D. (1993). A longitudinal study on the early development of leader-member exchanges. *Journal of Applied Psychology*, 78(4), 662-674.

Likert, R. (1961). *New patterns of management*. New York: McGraw-Hill.

Livingston, J. S. (1969). Pygmalion in management. *Harvard Business Review*, 47(4), 81-89

Locke, E.A., Frederick, E., Lee, C., & Bobko, P. (1984). The effect of self-efficacy, goals, and task strategies on task performance. *Journal of Applied Psychology*, 69, 241-251.

Locke, E. A., Saari, L. M., Shaw, K. N. & Latham, G. (1981). Goal setting and task performance: 1969-1980. *Psychological Bulletin*, 90, 125-152.

Lundberg, C. C. (1975). The effect of managerial expectation and task scope on team productivity and satisfaction: A laboratory study. *Journal of Business Research*, 3(3), 189-198.

Markin, J. R., & Lillis, M. C. (1975). Sales managers get what they expect. *Business Horizon* (June), 51-58.

Mase, H. S. (1970). *The effects of interpersonal expectancy upon learning achievement in an industrial setting*. Unpublished doctoral dissertation, City University of New York.

McClelland, D. C., & Boyatzis, R. (1982). Leadership motive pattern and long-term success in management. *Journal of Applied Psychology*, 67, 737-743.

McHugh, M., & Brennan, S. (1992). Organizational development and total stress management. *Leadership & Organization Development Journal*, 13(1), 27-32.

McNatt, D. B. (2000). Ancient Pygmalion joins contemporary management: A meta-analysis of the result. *Journal of Applied Psychology*, 85(2), 314-322.

Mendals, G. E., & Flanders, J. P. (1973). Teachers' expectations and pupil performance. *American Educational Journals*, 10, 36-42.

Merton, R. K. (1948). The self-fulfilling prophecy. *Antioch Review*, 8, 193-210.

Miller, V. D., & Jablin, F. M. (1991). Information seeking during organizational entry: Influence tactics, and a model of the process. *Academy of Management Review*, 16, 92-120.

Miller, D. T., & Turnbull, W. (1986). Expectancies and interpersonal processes. *Annual Review of Psychology*, 37, 233-256.

Mowday, R. T., Porter, L. W., Steers, R. M. (1982). *Employee-organization linkages: The psychology of commitment, absenteeism, and turnover*. New York: Academic Press.

Odiorne, G. S., *Management decision by objectives*. Englewood Cliffs, NJ: Prentice Hall.

O'Leary, V. E. (1977). *Toward understanding women*. Monterey, CA: Brooks/Cole.

Orpen, C. (1989). Self-esteem as a moderator of the effect of confidence in performance appraisals on managerial work attitudes. *Journal of Social Psychology*, 129(1), 133-144.

Oz, S., & Eden, D. (1994). Restraining the Golem: Boosting performance by changing the interpretation of low scores. *Journal of Applied Psychology*, 79, 744-754.

Pajares, F., & Kranzler, J. (1995). Self-efficacy beliefs and general mental ability in mathematical problem solving. *Contemporary Educational Psychology*, 20, 426-443.

Pajares, F., & Miller, M. D. (1994). Role of self-efficacy and self-concept beliefs in mathematical problem solving: A path analysis. *Journal of Educational Psychology*, 86, 193-203.

Peake, P. K., & Cervone, D. (1989). Sequence anchoring and self-efficacy: Primacy effects in the considerations of possibilities. *Social Cognition*, 7, 31-50.

Postman, N., & Weingartner, C. (1969). *Teaching as a subversive activity*. New York: Delacorte Press.

Quinones, M. A., Ford, J. K., & Teachout, M. S. (1995). The relationship between work experience and job performance: A conceptual and meta-analytic review. *Personnel Psychology*, 48(4), 887-893.

Reynolds, D. (1999). Motivating needs of managers in the on-site foodservice segment. In K. Chon (Ed.), *The Practice of graduate research in hospitality and tourism* New York: The Haworth Press, Inc., 129-142.

Reynolds, D. (2000). An exploratory investigation into behaviorally based success characteristics of foodservice managers. *Journal of Hospitality and Tourism Research*, 24(1), 127-138.

Reynolds, D., & Tabacchi, M. (1993). Burnout in full-service chain restaurants. *Cornell Hotel and Restaurant Administration Quarterly*, 34(2), 62-68.

Roach, B. (1986). Organizational decision makers: Different types for different levels. *Journal of Psychological Type*, 12, 16-24.

Rogers, C. R. (1959). A theory of therapy, personality, and interpersonal relationships, as developed in the client-centered framework. In S. Kock (Ed.), *Psychology: A study of a science. Formulations of the personal and the social context*. New York: McGraw-Hill, (3) 184-256.

Rosenthal, R. (1969). Interpersonal expectations: Effects of the experimenter's hypothesis. In R. Rosenthal & R. L. Rosnow (Eds.), *Artifact in behavioral research*. New York: Academic Press.

Rosenthal, R. (1971). The silent language of classrooms and laboratories. *Proceedings of the Parapsychological Association*, 8, 95-116.

Rosenthal, R. (1973). The mediation of Pygmalion effects: A four factor "theory." *Papua New Guinea Journal of Education*, 9, 1-12.

Rosenthal, R. (1974). *On the social psychology of the self-fulfilling prophecy: Further evidence for Pygmalion effects and their mediating mechanisms*. New York: MSS Modular Publications, Module 53.

Rosenthal, R. (1981). Pavlov's mice, Pfungst's horse, and Pygmalion's PONS: Some models for the study of interpersonal effects. In T. A. Sebeok & R. Rosenthal (Eds.), The Clever Hans phenomenon: Communication with horses, whales, apes, and people. *Annuals of the New York Academy of Sciences*. New York: New York Academy of Sciences, 364.

Rosenthal, R. (1989). *Experimenter expectancy, covert communication, and meta-analytic methods*. American Psychological Association meeting, August 14.

Rosenthal, R. (1991). Teacher expectancy effects: A brief update 25 years after the Pygmalion experiment. *Journal of Research in Education*, 1, 3-12.

Rosenthal, R. (1993). Interpersonal expectation: Some antecedents and some consequences. In P. D. Blanck (Ed.), *Interpersonal expectations* Cambridge, England: Cambridge University Press, 1-24.

Rosenthal, R. (1994). Interpersonal-expectancy effects: A 30-year perspective. *Current Directions in Psychological Science*, 3, 176-179.

Rosenthal, R., & Jacobson, L. (1968). *Pygmalion in the classroom: Teacher expectation and pupils' intellectual development*. New York: Holt, Rinehart & Winston.

Rosenthal, R., & Rubin, D. B. (1978). Interpersonal expectancy effects: The first 345 studies. *Behavioral and Brain Sciences*, 3, 377-415.

Rothbart, M., Dalfen, S., & Barrett, R. (1971). Effects of teacher's expectancy on student-teacher interaction. *Journal of Educational Psychology*, 62, 49-54.

Rubovits, P. S., & Maehr, M. L. (1973). Pygmalion Black and White. *Journal of Personality and Social Psychology*, 25, 210-218.

Saks, A. M., & Waldman, D. A. (1998). The relationship between age and job performance evaluations for entry-level professionals. *Journal of Organizational Behavior*, 19(4), 409-419.

Saunders, K. T., & Saunders, P. (1999). The influence of instructor gender on learning and instructor ratings. *Atlantic Economic Journal*, 27(4), 460-473.

Seton-Williams, M. V. (1993). *Greek legends and stories*. London: Rubicon Press.

Shimko, B. W. (1990). The McPygmalion effect. *Training and Development Journal*, 44(6), 64-70.

Shore, L. M., & Thornton, G. C., III. (1986). Effects of gender on self-and supervisory ratings. *Academy of Management Journal*, 29(1), 115-129.

Smith, P. B., Misumi, J., Tayeb, M., Peterson, M., & Bond, M. (1989). On the generality of leadership style measures across cultures. *Journal of Occupational Psychology*, 62(2), 97-110.

Smits, S., McLean, E., & Tanner, J. (1993). Managing high-achieving information system professionals. *Journal of Management Information Systems*, 9, 103-117.

Snow, R. E. (1969). Unfinished Pygmalion. *Contemporary Psychology*, 14, 197-199.

Solomon, R. L. (1949). An extension of control group design. *Psychological Bulletin* 46, 137-150.

Sproul, N. L. (1995). *Handbook of research methods: A guide for practitioners and students in the social sciences* (2nd ed.). London: The Scarecrow Press, Inc.

Stahl, M., & Harrell, A. (1982). Evolution and validation of a behavioral decision theory measurement approach to achievement, power, and affiliation. *Journal of Applied Psychology*, 67, 744-751.

Stajkovic, A.D., & Luthans, F. (1998a). Self-efficacy and work-related performance: A meta-analysis. *Psychological Bulletin*, 124(2), 240-261.

Stajkovic, A.D., & Luthans, F. (1998b). Social cognitive theory and self-efficacy: Going beyond traditional motivational and behavioral approaches. *Organizational Dynamics*, 26(4), 62-74.

Stumpf, S. A., & London, M. (1981). Capturing rater policies in evaluating candidates for promotion. *Academy of Management Journal*, 24, 752-766.

Sutton, C. D. (1986). *Pygmalion goes to work: The effects of supervisor expectations in a retail setting*. Unpublished doctoral dissertation, Texas A&M University.

Sutton, C. D., & Woodman, R. W. (1989). Pygmalion goes to work: The effects of supervisor expectations in a retail setting. *Journal of Applied Psychology*, 74, 943-950.

Taylor, M. C. (1993). Comment: Never-ending nets of moderators and mediators. In P. D. Blanck (Ed.), *Interpersonal expectations*. Cambridge, England: Cambridge University Press, 454-474.

Taylor, M. S., Audia, G., & Gupta, A. K. (1996). The effect of lengthening job tenure on managers' organizational commitment and turnover. *Organization Science*, 7(6), 632-648.

Telch, M. J., Bandura, A., Vinciguerra, P., Agras, A., & Stout, A. L. (1982). Social demand for consistency and congruence between self-efficacy and performance. *Behavior Therapy*, 13, 694-701.

Thorndike, R. L. (1968). Review of *Pygmalion in the Classroom* by R. Rosenthal and L. Jacobson. *American Educational Research Journal*, 5, 708-711.

Tolkien, J. R. R. (1954). *The fellowship of the ring* (*Part I of Lord of the rings*). Boston: Houghton Mifflin Company.

Van Velser, E. (1987). Database analysis; Leadership development program, Center for Creative Leadership. Paper presented at the Conference on Psychological Measures and Leadership, San Antonio, TX, as reported in Gardner, W. L. & Martinko, M. J. (1996). Using the Myers-Briggs Type indicator to study managers: A literature review and research agenda. *Journal of Management*, 22(1), 45-83.

Vrugt, A. (1990). Negative attitudes, nonverbal behavior and self-fulfilling prophecy in simulated therapy interviews. *Journal of Nonverbal Behavior*, 14, 77-86.

Watson, J., & Barone, S. (1976). The self-concept, personal values, and motivational orientations of black and white managers. *Academy of Management Journal*, 19, 36-45.

Weick, K. E. (1979). *The social psychology of organizing* (2nd ed.). Reading, MA: Addison-Wesley.

Weiner, B. (1985). An attributional theory of achievement, motivation and emotion. *Psychological Review*, 92, 548-573.

Wilson, D. K., Wallston, K. A., & King, J. E. (1990). Effects of contract framing, motivation to quit, and self-efficacy on smoking reduction. *Journal of Applied Psychology*, 1, 531-547.

Wood, R. E., & Bandura, A. (1989). Social cognitive theory of organizational management. *Academy of Management Review*, 14, 361-384.

Wood, R. E., & Locke, E. A. (1987). The relation of self-efficacy and grade goals to academic performance. *Educational and Psychological Measurement*, 47, 1013-1024.

Woodruff, S. L., & Cashman, J. F. (1993). Task, domain, and general efficacy: A reexamination of the self-efficacy scale. *Psychological Reports*, 72, 423-432.

Woolridge, B. (1994). Changing demographics of the workforce: Implications for the use of technology as a productivity improvement strategy. *Public Productivity & Management Review*, 17(4), 371-385.

Wylie, R. C. (1974). *The self-concept: A review of methodological considerations and measuring instruments* (rev. ed.). Lincoln, NE: University of Nebraska Press.

Yukl, G. (2000, February). *Transformational leadership: Does it matter in the year 2000*? Panel discussion including B. Avolio, J. Beyer, R. House, & T Judge, conducted at Cornell University.

Zalesny, M. D. (1990). Rater confidence and social influence in performance appraisals. *Journal of Applied Psychology*, 75(3), 274-289.

RESEARCH NOTES

Quick Service Chain Restaurant Managers: Temperament and Profitability

Nancy Swanger

SUMMARY. The purpose of this study was to describe the effect of manager temperament on unit profitability for quick service chain restaurants. A questionnaire incorporating the Keirsey Temperament Sorter was used to determine the temperament type of each unit manager, demographic data, and unit profitability. Of the 195 managers responding, only 69 included an answer to the question on unit net profit/loss. Results indicated that gender was significant in relation to unit profitability. Women managers reported profits 5.79% above the overall mean (Multiple Classification Analysis). Of the three most frequently occurring temperament types, almost 69% of the managers in the high profit range were ESTJ (extraverted, sensing, thinking, judging). Although there appeared to be a difference among the mean values, it was not significant (Chi-square). Further stud-

Nancy Swanger, EdD, is a Faculty Member, Department of Hospitality Management, College of Business Management, Washington State University, Todd Hall, Pullman, WA.

[Haworth co-indexing entry note]: "Quick Service Chain Restaurant Managers: Temperament and Profitability." Swanger, Nancy. Co-published simultaneously in *Journal of Restaurant & Foodservice Marketing* (The Haworth Hospitality Press, an imprint of The Haworth Press, Inc.) Vol. 4, No. 4, 2001, pp. 303-319; and: *Quick Service Restaurants, Franchising and Multi-Unit Chain Management* (ed: H. G. Parsa and Francis A. Kwansa) The Haworth Hospitality Press, an imprint of The Haworth Press, Inc., 2001, pp. 303-319. Single or multiple copies of this article are available for a fee from The Haworth Document Delivery Service [1-800-HAWORTH, 9:00 a.m. - 5:00 p.m. (EST). E-mail address: getinfo@haworthpressinc.com].

ies are recommended to determine the relationship between temperament and profitability.

KEYWORDS. Manager temperament, profitability, quick service restaurants, restaurant manager

As quick service chain restaurant (QSR) managers move into the 21st century, they face an unprecedented challenge: too many food outlets with increasingly fewer patrons per unit. Competition is fierce; managers will be expected to do more with less in order to maintain acceptable levels of profitability. As the demands and expectations of owners, guests, and employees increase, the manager of the 21st century will have to be versatile enough to adequately answer the call on all fronts. Certainly, with enough resources like time and money, companies could train their managers to become better competitors for the next generation of QSR operations. Unfortunately, both time and money have already become very scarce resources. Is it possible for owners to predetermine which managerial candidates are more likely to run profitable operations by discovering their temperament type through a simple, inexpensive test that could be used as a part of their selection process? In this study, an attempt has been made to answer this question.

According to Maynard and Champoux (1996), the turnover rate in the restaurant industry at the management level frequently runs between 50-100% annually. It was intended that results from the current study would provide additional information for future studies in lowering turnover levels of quick service chain restaurant managers.

LITERATURE REVIEW

Profitability is a very sensitive, proprietary issue; managers are reluctant to provide their unit's profit percentage. Franchisers are prohibited from providing possible net profit or loss information to prospective franchisees in company disclosure statements. Established franchise operators rarely share income statements with others for a variety of reasons; of which, the possible attraction of unwanted competition may head the list. A quick glance of the hospitality literature reveals that very little research has been conducted and published relating temperament of quick service chain restaurant managers and their unit's profitability. Hence, the current study focuses on relationship between personality type of QSR chain restaurant managers and their unit (restaurant) performance with the aid of Keirsey Temperament Sorter.

Keirsey Temperament Sorter

The Keirsey Temperament Sorter is a self-scoring instrument that many researchers have found to be very similar to the Myers-Briggs Type Indicator (MBTI). Most recently, studies by Quinn, Lewis, and Fischer (1992) and Tucker and Gillespie (1993) found that both instruments measured the same constructs and either could be used to determine a person's personality type. Both involve choices between *extraversive* or *introversive* attitudes, *sensation* or *intuition, feeling* or *thinking,* and *judging* or *perceiving* functions to describe and differentiate categories of people according to the way they prefer to use their minds (Murray, 1990).

Of the two instruments, the MBTI is the most widely known and the most used. It also claims more research publications than the Keirsey. However, there are some differences between MBIT and Keirsey Sorter. The MBTI is a Type B instrument, meaning only certified personnel can administer it, and there is a per-test fee that must be paid. Obviously, if used on a large sample, the dollar amount could be prohibitive. On the other hand, the only fee involved in using the Keirsey Temperament Sorter was the price of the book ($11.95) on the subject, *Please Understand Me* by David Keirsey and Marilyn Bates. The book includes the research instrument, procedures for its use and scoring, and detailed descriptions of each possible type. This information, including the instrument, is also available on line at no charge. Conceptually both MBTI and Keirsey Sorter are very similar, and both were found to be effective in measuring personality traits.

The research of Katherine Briggs and Isabel Briggs Myers begun in the 1950s gave practical application to the theoretical work on type done by Carl Jung during the 1920s, which had essentially been abandoned (Tieger & Barron-Tieger, 1995). The descriptions of the personality types given by Keirsey and Bates in the 1980s are a bit different from those given by Jung, but are essentially the same as those used by Myers and Briggs.

Carl Jung, a Swiss psychologist, distinguished personality types by the extravert and the introvert with four distinct functions.

> For the extrovert, the object is interesting and attractive a priori, as is the subject, or psychic reality, for the introvert. We could therefore use the expression "numinal accent" for this fact, by which I mean that for the

> extrovert the quality of positive significance and value attaches primarily to the object, so that it plays the predominant, determining, and decisive role in all psychic processes from the start, just as the subject does for the introvert.
>
> But the numinal accent does not decide only between subject and object; it also selects the conscious function of which the individual makes the principal use. I distinguish four functions: thinking, feeling, sensation, and intuition. (Jung, 1936, translated 1971)

Jung, however, did not have the functions of judging and perceiving as part of his theory. Isabel Myers and Katherine Briggs began looking for a practical application for Jung's theory during World War II. Their interest stemmed from the desire to find a way to better match women to jobs for the war effort. As no instrument could be found to aid in the process, the two women created one that eventually became the Myers-Briggs Type Indicator (Myers & Myers, 1995).

Myers and Briggs, and later Keirsey and Bates, used the following dimensions to type personality:

Extroversion versus Introversion: This preference is based on the way people draw their energy. Extroverts (E) prefer to be around people and renew their energy when doing so. Introverts (I) prefer solitude when their energy levels are low and need quiet time to rejuvenate. About 75 percent of the general population of the United States is Extroverted, and 25 percent of the population is Introverted (Tieger & Barron-Tieger, 1995). Tieger and Barron-Tieger (1995) note this Extroverted population may help account for this country's fondness of fast food–Extroverts are often on the run. Keirsey and Bates (1984) report the main word that differentiates an Extrovert from an Introvert is "sociability" as opposed to "territoriality." Other words used by Keirsey and Bates (1984) to cue the preference of Extroversion over Introversion included: breadth versus depth, external versus internal, extensive versus intensive, and interaction versus concentration.

Intuition versus Sensation: This preference in based on the way people process information. The Sensors (S) prefer to rely on their five senses to get accurate information. The Intuitives (N) prefer to rely on their "sixth sense" or gut feeling when processing information. About 75 percent of the population prefer Sensation, while 25 percent of the population prefer Intuition (Tieger & Barron-Tieger, 1995). Keirsey and Bates (1984) provided cue words to indicate a preference of Sensation over Intuition. Those words included experience versus hunches, past versus future, realistic versus speculative, and perspiration versus inspiration.

Thinking versus Feeling: This preference is based on the ways people make choices or decisions. The Thinkers (T) prefer impersonal, objective judgments and the Feelers (F) prefer value judgments. Tieger and Barron-Tieger (1995) reported that the population is evenly split between Thinkers and Feelers; however, this is the only type dimension in which clear gender differences show up. About two-thirds of men prefer Thinking, and about two-thirds of women prefer Feeling. The cue words provided by Keirsey and Bates (1984) to show preference of Thinking over Feeling include: objective versus subjective, principles versus values, firmness versus persuasion, and critique versus appreciate.

Judging versus Perceiving: This preference is based on the way people prefer to live their lives. The Judgers (J) prefer orderliness, structure and routine. The Perceivers (P) prefer openness, options and spontaneity. Again, the population is evenly split on this dimension (Tieger & Barron-Tieger, 1995). Keirsey and Bates (1984) provided the following cue words to show preference for Judging over Perceiving: settled versus pending, fixed versus flexible, completed versus emergent, and decisive versus tentative.

Mixed Types: An X appears in the temperament type when the respondent's selections within the categories (E/I, S/N, T/F, and J/P) are equal. This indicates a mixed type. Mixed Typers do not belong any one single type.

Keirsey (1997) has given each of the sixteen types a one-word descriptor similar to the ones used to describe the different management temperaments. They are: ESFP-Performer, ISFP-Composer, ESTP-Promoter, ISTP-Operator, ESFJ-Provider, ISFJ-Protector, ESTJ-Supervisor, ISTJ-Inspector, ENFP-Advocate, INFP-Conciliator, ENFJ-Teacher, INFJ-Counselor, ENTP-Inventor, INTP-Architect, ENTJ-Fieldmarshal, and INTJ-Masterplanner.

Different preferences among these dimensions yield sixteen personality types. Underlying these sixteen personality types are four temperaments as indicated by Hippocrates, Spranger and others (Keirsey & Bates, 1984). These four temperaments are known as Guardians, Idealists, Rationals, and Artisans (Keirsey, 1997). Keirsey's descriptions of these temperaments are as follows:

Guardians (SJ): The temperament of the Guardians is driven by the need for status and this group comprises between 40-45 percent of the population. The types that are included in this temperament are ISTJ, ESTJ, ISFJ, and ESFJ.

At work (Tieger & Barron-Tieger, 1995), the SJs need to belong, to serve, and to do the right thing. They value stability, orderliness, cooperation, consistency, and reliability, and they tend to be serious and hardworking. SJs demand a great deal of themselves on the job and expect the same of others. In a quick service restaurant, the SJs would make good managers because of their strong needs to serve others, enforce standards, assume responsibility, and to do the right thing at the right time. Keirsey (1997) describes the SJ manager as the Stabilizer.

Idealists (NF): The temperament of the Idealists is driven by the need for meaning and they comprise between 8-10 percent of the population. The types included in this temperament are INFP, ENFP, INFJ, and ENFJ.

At work, NFs enjoy using their natural ability to understand and connect with other people. They are naturally empathetic and focus on the needs of the people involved in their work (Tieger & Barron-Tieger, 1995). In a quick service restaurant, the NFs would make good crew trainers because of their desire to develop the good in everyone and communicate with people. The NFs are also born entrepreneurs; entrepreneurs founded the restaurant industry. Keirsey (1997) dubs the NF manager the Catalyst.

Rationals (NT): The temperament of the Rationals is driven by the need for power and they make up between 5-7 percent of the population. Types included in this temperament are the INTPs, the ENTPs, INTJs, and ENTJs.

At work, NTs enjoy using their abilities to see possibilities and analyze them logically to solve problems. They are interested in constantly acquiring knowledge, either for its own sake of for a strategic purpose (Tieger & Barron-Tieger, 1995). The NT would be most comfortable working at the corporate level of a quick service chain restaurant because of the need to develop strategic plans, focus on the future, and be a visionary. In fact, Keirsey (1997) refers to the NT manager as the Visionary.

Artisans (SP): The temperament of the Artisans is driven by the need for freedom and they comprise 35-40 percent of the population. This temperament includes the ISFP, ESFP, ISTP, and ESTP types. Tieger and Barron-Tieger (1995) describe the SPs at work as needing to be active and free to act on their impulses. They focus on what can be accomplished in the here and now. SPs value heroic deeds and masterful acts and like moving from one challenge to the next. The SPs might be drawn to a career in quick service chain restaurants because of their need to vary their work patterns each day, because they have amazing endurance, and because they like to have fun and entertain. The Negotiator is the term used by Keirsey (1997) to describe the SP manager.

According to Watson (1990), the Keirsey Temperament Sorter has not been tested as thoroughly as the MBTI for validity and reliability. But when the current authors contacted David Keirsey, son of late Keirsey, regarding validity and reliability of the Keirsey Sorter, he responded as follows:

> . . . Informally, the Keirsey Temperament Sorter seems as reliable as the Myers-Briggs, both about .25 error. There is a study correlating the Keirsey Temperament Sorter with the Myers-Briggs, and informally my father found the correlation has been shown to be about .75.
>
> He further noted that "A study confirms the stated correlation, and the reference is 'Correlations among three measures of personality

type,' Irving F. Tucker, Bonnie Gillespie, *Perceptual and Motor Skills,* Oct. 1993, v77, n2, p650." (D. Keirsey, personal communication, February 25, 1998)

The Myers-Briggs Type Indicator: Atlas of Type Tables (1986) provided a breakdown by personality type within occupations. According to MBTI, in the occupational category, "Managers: Restaurant, Cafeteria, Bar and Food Service," five types were found to be most prominent: ESTJ–27.56%, ISTJ–13.46%, ESFJ and ENTJ–9.62% each, and ISFJ–8.87%. In the restaurant industry, like any other industry, some managers perform better than others. Some restaurants are more profitable than others. Thus, MBTI personality types and performance of a restaurant manager are related–the personality types of the managers in the more profitable restaurants may be better suited to the profession than the manager personality types in the less profitable restaurants.

Quick Service Chain Restaurant Profitability

Withiam (1997) cited figures from the 1997 Public Chain Restaurant Company Report showing a profit margin of 14.7% on 1996 sales for McDonald's–$10.7 billion in revenues and net income of $1.5 billion. The National Restaurant Association (1997) released figures from a study conducted with the accounting firm of Deloitte & Touche on 1995 sales information of 2,200 restaurants that found quick service restaurants to show an average pretax profit of 9%. McDonald's performed well above the average, by comparison, to the profit percentage cited in the Deloitte and Touche study.

Of the aggregate $112.5 billion, as reported in *Nation's Restaurant News,* for the year ending December 31, 1996, quick service chains accounted for just over 63% of the sales or nearly $71 billion. Classifications of restaurants included sandwich, pizza, chicken, snack, and seafood.

Mone and Umbreit (as cited in Shingler & Ludwick, 1995) state, "At the single unit level, restaurants are examples of consumer service organizations requiring managers to implement a set of standardized operating procedures to achieve predetermined goals." These goals are often centered on sales volume, food/paper/labor costs, and profit.

Many of the chain operations franchise their business format. This allows people (franchisees) to use the name, logo, recipes, system, products, and marketing of the particular chain. Why do people buy franchises rather than run independent businesses? The U.S. Department of Commerce (as cited in George, 1995) reports that 38% of all new businesses fail within the first year of operation and 77% fail within the first five years of operation. The most frequently named reason for failure is the combination of a lack of a sound business control system and poor management skills. Buying into a proven concept, such as a chain operation, greatly reduces the risk of failure as the system has been previously tested and refined. Bond (as cited in George, 1995) suggests that only between 4.4% and 5% of all franchise businesses fail within the first year.

A survey conducted by Rice (1997) found the unit manager emerging as being the most critical element in employee satisfaction. Unit managers normally earn a salary in the $25,000 to $40,000 range and are eligible for bonus compensation equal to 10% to 35% of their base salary (Shingler & Ludwick, 1995).

Although the number of qualified applicants for salaried positions in the restaurant industry has declined, operators have become more stringent in their selection process. The National Restaurant Association (1996) stated that "although there is a dwindling supply of labor, the consumer increasingly expects better service, and better service requires better employees." Rice (1997) projected there will be a need for about 160,000 new unit managers by 2005; unit managers have the second highest anticipated percentage growth. The Research and Analysis Bureau of the Idaho Department of Employment shows food service managers/fast food positions on the lists of fast growing occupations and high demand occupations through the year 2005.

According to Bedford and Ineson (1995), testing for management trainee selection is common and occurs during one of three stages of the hiring process: screening applicants for further investigation, screening selected applicants prior to the final assessment, and for making or supporting final decisions. Bedford and Ineson (1995) state that personality tests may be more time consuming and costly to implement for selection purposes than cognitive tests, yet perhaps should be given greater consideration for selection in an industry where personality traits are believed to be crucial to successful performance.

Personality Type and Performance

Personality and performance can be measured in many ways. With regard to personality, some of the personality traits which were studied earlier included positive and negative affectivity, Type A and Type B personalities, the "Big Five" (extraversion, agreeableness, conscientiousness, emotional stability, and openness to experience), charisma, and competitiveness. Some of the instruments used to measure personality included the NEO Five-Factor Inventory, the Eysenck Personality Profiler, the Hogan Personality Inventory, and the Jackson Personality Inventory. One study (Nordvik and Brovold, 1998) compared the personality dimensions of the Myers-Briggs Type Indicator with the respondents' preferences for leadership tasks. They found some variations among leadership traits. Similar variations were found when performance was considered. Studies generally measured performance in terms of individual, team, or organizational results. It appears as if the effect personality has on performance may depend on the instrument being used, which traits are being defined, and what level of performance is being measured.

Recent literature indicates that there is some general support, in recent years, that personality does indeed have an effect on performance. Furnham, Forde, and Ferrari (1999) conclude, "personality factors appeared to account for between 20 and 30% of the variance in work performance." According to Pervin (as cited in George, 1992), "for each individual there are environments (interpersonal and noninterpersonal) which more or less match the characteristics of his [or her] personality. A 'match' or 'best-fit' of individual to environment is viewed as expressing itself in high performance, satisfaction, and little stress in the system whereas a 'lack of fit' is viewed as resulting in decreased performance, dissatisfaction, and stress in the system." Huston and Huston (1995) state, "In any profession, certain personality types are better suited for certain roles than others." Hogan, Hogan, and Roberts (1996) conclude "that (a) well-constructed measures of normal personality are valid predictors of performance in virtually all occupations, (b) they do not result in adverse impact for job applicants from minority groups, and (c) using well-developed personality measures for pre-employment screening is a way to promote social justice and increase organizational productivity." This leads to the following research hypotheses, stated in null form:

H1: In quick service restaurant chains, performance of a restaurant unit is not related to the personality type of the unit manager.

H2: In quick service restaurant chains, performance of a restaurant unit is not related to the demographic variables of the unit manager.

Zemke (1992) includes the following in his work: "Allen L. Hammer is a researcher with Consulting Psychologists Press, Inc. of Palo Alto, CA, the publisher of the MBTI, and editor-in-chief of the company's interpretation manual. 'The underlying concept for the use of instruments in career counseling since the 1930s is matching people with a job that is congruent with their interests and preferences. The MBTI does that as well or better than any instrument on the market,' Hammer says." As mentioned earlier, the MBTI and the Keirsey Temperament Sorter have been determined to measure the same constructs.

METHODOLOGY

The subjects in this study were 326 quick service chain restaurant managers in Idaho. All quick service chain restaurants, obtained from the most current list of establishments in each of the health districts in Idaho and from Survey Sampling, Inc., were identified and the entire population was surveyed. Telephone calls were made to all the restaurants on the list to obtain the manager's name. The unit manager was then contacted, via mail, to participate in the study.

The subjects were informed that the purpose of the study was to determine the relationship between individual temperament and unit profitability. Participation was voluntary; thus, the completion of the questionnaire constituted informed consent. Managers were assured that all responses would be kept anonymous and confidential.

A survey was used in this study that incorporated the Keirsey Temperament Sorter and some general demographic questions. Each of the sixteen types (INFP, ENFP, INFJ, ENFJ, ISFP, ESFP, ISFJ, ESFJ, INTP, ENTP, INTJ, ENTJ, ISTP, ESTP, ISTJ, and ESTJ) and any additional mixed types categorized responses to the Keirsey Temperament Sorter, as well.

All subjects were given the same survey and asked to respond to all questions. Completion required about 20 minutes. The Social Survey Research Unit at the University of Idaho was used for design of the questionnaire and for data collection and entry. A pilot study, as recommended by Thomas and Nelson (1996), was conducted in two neighboring counties in late December 1997. A questionnaire with cover letter was mailed to all 326 names on January 14, 1998. A postcard reminder was mailed January 21 and a second questionnaire with cover letter was mailed February 4. On February 18, managers who had not yet responded were called on the phone and reminded to return their questionnaire. A total of 48 questionnaires were mailed to managers who requested another one on February 19. There were 195 completed surveys with a response rate of 59.8%.

Unit profitability was the dependent variable. Three ranges of unit profitability (low, medium, and high) were categorized after data collection was completed with approximately one-third of the valid responses in each level. The independent variable, manager temperament, was measured with Keirsey Sorter instrument.

A multiple classification analysis (MCA) was used to determine the relationship between temperament type and profitability. Chi-square was used to analyze the relationship between the three most frequently occurring temperament types and the three ranges of profitability.

The following independent variables and categories were established prior to running the MCA:

1. Gender (2 categories with male/female)
2. Years Total Restaurant Experience (4 categories with approximately 25% of the valid responses in each)
3. Years With Current Employer (4 categories with approximately 25% of the valid responses in each)
4. Overall Job Satisfaction (5 categories using Likert-scale choices from the questionnaire)
5. Temperament Type (5 categories using the 4 most frequently occurring types and one labeled "other" to include all remaining types)

RESULTS AND DISCUSSION

The obtained results indicated that a typical QSR manager in Idaho is a 33-year-old Caucasian married male with some college or technical schooling, earning between $20,000 and $29,999. This manager has almost 10.5 years of restaurant experience and has been with their current employer for over 6 years. Five temperament types represent 80% of the population: ESTJ (28%), ESFJ (24%), ISTJ (11%), ISFJ (9%), and XSFJ (8%). Though the percentages varied a bit, the four most frequently occurring types among Idaho restaurant managers were consistent with the four most frequently occurring types presented by the *Myers-Briggs Type Indicator: Atlas of Type Tables* for that occupation.

An analysis of variance, using multiple classification analysis, showed that of this study's variables potentially influencing unit profitability, only gender was significant, $F(1,62) = 3.69, p < .10$. As shown in Table 1, women managers were in restaurants producing above the mean for profitability. This provides partial rejection of the Hypothesis 2 stating no differences between demographics and profitability.

Chi-square was also used to test the significance of the differences between levels of education and levels of income. Managers with an associate's degree or higher were more likely to be in the higher income groups

The other factors (years of restaurant experience, years with current employer, overall job satisfaction, and temperament type) were not statistically significant. However, of the personality types analyzed, the ESTJ temperament type performed above the mean. Table 2 contains the ANOVA table and Table 3 contains the Multiple Classification Analysis.

Chi-square was used to test the significance of differences in personality type on the level (low, medium, and high) of unit profitability. Of the three most frequently occurring types (ESFJ, ESTJ, and ISTJ), 68.8% of the managers in the high profit category were ESTJ. Although there appeared to be a difference, it was not significant. Table 4 contains a summary table of the Chi-square analysis. The posited hypothesis 1 could not be rejected. Hypothesis One states that there is no difference between personality types and unit profitability.

TABLE 1. Unit Profit/Loss Percentage by Gender in Comparison to the Grand Mean

	n	*M*
Female	16	21.99%
Male	47	14.23%

Profitability Grand Mean = 16.2%

TABLE 2. Analysis of Variance for Restaurant Profitability

Source	*df*	*SS*	*MS*	*F*
Gender	1	622.42	622.42	3.69*
Years Total Restaurant Experience	3	112.41	37.47	.22
Years with Current Employer	3	120.68	40.23	.24
Overall Job Satisfaction	4	395.07	98.77	.59
Temperament Type	4	644.40	161.10	.96
Error	47	7931.50	168.76	
Total	62	10,293.51	166.02	

*$p < .10$.

Of the 195 respondents, 74% were managers, 19% were owner/managers, 6% were area/district managers, and 1% were owners. Regarding type of ownership, 77% of the businesses were franchised independents, 11% were company owned, 6% were joint venture, and 6% reported varying answers ($n = 171$).

Only 39% of the respondents included an answer to the net profit/loss question. Of the information obtained, there is some question as to its accuracy. Were the profit percentages reported actual, estimated, or created? The outliers that reported 100% are not included in data analysis. The National Restaurant Association released figures, as mentioned earlier, that showed an average profitability of 9% for quick service restaurants. The study completed by this researcher in Idaho found average profitability to be 16.2%.

Do Idaho's quick service restaurants really outperform the rest of the country on average, or were the percentages misreported? Of the non-respondents to the profitability question, it was either left blank, answered with, "I don't know; the bookkeeper/owner has that information," or answered with, "I can't give this information out." Only five managers reported a net loss. The re-

TABLE 3. Multiple Classification Analysis for Restaurant Profitability

Variable/Category	*n*	Deviation from Grand *M**	β**
Gender			0.26
Male	47	−1.97	
Female	16	5.79	
Years Total Restaurant Experience			0.12
0.00-4.00 years		7	−1.86
4.01-9.00 years		18	1.29
9.01-15.00 years		22	1.14
15.01-32.00 years		16	−2.21
Years with Current Employer			0.13
0.00-2.00 years		8	1.18
2.01-4.00 years		16	0.41
4.01-9.00 years		20	−2.28
9.01-32.00 years		19	1.56
Overall Job Satisfaction			0.21
Poor		1	−20.25
Fair		5	3.16
Average		12	−0.53
Good		26	0.62
Excellent	19	−.28	
Temperament Type			0.27
ESFJ	13	−5.16	
ESTJ	22	2.26	
ISTJ	10	−0.14	
ISFJ	2	−10.47	
Other	16	2.47	
R^2			0.23

*Restaurant Profit/Loss Grand *M* = 16.20. Deviations from the Grand *M* have been adjusted for the independent variables.

**Statistical significance is closely related to sample size. Due to the small sample size in this study, emphasis will be on relative importance using β coefficients rather than statistical significance.

TABLE 4. Relationship Between Most Frequently Occurring Temperament Types and Levels of Unit Profitability

Type	−5.00% to 9.00%	9.01% to 19.50%	19.51% to 45.00%
ESTJ	33.3%	35.7%	68.8%
	$n = 6$	$n = 5$	$n = 11$
ESFJ	38.9%	42.9%	12.5%
	$n = 7$	$n = 6$	$n = 2$
ISTJ	27.8%	21.4%	18.8%
	$n = 5$	$n = 3$	$n = 3$

$X^2 = 5.94$
Note. $n = 48$; $p > .10$.

searcher assumes that some managers may have been reluctant to report net loss and thus, left the profitability question blank.

In replicating this study, the researcher would use a method suggested by one of the reviewers of this article. The method involves asking managers for food costs and labor costs, as percentages, through the use of open-ended questions. These two percentages can be added together and subtracted from 1 to yield the gross profit percentage, as shown in the following formula:

$$\text{Gross Profit \%} = 1 - (\text{Food Cost \%} + \text{Labor Cost \%})$$

Many managers receive bonuses based on keeping food costs and labor costs in line, and thus are very aware of those percentages in relation to company goals. Reporting their food and labor cost percentages may seem less threatening to managers than reporting their unit's profit percentage, and the reported percentages are more likely to be accurate.

Getting the response rate to 59.8% was a real struggle. Additional follow-up phone calls were made after the usual mailings were complete. The Survey Unit personnel attributed that to the fact they were dealing with a population of very busy people. The researcher came to a different conclusion. When the researcher conducted the pilot study, 17 questionnaires were mailed out to quick service chain restaurant managers in two neighboring counties. Within about 2 weeks, 12 of the questionnaires, or 71%, were returned, using only one mailing and no follow up. The researcher used plain manila envelopes to mail the materials and included her name and home address on the mailing labels. The packet of materials sent from the Survey Unit went out on University of Idaho College of Agriculture stationery. Is it possible when managers received mail at their restaurant from the University of Idaho College of Agriculture, some chose to throw it away before opening–seeing no reason to have been contacted by the University of Idaho? One of the managers in the study contacted the researcher to apologize for having thrown away the first mailing because they thought it was "junk" mail. Why would the University of Idaho College of Agriculture be contacting restaurant managers?

Returning to the response rate from the pilot study, it may have been high because managers saw the name of a person and were curious as to the reason for the correspondence. Similar results were reported by Weaver et al. (1991) in their study comparing first class mail versus bulk mail. Because quick service chain restaurant managers deal with customers on a daily basis, they almost had to assume the person listed on the return label was a customer needing some attention. After discovering the letter was not a complaint, they may have been relieved it was only a questionnaire and filled it out. But when envelopes arrived in official university envelopes most likely they were discarded.

Further studies might be conducted with managers in other segments of the hospitality industry (hotels, full-service restaurants, institutional operations, clubs, casinos, etc.) to find out if the results of this study are comparable.

One of the main goals of the study was to identify a tool that restaurant owners could use as part of a hiring process to better match people to the position of manager. Although, no statistical significance resulted when comparing temperament type to profitability, a practical application may still be in place. As reported, almost 69% of the managers in the high level of profitability were of the ESTJ type. All other selection criteria being equal, if one of the applicants were to type as ESTJ, the restaurateurs would potentially hire that individual over the other candidates. Again, as a point of emphasis, use of the Keirsey Temperament Sorter is recommended only as a part of a complete selection process.

As indicated by the results of the Chi-square analysis, income level increases as education level increases. Hospitality educators need to position themselves and their programs so managers employed in the field have the opportunity to work on skill building or a college degree. In order to attract the working manager, educators must examine programs for flexibility and relevance. Restaurants can be open 24 hours per day, 365 days per year. To accommodate potential students from industry, educators must look to alternate delivery. On-campus offerings must include early morning, late evening, and weekend options. Off-campus offerings must include correspondence/independent study, directed study, and web-centric courses. Implementation of a portfolio program and/or a means for challenging coursework for students with extensive on-the-job experience is recommended. People who are working in the field expect relevancy of course material; they know what's real and what's not or what works and what doesn't. Therefore, professional seminars or relevant higher education degree programs could be more beneficial in enhancing the profitability of restaurant units.

There are many factors that contribute to the profitability of a quick service chain restaurant–variables beyond the scope of this study. However, the unit manager will always be the most important part of the success equation, in this researcher's opinion. Don Smith, Professor Emeritus at Washington State University, holds a fundamental belief that "the coach makes the difference" (personal communication via e-mail, February 24, 1998). This sports analogy for restaurant managers sums up the importance of having the right person for the job. The personality the manager brings to the operation cannot be changed or trained, yet it can be tested.

REFERENCES

Bedford, R. & Ineson, E. (1995). Psychometric testing for management selection in the United Kingdom hotel and catering industry. *Hospitality Research Journal, 19* (3), 83-94.

Dillman, D. (1978). *Mail and telephone surveys: The total design method.* New York: John Wiley & Sons.

Furnham, A., Forde, L., & Ferrarri, K. (1999). Personality and work motivation. Personality and Individual Differences, 26 (6), 1035-1043.

George, J. (1992). The role of personality in organizational life: issues and evidence. *Journal of Management*, 18 (2), 185-213.

George, T. (1995). Franchising and the hospitality industry. In R. Brymer (Ed.), *Hospitality management: An introduction to the industry* (pp. 109-122). Dubuque, IA: Kendall/Hunt.

Hogan, R., Hogan, J., & Roberts, B. (1996). Personality measurement and employment decisions: questions and answers. *The American Psychologist*, 51 (5), 469-477.

Huston, J., & Huston, T. (1995). How learning style and personality type can affect performance. *Health Care Supervision*, 13 (4), 38-45.

Jung, C. G. (1971). Psychological types. Bollingen Series XX. *The Collected Works of C. G. Jung* (Vol. 6). Princeton, NJ: Princeton University Press.

Keirsey, D. (1997). *The Four Temperaments* [On-line]. Available Internet: http://keirsey.com/personality/sj.html, http://keirsey.com/personality/nf.html, http://keirsey.com/personality/nt.html, http://keirsey.com/personality/sp.html

Keirsey, D. & Bates, M. (1984). *Please understand me: Character and temperament types* (Rev. ed.). Del Mar, CA: Prometheus Nemesis.

Macdaid, G., McCaulley, M., & Kainz, R. (1986). *Myers-Briggs Type Indicator: Atlas of type tables.* Gainsville, FL: Center for Applications of Psychological Type, Inc.

Maynard, B., and Champoux, T. (1996). *Heart, soul and spirit: Bold strategies for transforming your organization.* Redmond, WA: Success Research, Inc.

Murray, J. (1990). Review of research on the Myers-Briggs Type Indicator. *Perceptual and Motor Skills*, 70, 1187-1202.

Myers, I., and Myers, P. (1995). *Gifts differing: Understanding personality type* (Rev. ed.). Palo Alto, CA: Davies-Black Publishing.

National Restaurant Association. (1997). *Restaurant Industry Operations Report–1996* [On-line]. Available Internet: http://www.restaurant.org/research/pocket/index.htm

National Restaurant Association. (1996, December). *1997 Restaurant industry forecast.*

Nordvik, H. & Brovold, H. (1998). Personality traits in leadership tasks. *Scandinavian Journal of Psychology*, 39 (2), 61-64.

Quinn, M., Lewis, R., and Fischer, K. (1992). A cross-correlation of the Myers-Briggs and Keirsey instruments. *Journal of College Student Development*, 33, 279-280.

Rice, G. (1997, January). *Industry of choice: A report on foodservice employees.* (Available from The Educational Foundation of the National Restaurant Association, 250 S. Wacker Drive, Suite 1400, Chicago, IL 60606.)

Shingler, G. & Ludwick, J. (1995). The quick service (fast food) segment. In R. Brymer (Ed.), *Hospitality management: An introduction to the industry* (pp. 417-422). Dubuque, IA: Kendall/Hunt.

Tieger, P. & Barron-Tieger, B. (1995). *Do What You Are* (2nd ed.) Boston, MA: Little, Brown and Company.

Thomas, J. & Nelson, J. (1996). *Research Methods in Physical Activity* (3rd ed.). Champaign, IL: Human Kinetics.

Tucker, I. & Gillespie, B. (1993). Correlations among three measures of personality type. *Perceptual and Motor Skills*, 77, 650.

Watson, L. (1990). An investigation into the relationships between job satisfaction, temperament type, and selected demographic variables among West Virginia vocational agriculture teachers (Doctoral dissertation, Virginia Polytechnic Institute and State University, Blacksburg, VA, USA 1990). *Dissertation Abstracts International-A, 52/01*, 58.

Weaver, P., E. Chiu and K. McCleary (1991). Can choice of letterhead affect your mail survey response rate?. *Hospitality and Tourism Educator, 3(2), 10-13.*

Withiam, G. (1997). The measure of a restaurant. *Cornell Hotel and Restaurant Administration Quarterly*, 38 (5), 15.

Zemke, R. (1992). Second thoughts about the MBTI. *Training*, 29 (4), 43-47.

Strategic Alliances: The Economics of Dual Branding in Restaurant Franchising

Yae Sock Roh
Cathrine Smestad

SUMMARY. The adaptation of dual brand franchising has become one of the most popular means of doing business in the restaurant industry. The purpose of the paper is to explore the economic significance of dual branding by examining various types of strategic alliances in restaurant franchising. In addition, the research treats the motivations for, and importance of, dual brand franchising from both franchisor and franchisee perspectives. Several dual brand franchising models are also identified. Theoretical economic implications as they relate to dual franchising are investigated. The research offers some discussion of possible future trends and prospects for dual branding in restaurant franchising.

KEYWORDS. Restaurants, franchising, dual branding, marketing, restaurant chains

Yae Sock Roh is Assistant Professor, 108-D Smith Hall, Department of Marketing and Hospitality Service Administration, College of Business, Central Michigan University.

Cathrine Smestad is affiliated with Whittemore School of Business and Economics, University of New Hampshire.

[Haworth co-indexing entry note]: "Strategic Alliances: The Economics of Dual Branding in Restaurant Franchising." Roh, Yae Sock, and Cathrine Smestad. Co-published simultaneously in *Journal of Restaurant & Foodservice Marketing* (The Haworth Hospitality Press, an imprint of The Haworth Press, Inc.) Vol. 4, No. 4, 2001, pp. 321-332; and: *Quick Service Restaurants, Franchising and Multi-Unit Chain Management* (ed: H. G. Parsa and Francis A. Kwansa) The Haworth Hospitality Press, an imprint of The Haworth Press, Inc., 2001, pp. 321-332. Single or multiple copies of this article are available for a fee from The Haworth Document Delivery Service [1-800-HAWORTH, 9:00 a.m. - 5:00 p.m. (EST). E-mail address: getinfo@haworthpressinc.com].

Franchising restaurants have a reputation for entering the marketplace more successfully and with fewer search costs than independent restaurants (Young et al., 1999). However, recent statistics show that Quick Service franchises have continued to lag in sales and are challenged by tremendous competition, high operating costs, and saturation of similar concepts in a given territory. Thus, restaurant franchisees often seek alternative ways to strengthen their market position to appeal to consumers. The motivation behind dual branding stems from the desire to discover an alternative way to solve these recurrent problems in the fast-food industry. Dual branding franchising often refers to co-branding, or entailing the multiple concepts of different franchise brands under one roof.

In an effort to reach target markets, franchisors have developed various strategies such as quality and service improvements, multiple products, reduction of workforce, and engagement in international markets. Among these enhancements, dual branding strategies have gained widespread acceptance in many sectors of the franchising industry (Justis and Castrogiovanni, 1998). Miami Subs Corp., headquartered in Florida, was one of the first restaurant chains to experiment with dual branding in the late 1980s. The chain decided to add Baskin-Robbins ice cream and frozen Yogurt in order to increase their sales (*Amusement Business*, 1997). Since then, dual brand franchising has gone through an unprecedented growth and the use of dual branding in the restaurant franchise industry has become increasingly visible. Although no statistics are readily available for other industries, it is estimated that approximately 20% of all food franchises are co-branded (Shubart, 1999).

The primary benefit of a dual brand is not only to increase efficiency in the overall operations, but to appeal to diverse customers' needs. Although not yet supported by empirical evidence, many trade journals and industry practitioners have suggested that dual branding arrangements have resulted in reductions in capital investment and labor overheads and have created efficiencies relating to cross-trained employees (Van Houten, 1997). However, their arguments have not yet provided a foundation in economic theory and have not yet inquired as to why some multiple products work while others do not.

The primary object of the paper is to provide such theoretical grounding for dual branding in restaurant franchising. The second section of the paper reviews current trends in strategic alliances between two or more different franchisees. Four models of dual branding and the differences among them are identified in the third section. The fourth section considers the economic implications of dual branding. The last section concludes with some discussion of possible future trends and prospects for dual branding in restaurant franchising.

EXAMPLES OF STRATEGIC ALLIANCE

Recently, the franchising industry has gained experience with various forms of strategic alliances. Co-branding is a form of partnering that enables the competitive advantages associated with two (or more) brands to be combined. It is increasingly seen in well known Quick Service franchisees which open their units in non-traditional locations, such as discount department stores or gas stations. The following section illustrates some strategic alliances of restaurant franchising with other products and services.

Multiple Quick Service Restaurants

Fast-food restaurants can dual brand with another products in various ways. However, in order to be successful, it is necessary that the two or more food items complement each other rather than be in competition with one another. This prerequisite is a crucial factor because the combination enables each business to bring in customers at different times of the day. As the consumers prefer variety of selection over a single item, dual branding can decrease the amount of off peak-meal periods.

For example, after having gone through a testing period, Arby's franchisor made T.J. Cinnamons brand available to all their franchisees systemwide. As a result, the cinnamon rolls augment sales in the morning hours for breakfast and for snacking during the day without overlapping Arby's sales during lunch and dinner time (Hamstra, 1997).

Gas Station/Convenience Stores and Quick Service Restaurants

Food service franchisors are also co-branding with non-food industries as well. Gas stations and convenience stores are partnering with fast-food restaurants to buy or lease prime retail property and operate together by dual branding their businesses. It is not unusual to find that TCBY offering frozen yogurt and ice cream products at gas stations or retail stores. According to the National Convenience Stores Association, Quick Service makes up almost 14% of c-stores merchandise sales (Lowe, 1997).

Recently, the results of an annual survey about branded fast-food programs in the petroleum and convenience store industry indicated that eight out of ten stores increased their overall sales at a growth rate of 15 percent (Killeen et al., 1997).

Quick Service Restaurants and Large Discount Stores

The saturated market has prompted fast-food restaurants to search vigorously for new locations to operate its stores in. Interestingly enough, large discount stores such as Wal-Mart, Target, Home Depot, and K-mart have become popular locations. By the middle of the 1990's, McDonald's had already opened restaurants inside more than 700 Wal-Marts, and Little Caesar's Pizza and K-mart had developed over 600 jointly-franchised restaurants. These expansions led to an immense increase in consumption levels and a rapid rise in sales (Young et al., 1997).

Though this partnering can be a beneficial alliance for both partners, major disagreements are likely to occur between the Quick Service franchisee and the discount stores about where the food outlet should be placed. Clearly, Quick Service Restaurant companies prefer their venue to be located in the front of the store near the entrance, so that the store is visible to customers. On the other hand, discount stores want food courts located in the back of the store to limit food smells in the store at large. This tension can be a controversial issue between the two partners and should be solved prior to forming the alliance.

Restaurants and Hotels

Hotel restaurants have a reputation among consumers as providing poor value by offering indifferent service and inferior food. They have also been known to be pricey, due to the high cost structure of hotel restaurants (Strate and Rappole, 1997; Shubart, 1999). Thus co-branding arrangements allow restaurants to locate in the lodging franchise's prime locations and to cater not only for room service, but also for local markets.

Choice Hotel International has instituted its Pizza Hut arrangements across all its brands while the Metromedia Steakhouse Co. agreement focuses development alongside Comfort Inns. O'Rourke, V.P. of Marketing and Promotions for Choice Hotel International suggests that co-branding not only makes a limited-service hotel a better value to the discerning customer, but also brings other benefits, such as increased advertising revenue. For example. Choice hotels allowed Pizza Hut to do a joint promotion (Andorka, 1995)

A Pennsylvania Holiday Inn is an excellent example of an unsuccessful hotel that improved its performance through dual branding. When the hotel operated its own restaurant, sales were merely $450,000 annually, forcing the hotel to reconsider what type of a restaurant would bring about greater sales. The hotel decided to place a franchised restaurant in place of the regular hotel restaurant and watched its sales grow to $4 million in its first year of operations (Parseghian, 1996).

Dual branding has not only generated significant increases in restaurant sales; it has also brought about a significant rise in hotel occupancy. A manager of a Best Western hotel reported that hotel occupancy increased as much as 10 percent when it changed over from a self-operated restaurant to a franchised Country Kitchen (Boone, 1997).

MODELS OF FRANCHISING

The concept and application of dual branding may take various forms. The most common methods are the following: franchisee within a local retail store, a single franchisee with multiple products, multiple franchisees with multiple products all under the same roof, or a portfolio of franchising systems.

Franchisee Within a Local Retail Store

A common way to build a dual branding is to create a franchisee contract, in which a franchise is placed inside of a local retail store (see Figure 1). In these situations, the franchisee leases space from the retail store and pays rent to the leaser, along with paying the usual franchise fees and royalties to its franchisor (Justis and Castrogiovanni, 1998). This situation is very profitable to both the retail store and the franchisee as one attracts customers for the benefit of the other: in essence, they help each other either by attracting more customers or keeping the customers in the store for a longer period of time. This type of relationship is known as a Franshisor Contract (Young et al., 1997)

Single Franchisee with Multiple Products

A single franchisee is a dual branding relationship in which there is only one franchisee but it has the rights to operate franchises from multiple franchisors (see Figure 2) (Justis and Castrogiovanni, 1998). This type of co-branding has many benefits. One important benefit is that only one manager or owner operates the store. Less conflict is likely to occur among the franchises as this individual makes decisions about such issues as signage, location of the store, advertisements, and staffing. This one franchisee does, however, have to pay royalty and franchise fees to all his or her franchisors. He also has to pay the entire overhead, real estate, and development costs himself. However, the operation should bring in more customers and increase its sales

FIGURE 1. Franchisee Within Local Retail Store

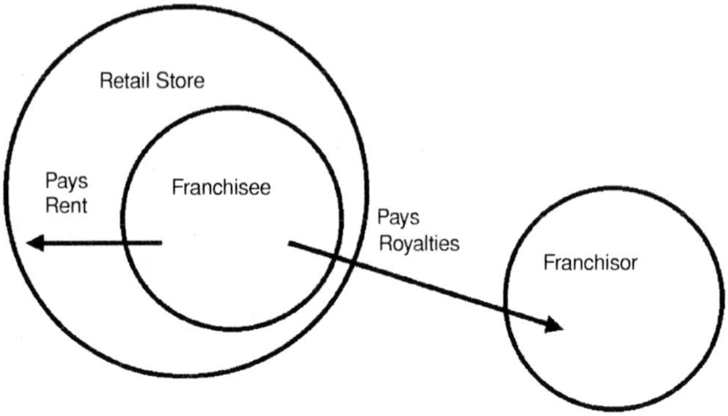

FIGURE 2. Single Franchisee with Multiple Franchises

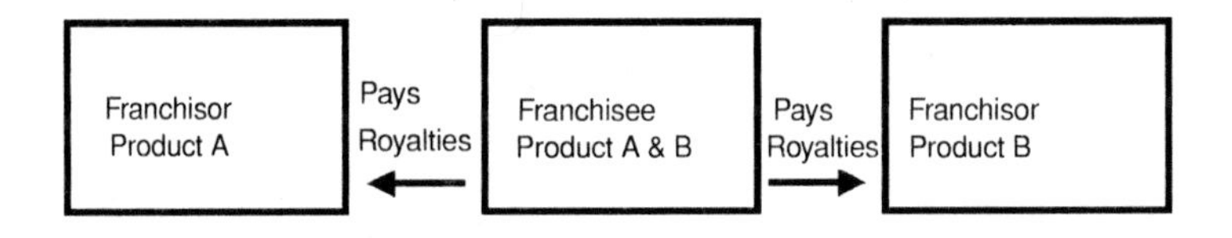

so that the fees and costs he has to pay are more easily covered than if he were only franchising one free-standing store.

Dual Franchisees with Dual Products

The most common and classic example of dual branding is a single store which offers two food concepts. A type of dual branding that occurs when two local franchisees choose to place their businesses side-by side in the same building (see Figure 3), dual franchising, can make both franchises much more efficient because it enables them to operate with significantly lower costs than stand alone businesses.

In this type of dual branding agreement, both franchisees pay royalty and franchise fees to their respective franchisors. They also receive managerial assistance in operating and quality standards from their franchisors. In this alliance, all overhead, real estate development, parking and upkeep costs are shared by both franchisees. Some alliances also share costs associated with staffing, equipment, and managers.

FIGURE 3. Two Franchisees Operating Their Own Franchises

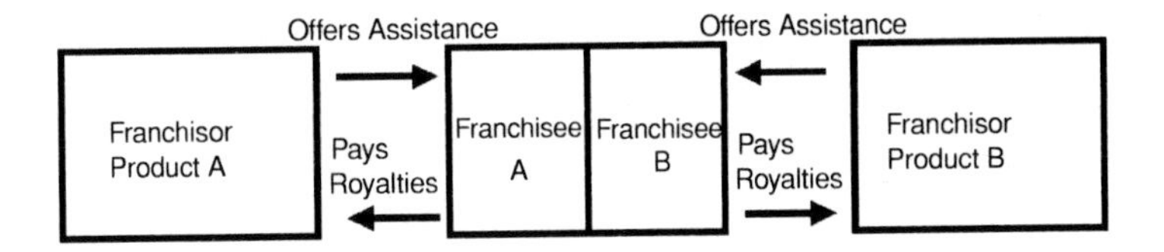

Portfolio Franchising System

A portfolio franchising system grants a single franchisee the rights to operate several franchises on the same site (Justis and Castrogiovanni, 1998) (see Figure 4). Portfolio franchising systems are frequently found at the non-traditional sites such as sports stadiums, bus and train stations, interstate travel plazas, airport terminals and college campuses.

Portfolio franchising enables the franchisee to save on various costs while bringing in a greater volume of customers, increasing the operation's total sales. However, the franchisee must pay franchise and royalty fees to each of the franchisors who are involved in the portfolio franchise system.

ECONOMIC ISSUES

This section of the paper investigates the economic significance of co-branding. As previously indicated, by and large, dual branding franchising restaurants deal with pairings of two menus that are complementary to each other. The success of cobranding depends on the following economic elements: capital intensity, reduction of costs and risk of operations.

Capital Intensity

Most business firms prefer to support a high level of sales or operating cash flow with a small amount of assets (i.e., efficient utilization of assets). However, restaurant properties are fixed asset intensive. The proportion of fixed assets, such as land, buildings, kitchen equipment, and furniture are substantially higher than those of manufacturing companies. Thus, how to make efficient use of existing fixed assets has been a challenging task for the industry practitioners.

Capital intensity is the level of investment in the fixed assets needed to generate sales. The lower the capital intensity, the more efficient use of capital. An

FIGURE 4. Franchisee Portfolio

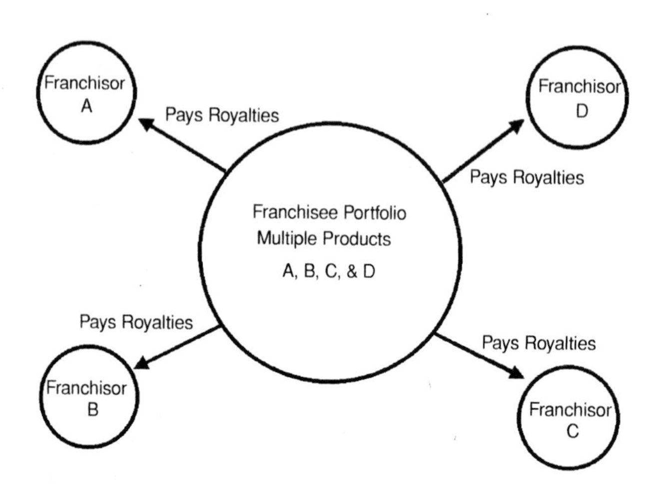

example of a strategy that reduced capital intensity is the development of breakfast items by McDonald's in the early 1970s. Prior to offering breakfast, McDonald's realized that their facilities and equipment sat idle during the early morning hours. With minimal investment in labor and raw materials, McDonald's restaurants were able to generate additional revenue. Thus, the level of capital investment employed to produce more goods and service took fewer dollars for McDonald's; it is estimated that one quarter of all breakfasts now eaten outside the home are consumed at McDonald's.

Several successful cobranding examples also demonstrate efficient use of capital investment. Cobranding between Kentucky Fried Chicken (KFC) and Taco Bell resulted KFC attracting a stronger dinner crowd, while Taco Bell did its best business serving lunch and late night snacks. Giving diners an extra menu increased KFC's salea whopping 10%.

Although it would cost as much as $90,000 per store to add Taco Bell to the existing KFC store, KFC perceived that the benefits of cobranding would exceed the costs of conversion (i.e., positive net present value) (Fischer, 1995). Similarly, in Canada, Wendy's International Inc. operates hamburger and doughnut outlets. The Horton's menu draws many customers early in the morning when the Wendy's menu is not available.

Also, several of Arby's dual branding test markets have proved fruitful. During the past several years, the chain has converted 15 stores in four markets to co-brand units with partners such as Long John Silver's, Sbarro's, Mrs. Winner's Chicken and Green Burrito Restaurants. By combining Arby's lunch menu with its partners' dinner focus meals, Arby's boosted average units volume at the dual brand locations by 50%.

Even if cobranding opportunities are not available, franchisees can seek a location near a complementary franchisee. For instance, Starbucks Coffee Co. has strategically positioned some of its stores next to bagel shops to provide customers with an easy breakfast combination.

Reduced Costs

Dual brand franchising helps franchisees reduce the costs of daily business because it allows multiple partners to share fixed costs, including equipment, management and real estate expenses. The marginal cost of adding the second franchise store to the existing store is less than the marginal cost of opening an independent establishment alone. In fact, an average franchisee is usually able to set up units for 40% of the capital costs of setting up free standing ones because the franchise units are able to share common fixtures, equipment and space. Also, Wendy's and Horton's have found that they save about 25% on building and site costs by each occupying half the space.

Some franchisees have also found that labor costs are 60-80% of what they would have been if the franchise unit had been staffed separately (Justis and Castrogiovanni, 1998). For example, two or more food chains can operate autonomously, but share the site and a single manager. Also, the dual branding between T.J Cinnamons and Arby's required barely any extra staffing which makes for considerable efficiency in labor costs. The Arby's employees were taught how to make the cinnamon rolls themselves (Hamstra, 1997).

McDonald's also found that dual branding with gas stations offers convenience for the customers and reduces costs of running a business. For example, when expanding on a current operation, McDonald's uses "satellites" instead of a regular restaurant. These satellites are smaller versions of the regular stores offering limited, best selling menu items. However, these satellites generate the same sales volume and are the same size as full size restaurants (Burns, 1995). Through the process of developing satellite stores, McDonald's saved its development costs by one third of that of its regular stores simply by standardizing its equipment and designs. What makes these sites more successful than other regular sites are that they share a multitude of costs with the other side-by-side operations in prime retail locations in highly saturated areas or smaller towns.

An additional cost advantage of cobranding is to get volume discounts through existing franchise suppliers. From a consumer's point of view, co-branding also helps customers minimize search costs.

Risk Reduction

Dual branding also helps restaurant firms reduce operational risk. For example, the addition of pizza (chicken) to coffee (donuts) diversifies the risk of each operation. During the breakfast periods, sales for coffee (donuts) tend to be brisk; Pizza (chicken) on the other hand, tends to be most in demand during lunch and dinner periods. Thus, the sales of the specific foods are inversely or negatively related depending on the time of day. When the return from coffee (donuts) operations is high, the return from chicken (pizza) sales will presumably be low. As a result, the combined returns are more stable and therefore less risky. The portfolio effect of reduced variability of sales occurs because there is a negative correlation between complementary items. Indeed, the recent strategic alliance between TCBY, Wall Street Deli, A&W Restaurants and other national chains allowed them to smooth out their seasonal fluctuations (*Amusement Business*, 1997).

The authors conducted an informal interview with a local franchisor who is currently engaged in dual brand practice. In order to provide promised confidentiality, the name of the restaurant is undisclosed. The operator stated that the restaurant is considerably busier than before due to the wide variety of the menu. According to Calvin Heskell, the CEO of Franchise Solutions, a leading supplier of information based on services to the franchise industry, a significant number of prospective franchisees are trying to pursue dual brand franchise concept rather than single brand. He pointed out that potential franchisees are already aware of less risk when they offer multiple products.

FUTURE AND PROSPECTS

The purpose of this paper was to provide an analysis of the economic implications of dual branding in restaurant franchising. In particular, the paper examined different types of dual branding alliances by illustrating various models.

While dual branding has been used as a strategic management tool for certain businesses, forming synergistic alliances with other brands can require extra caution. For example, cobranding between a Quick Service hamburger restaurant, Carl's Jr., and Long John Silver's Restaurant Inc., a fish chain, have been notably less successful than other instances of cobranding because the Long John Silver's name wasn't well enough established in Southern California markets (Tannenbaun, 1996). Thus, prior to executing dual brand concepts, it is important to test the brands' compatibility and complementarily. The prospective partners must determine the duration of the test period, the number of units to be tested and future competition of the test markets. Such pilot tests should reduce the future risk of operations.

Industry expertise has argued that cobranding has clear logical appeal and several advantages over conventional franchising systems. However, there are many unanswered questions and possible challenges when prospective franchisees deal with heterogeneous products (Shubart, 1999). For example, required labor for one branded concept may be significantly different from that of the other. Consumers' acceptance of the brand may also be variable. For this reason, some franchisors may be reluctant to engage in dual franchising because it may jeopardize the uniqueness of each franchise system. Consequently, the conversion can be impractical or otherwise uneconomical for a certain franchise.

While the practice of dual branding prevails in the fast restaurant industry, it has not reached a stage of maturity from a technological point of view. In the near future, firms must resolve the series of technological problems that may arise when two or more franchisors are involved in dual franchising. In particular, these problems are more likely to occur when two unaffiliated brands co-brand rather than when one parent with its own brands are linked together. Elaborate formulas will need to be devised for dividing the royalty fees from each product. The point-of sale cash register or computer systems should be able to accurately segregate such product sales information (Ritchie, 1998).

Managerial problems are an another important issue for the future of dual branding franchising. Managerial interests are not necessarily aligned in dual branding when each franchisee has its own manager. Each manager has his/her own managerial philosophy and strategy that may affect both units. It is therefore vital for managers to make clear decisions that are beneficial for both franchisees (Justis and Castrogiovanni, 1998). Many dual branding partners prefer to have a single manager in order to minimize the chances of conflicts of interest in the dual franchising system.

In sum, given that the concept of co-branding franchising is a relatively new phenomenon, only descriptive studies or case studies are available at present. Although it is suggested that such strategic alliances can reduce costs and operational risks, and create efficiencies relating to cross-trained employees, persuasive empirical evidence is completely absent. In other words, the concept of co-branding is still considered a recent trend and the literature examining such relationships is extremely limited (Young et al.,1999). Admittedly, the research is a challenging topic to tackle using any empirical methodology because the practitioners in the area are reticent about sharing their operating information. However, a future avenue of research would be more deep and through investigation along with empirical studies to test certain hypothesis on the economic significance of dual branding.

REFERENCES

Amusement Business (1997). Dual Branding Trend Better Than One: Boosts Sales, Peak Times and Volume. 109 (17), 16.

Andorka, F. (1995). High-Recognition Restaurants. *Hotel & Motel Management*, 210 (16), 43-44.

Boone, J. (1997). Hotel-Restaurant Co-Branding–A Preliminary Study. *Cornell Hotel and Restaurant Administration Quarterly*, 38 (5), 34-43.

Burns, G. (1995). French-Fries With That Quart of Oil? *Business Week, (November 27), 3452, 86-87.*

Fischer, D. (1995). The New Meal Deals. *U.S. News and World Report*, 119 (17), 66.

Hamstra, M. (1997). Arby's Rolls T.J. Cinnamons Into Dual-Branded Stores. *Nation's Restaurant News*, 31 (27), 5-6.

Justis, R. and G. Castrogiovanni (1998). Co-Branding: A Franchise Growth Strategy. *Conference Proceedings of the Society of Franchising*, Las Vegas, NV.

Killeen, B., J. Tarnowski, and J.Callanan (1997). 1997 Branded Fast-Food Survey. *The Journal of Petroleum Marketing*, (September), 43-50.

Lowe, K. (1997). QSR's and C-stores: Perfect Partners? *Restaurants and Institutions*, 107 (23), 90-92.

Parseghian, P. (1996). Branding Offers Hotels Opportunity to Increase Food Sales. *Nation*'s Restaurant News, 30 (23), 96.

Ritchie, H. (1998). Dual Branding. *Food Service Director*, 11 (4), 174-175.

Strate, Robert, and Clinton Rappole (1997). Strategic Alliances Between Hotels and Restaurants. *Cornell Hotel and Restaurant Administration Quarterly*, 38 (3), 50-61.

Shubart, E. (1999). Cobranding, *Franchise Times*, 5(10), 40-41.

Tannenbaun, J. (1996). Franchises Try New Fast-Food Combos, *Wall Street Journal*, October 16, B1.

Van Houten, B. (1998): Casual Twins–Double Unit Gives New Meaning to Co-Branding. *Restaurant Business*, (December 15), 17-18.

Young, J., A. Paswan, J. Buch, and L. Ashby (1997). Fast-Food Franchisors and Discount Retailers: A tale of Two Alliances and Beyond. *Conference Proceedings of the Society of Franchising*, Las Vegas, NV.

Young, J., C. Hoggatt, and A. Paswan (1999). Co-branding Relationships: Franchisors Partnering With Other Franchisors. *Proceedings of the Society of Franchising*, *Society of Franchising*. Miami, FL.

Index